Science: Philosophy, History and Education

Series Editors
Kostas Kampourakis, University of Geneva, Geneva, Switzerland

Editorial Board
Fouad Abd-El-Khalick, The University of North Carolina at Chapel Hill, Chapel Hill, USA
María Pilar Jiménez Aleixandre, University of Santiago de Compostela, Santiago de Compostela, Spain
Theodore Arabatzis, University of Athens, Athens, Greece
Sibel Erduran, University of Oxford, Oxford, UK
Martin Kusch, University of Vienna, Vienna, Austria
Norman G. Lederman, Illinois Institute of Technology, Chicago, USA
Alan C. Love, University of Minnesota - Twin Cities, Minneapolis, USA
Michael Matthews, University of New South Wales, Sydney, Australia
Andreas Müller, University of Geneva, Geneva, Switzerland
Ross Nehm, Stony Brook University (SUNY), Stony Brook, USA
Stathis Psillos, University of Athens, Athens, Greece
Michael Reiss, UCL Institute of Education, London, UK
Thomas Reydon, Leibniz Universität Hannover, Hannover, Germany
Bruno J. Strasser, University of Geneva, Geneva, Switzerland
Marcel Weber, University of Geneva, Geneva, Switzerland
Alice Siu Ling Wong, The University of Hong Kong, Hong Kong, China

Scope of the Series

This book series serves as a venue for the exchange of the complementary perspectives of science educators and HPS scholars. History and philosophy of science (HPS) contributes a lot to science education and there is currently an increased interest for exploring this relationship further. Science educators have started delving into the details of HPS scholarship, often in collaboration with HPS scholars. In addition, and perhaps most importantly, HPS scholars have come to realize that they have a lot to contribute to science education, predominantly in two domains: a) understanding concepts and b) understanding the nature of science. In order to teach about central science concepts such as "force", "adaptation", "electron" etc, the contribution of HPS scholars is fundamental in answering questions such as: a) When was the concept created or coined? What was its initial meaning and how different is it today? Accordingly, in order to teach about the nature of science the contribution of HPS scholar is crucial in clarifying the characteristics of scientific knowledge and in presenting exemplar cases from the history of science that provide an authentic image of how science has been done. The series aims to publish authoritative and comprehensive books and to establish that HPS-informed science education should be the norm and not some special case. This series complements the journal Science & Education http://www.springer.com/journal/11191 Book Proposals should be sent to the Publishing Editor at Claudia.Acuna@springer.com.

More information about this series at http://www.springer.com/series/13387

Michael R. Matthews

Feng Shui: Teaching About Science and Pseudoscience

Michael R. Matthews
University of New South Wales
Sydney, NSW, Australia

ISSN 2520-8594   ISSN 2520-8608 (electronic)
Science: Philosophy, History and Education
ISBN 978-3-030-18821-4   ISBN 978-3-030-18822-1 (eBook)
https://doi.org/10.1007/978-3-030-18822-1

© Springer Nature Switzerland AG 2019
This work is subject to copyright. All rights are reserved by the Publisher, whether the whole or part of the material is concerned, specifically the rights of translation, reprinting, reuse of illustrations, recitation, broadcasting, reproduction on microfilms or in any other physical way, and transmission or information storage and retrieval, electronic adaptation, computer software, or by similar or dissimilar methodology now known or hereafter developed.
The use of general descriptive names, registered names, trademarks, service marks, etc. in this publication does not imply, even in the absence of a specific statement, that such names are exempt from the relevant protective laws and regulations and therefore free for general use.
The publisher, the authors, and the editors are safe to assume that the advice and information in this book are believed to be true and accurate at the date of publication. Neither the publisher nor the authors or the editors give a warranty, express or implied, with respect to the material contained herein or for any errors or omissions that may have been made. The publisher remains neutral with regard to jurisdictional claims in published maps and institutional affiliations.

This Springer imprint is published by the registered company Springer Nature Switzerland AG.
The registered company address is: Gewerbestrasse 11, 6330 Cham, Switzerland

This book is dedicated to my friend and teacher, Mario Bunge.
I am among many, in countless countries around the world, who admire his wide-ranging, cross-disciplinary, serious scholarship, his insistence that science and philosophy need to mutually engage with each other, his lucidity in argument and clarity of writing and his lifelong commitment to Enlightenment values.
The book is published in his centenary year. Hopefully, he will be blessed with more birthdays and precious time with family, friends and books.

Foreword

Michael Matthews is the editor of the substantial three-volume and highly regarded, *International Handbook of Research in History, Philosophy and Science Teaching*, which was published by Springer in 2014. A few years later, Michael edited the volume *History, Philosophy and Science Teaching: New Perspectives*, which nicely complements the abovementioned handbook and which points to new directions in HPS and science teaching research. For those who might have thought that Michael had said his last words and that he now enjoys retirement, the present volume shows that this is not the case. Not only is Michael back, but with an extraordinary work that opens a new era in this field.

The topic of this volume is feng shui: a worldview and practice that, as Michael informs us in his introduction, '…has been, and is, to Asia somewhat like Christian belief was to the European medieval world. Like medieval Christianity, feng shui impinges on daily and social life; it informs people's understanding of their place in the cosmos'. I must admit that I had a little idea what feng shui was before reading the book, and I had heard only a little about it. I could therefore not even imagine that there might be a connection between feng shui and science teaching. But as the title of the present volume, *Feng Shui: Teaching About Science and Pseudoscience*, indicates, there is indeed a connection, which is actually a very important one.

In this book, Michael Matthews makes the case that important educational responsibilities are met, and opportunities arise, if feng shui is included in science teaching programmes. Feng shui is a deeply entrenched, three-millennia-old system of Asian beliefs and practices about nature, architecture, health and divination. It is part of a comprehensive and ancient worldview built around the belief in chi (qi), the putative universal energy or life-force that animates all existence—the cosmos, the solar system, the Earth and the human bodies. Harmonious living requires building in accord with local chi streams, good health requires replenishment and manipulation of internal chi flow and a beneficent afterlife is enhanced when buried in conformity with local chi directions. Traditional Chinese Medicine is based on the proper manipulation of internal chi by acupuncture, tai chi and qigong exercises and herbal dietary supplements.

This book provides a richly documented account of the historical, cultural, philosophical and practical dimensions of feng shui. It argues that educational systems have a responsibility to examine the claims of feng shui and that this provides opportunities for students to better learn about the key features of the nature of science, the demarcation of science and non-science, the characteristics of pseudoscience and the engagement of science with diverse worldviews. The author argues that although feng shui presents itself as a science or an alternate science, it is actually a pseudoscience.

Therefore, the book provides an in-depth look in an important topic that could be used to teach about the nature of pseudoscience. I found learning about feng shui fascinating, and I considered the discussion of its being pseudoscience illuminating. Michael Matthews should be commended for producing another thought-provoking and far-reaching book. For Asian educators, he has provided an engaging account of the history and philosophy of a major Asian worldview that can enrich science instruction. And he has successfully shifted the attention of the Western-dominated international science education community to a non-Western tradition that is worth learning about and that can illuminate the fundamental questions about the nature of science and its connection to culture.

University of Geneva   Kostas Kampourakis
Geneva, Switzerland   Series Editor

Preface and Acknowledgements

This book is a contribution to the 'History, Philosophy and Science Teaching' (HPS&ST) research programme. That programme, stretching back to at least Ernst Mach's educational writings in the 19th century, has been concerned with the utilization of history and philosophy of science in theoretical, curricula and pedagogical issues in science teaching (Matthews 1988, 2014, 2015a, b). The book is ambitious in the wide range of historical, philosophical, cultural, political, scientific and educational issues discussed—perhaps too ambitious. Clearly, much more finesse, nuance, detail and scholarship are needed beyond what I can provide. Perhaps, this will be given by others. Hopefully, there is enough in the book to give researchers, teachers and students assistance in their own study of the neglected subject of feng shui and of the more general issues of science, philosophy, culture, social psychology, politics and education into which research on feng shui necessarily lead.

Reading about feng shui and associated qigong (*chi kung*) has occupied me for the past number of years. The practice of feng shui appeals to a 3000-year-old Asian worldview whose elements are life-force, all-pervasive subtle energies and cosmic harmonization. Composites of this worldview are gaining more and more traction in the West, especially where environmentalism, postmodernism and New Age sentiments are strong. Chi (*qi*) is its most distinctive and recognized ontological element, along with the universality of a yin-yang dialectic or interactive mechanism for all natural and human processes. Aspects of the worldview are incorporated in all significant Asian religions.

There are on-going Asian debates about the 'authenticity' of feng shui and of its claims to the mantle of knowledge. Comparable debates have long occurred in the West. Since even before the Reformation, the West has been familiar with debate about the corruption of religion. Protestants are appalled that Christianity should be identified with the baroque practices and beliefs of the Roman Catholic church; mainstream Protestants are dismayed that their faith is associated with the venality, shallowness and theatrics of televangelists. The same transitions and debates occur in Islam and Judaism. In the Abrahamic religions, there has always been a debate about legitimate and illegitimate development of doctrine and practice. A central historical/philosophical question in discussion of feng shui is whether its practice and beliefs are a legitimate expression of the basic Asian tradition or a corruption of it. Is contemporary feng shui practice consistent or inconsistent with the core worldview? Is it a corruption? The Asian traditions are fundamentally naturalistic, not supernatural; so the Abrahamic question of the existence of God does not really arise from within, but the same dynamic is replaced by questions about the existence of chi and its properties.

I have talked on the subject at different Asian HPS&ST conferences while visiting different Asian universities and have been struck why something so omnipresent and important in Asian culture, and now in numerous other cultures, is not more discussed in philosophy of science and in science education.

The subject well fits my own intellectual interests and scholarly background in science, philosophy, history of science, philosophy of science, philosophy of education and science education. All these fields contribute to better examination of feng shui and its potentially productive role in education. The productivity is best exploited in coordinated cross-disciplinary teaching where science, history, philosophy, religion, cultural studies and economics might all contribute to the fuller exposition and appraisal of the topic (see Fig. 11.3). The productive discussion of most serious topics in science education needs to be informed on the one hand by the history and philosophy of science and on the other by philosophy of education. The pressing topics are mainly normative: How are curriculum aims *justified*? What *should* be taught in science programmes? What *ethical constraints* are there on how science is taught? What is a *scientific outlook* or *habit of mind*? *Should* adoption of such an outlook be an aim of science instruction? Is science *compromised* by racial, gender, political, economic or cultural biases? What *should* happen in science programmes when the science clashes with deep-seated cultural and religious convictions of students? These questions lead straight to both philosophy of education and history and philosophy of science (Matthews 2019b).

This book makes seven basic claims that follow each other in a stepwise manner. These are the following:

1. Science needs to be taught in conjunction with the history and philosophy of science (HPS) as it is only in this way that proper student understanding and appreciation of science—its methodology, ontology, epistemology and interrelations with culture—can be acquired.
2. Science education has to fulfil disciplinary, personal, social and cultural purposes. Each goal requires the formation and utilization of a scientific habit of mind.
3. Accepting the social and cultural purposes of science education means that teaching about feng shui is an educational obligation in societies where the practice and beliefs are established; to not do so is educationally irresponsible.
4. Such teaching about feng shui, or its local surrogates (astrology and other such belief systems) where feng shui is not established in the culture, allows for the clarification of many central issues in HPS, specifically concerning the nature of science (NOS), a topic that appears in nearly all national and provincial science curricula.

5. Chi, the central ontological element of feng shui, and its related worldview do not exist. There are no scientific grounds to warrant its existence nor have reasonable and convincing non-scientific grounds been advanced.
6. The distinction between science and pseudoscience can be consistently and intelligently maintained; and the elaboration of feng shui by its sophisticated proponents constitutes a pseudoscience.
7. The history and philosophy of science does establish that science depends upon and thrives in an open society where freedom of belief, religion, politics, association and communication are all valued and promoted. There is a historical and philosophical concordance between science and liberalism.

China is examined as a test case of the final claim. It is a sober coincidence that the book is published in the 100-year anniversary of the 1919 May Fourth uprising that appealed to Mr. Science and Mr. Democracy and the 30-year anniversary of the Tiananmen Square demonstrations and killings. Each was a momentous social upheaval fuelled by ideas of science and its dependence on intellectual, social and political liberty, convictions that had traction among university science and philosophy faculty and students. Hopefully, the book contributes to the ongoing articulation and defence of this conviction.

I am fortunate to have spent some years at The University of Sydney as a full-time and then part-time student, completing degrees in science, psychology, philosophy, HPS and education. All of this was a beneficial preparation for my own high school science teaching and for teaching philosophy of education and science education at the Sydney Teachers' College, the University of Auckland and the University of New South Wales. The early University of Sydney years gave me a cross-disciplinary, HPS orientation to theoretical, curricular and pedagogical issues in science education. Such an orientation is apparent in the contents of this book. I am especially grateful for the teaching of the marvellously qualified staff in philosophy and in philosophy of education at the University of Sydney. As with so many, I have a special intellectual debt to Wallis Suchting, a teacher and friend.

This rewarding educational experience was continued during my first sabbatical leave (1978) at the Centre for History and Philosophy of Science at Boston University where I had the great good fortune to attend classes of Robert Cohen, Marx Wartofsky and Abner Shimony. The Boston period enabled me to experience the vibrant cross-disciplinary 'BU Style' in HPS research and debate that has been manifested in volume after volume of *Boston Studies in Philosophy of Science* since its initial appearance in 1963. That intellectual apprenticeship at Sydney University and Boston University has served me well (Matthews 2009). Not all science education students are so well served by their institutions and teachers, something that can be mitigated a little by including more philosophical studies in education programmes (Matthews 2015a, chap.12).

For 30 years, I have been involved in the International History, Philosophy and Science Teaching Group, and for much of that time, I edited the journal *Science Education and Culture: The Contribution of History and Philosophy of Science* (Matthews 2015b). Group membership and journal editorship put me in contact with scores of inspiring philosophers, historians, scientists and educators. Many of these are contributors to the 3-volume, 76-chapter *International Handbook of Research in History, Philosophy and Science Teaching* (Matthews 2014). I owe a great deal to all of these colleagues and friends.

This book's origin lies in work done over many years for a chapter on feng shui in the Springer anthology *History, Philosophy and Science Teaching: New Perspectives* (Matthews 2018), which was shepherded through Springer by their education editors Bernadette Ohmer and Marianna Pascale and published in the Springer book series edited by Kostas Kampourakis. I am grateful to all three for their encouragement and support of that venture.

The anthology chapter dealt with some of the matters canvassed here, but it was clear to many that an adequate treatment of feng shui belief and practice required many more matters to be investigated and for these to be done with more sophistication. Pleasingly, this opportunity was provided by Claudia Acuna (Education Editor for Springer) and Kostas Kampourakis (book series editor), who after reviews of the book proposal issued a Springer contract for this larger work. All authors will attest that researching and writing while having a contract is altogether different, and so much more satisfying, than researching and writing in hope of obtaining a contract. Reviews still need to be faced, but a contract gives an assurance that the research is not pointless and that the resulting manuscript is going to be considered for publication. This is an enormous psychological relief and encouragement for pursuing newfound avenues of research in the topic.

For 40 years since 1975, the University of New South Wales has supported and encouraged my research and teaching in philosophy of education, HPS and science education. It has been an excellent home for the bulk of my academic life allowing a 2-year furlough at the University of Auckland (1993–1994) where I first engaged with the philosophical and educational issues occasioned when constructivism is adopted as a theory of education (Matthews 1995).

I am grateful to a number of colleagues who have carefully read, commented on and copyedited different chapters of the book: Robert Nola (University of Auckland), Jim Mackenzie (University of Sydney), Robert Carson (Montana State University), Robyn Yucel (La Trobe University), Colin Gauld (formerly University of New South Wales), Paulo Maurício (University of Lisbon) and Yuanlin Gao (Tianjin University).

I am especially grateful to Paul McColl (a retired science teacher and tutor at the University of Melbourne) for patiently copyediting the entire book including revised versions, for making invaluable suggestions about ordering of the sections, for identifying confused portions of the argument and for ensuring that the considerable reference list is accurate and complete. At the very end, he importantly suggested a reorganization and restructuring of the final and penultimate chapters. This I have done with considerable improvement to the readability and argument of the book. All of Paul's work was done in a most timely manner, usually with 1-day turn-arounds, so enabling different Springer deadlines to be easily met.

The background reading, writing, rewriting many times and correcting the manuscript have taken hundreds and hundreds of hours over a number of years. I am fortunate that my wife, Julie House, and my daughters, Clare, Alice and Amelia, have been forbearing of the time spent in my study that could have been well spent outside of it.

The writings of many people have influenced the arguments of this book, and the names of most can be seen in the Reference list. But one, Mario Bunge, warrants a particular mention because this book is published in 2019, the year he hopefully celebrates his 100th birthday on September 21. He is a remarkable exemplar of a philosophical scientist, a scientific philosopher and a proponent of the Enlightenment tradition in philosophy and in education. For the five years that I have worked on this book, I have also been organising and editing a 41-chapter Festschrift for Mario (Matthews 2019a). It has been a very fruitful conjunction of tasks as is evident in the citations herein. To Mario, I gratefully dedicate the book.

Sydney, NSW, Australia   Michael R. Matthews
February 2019

References

Matthews, M. R. (1988). A role for history and philosophy in science teaching. *Educational Philosophy and Theory, 20*(2), 67–81.
Matthews, M. R. (1995). *Challenging New Zealand science education*. Palmerston North. Dunmore Press.
Matthews, M. R. (2009). The Philosophy of Education Society of Australasia (PESA) and My intellectual growing-up. *Educational Philosophy and Theory, 41*(7), 777–781.
Matthews, M. R. (Ed.). (2014). *International handbook of research in history, philosophy and science teaching* (3 Vols.). Dordrecht: Springer.
Matthews, M. R. (2015a). *Science teaching: The contribution of history and philosophy of science: 20th anniversary revised and enlarged edition*. New York: Routledge.
Matthews, M. R. (2015b). Reflections on 25-years of journal editorship. *Science & Education, 24*(5–6), 749–805.
Matthews, M. R. (2018). Feng shui: Educational responsibilities and opportunities. In M. R. Matthews (Ed.), *History, philosophy and science teaching: New perspectives* (pp. 3–41). Dordrecht: Springer.
Matthews, M. R. (Ed.). (2019a). *Mario Bunge: A centenary Festschrift*. Dordrecht: Springer.
Matthews, M. R. (2019b). The contribution of philosophy to science teacher education. In A. D. Colgan & B. Maxwell (Eds.), *Philosophical thinking in teacher education*. New York: Routledge.

Appraisals

In this remarkably detailed and informed study, Michael Matthews exposes and debunks the scientific pretensions of Feng Shui, and at the same time urges its usefulness for science education as a vehicle to teach science students about the nature of science and the place of testing and confirmation/falsification in it, the mutual interaction of science and culture, and much more. This book secures Feng Shui's place in the panoply of pseudosciences that science students need to confront, understand and evaluate. Matthews has produced another tour de force that will repay close study by students, scientists, and all those concerned to understand science, culture, and the science/culture nexus.

Harvey Siegel, Philosophy, University of Miami, USA

Even the most experienced – jaded? – authors sometimes encounter a book that astounds them and makes them realize how much is of seminal importance in their field and of huge educational worth. The present writer is the former and Michael Matthews seminal work on Feng Shui is the latter. With great erudition and even greater fluidity of style, Matthews introduces us to this now-world-wide belief system, analyzing its claims and discussing its place – science or pseudoscience – in the intellectual canon. As always, Matthews – a leading philosopher of science education – looks to the pedagogical message for teachers and their students. Truly, paradigm making.

Michael Ruse, Philosophy, Florida State University, USA

An excellent book. The history is fascinating. The analysis makes an important contribution to science literature.

James Alcock, Psychology, York University, Canada

The book is one of the best research works published on Feng Shui. It opens up vast horizons for viewing science in new perspectives. It is an outstanding contribution to the fields of the history of science, philosophy of science and science education.

Wang Youjun, Philosophy, Shanghai Normal University, China

By broadening the context of HPS&ST *research to Chinese culture, this book provides further relevant and useful information for science teaching and learning. The book is an important resource for science curriculum designers, teacher educators, researchers and teachers.*

Sibel Erduran, Education, University of Oxford, United Kingdom

This book is an introduction to the fruitful topic of feng shui for science educators and science teachers who have interests HPS and STS.

Chen-Yung Lin, Education, National Taiwan Normal University, Taiwan

A terrific book. It cogently explicates how fengshui is a pseudoscience, and why it is of momentous importance in teaching about science and pseudoscience in our time.

Bangping Ding, Education, Capital Normal University, China

Feng Shui is a peculiar phenomenon of knowledge and technics dealing with the relationship between humans and nature. It continues to survive while plenty of pre-modern knowledges have faded out. This book provides an in-depth study of Feng Shui in different periods, considering its philosophical, historical and educational dimensions; it shows how persuasive is its influence in social culture; and lays out its internal logic. Michael Matthews' excellent book provides a commendable study for those interested in theory, practice, and history of this pre-modern knowledge system especially from a perspective of the 'demarcation problem' between science and pseudoscience.

Yao Dazhi, History of Science, Chinese Academy of Sciences, China

Contents

Part I Feng Shui: Educational Responsibilities and Opportunities

1 Introduction... 3
   What Is Feng Shui?.. 5
   Aversion to Testing... 8
   Feng Shui in a Science Programme..................................... 9
   Conclusion... 11

2 The Cultural Contribution of Science Education.................... 13
   Science, Worldviews, and Education................................... 14
   Scientific Attitudes... 16
   A Role for History and Philosophy in Science Education............... 18
   Prevalence of Unscientific Beliefs................................... 19
   The Spectrum of Unwarranted Beliefs.................................. 23
   The Cultural Responsibility of School Science Programmes............. 26
   Religion and Superstition.. 28
   Superstition in Asia... 30
   Academic Neglect of Feng Shui.. 32
   The Scientific Habit of Mind... 34
   Scientific Literacy.. 35
   Francis Bacon and Critical Evidential Support........................ 38
   Conclusion... 39

Part II Feng Shui: Its Theory and Practice

3 Feng Shui and Chi.. 43
   Daoist Origins... 44
   Historical Development... 49
   Cosmology and Science.. 53
   Chi-fusion... 55

xvii

	Chi and Science.	57
	Chi-Based Worldviews	60
	Conclusion	62
4	**Feng Shui Practice**	65
	The Domain of Feng Shui.	65
	Form and Compass Schools	68
	Feng Shui in Hong Kong	72
	Feng Shui in Taiwan	75
	Feng Shui and Western Architecture and Construction	76
	Feng Shui on the Web.	78
	Divination	82
	I Ching or *Book of Changes*	84
	Conclusion	90
5	**Feng Shui and Traditional Chinese Medicine**	91
	Chi-Based Medicine	95
	Chinese Government's Promotion of Traditional Chinese Medicine	103
	Appraising Acupuncture	106
	Conclusion	111

Part III Feng Shui: A Historical-Philosophical Narrative

6	**Matteo Ricci: A Sixteenth-Century Appraisal of Feng Shui**	115
	The Jesuit Mission	116
	Ricci's China Travels and Journal.	120
	Astronomy.	123
	Observations on Feng Shui	126
	Contemporary Appraisal of Matteo Ricci.	130
	Conclusion	131
7	**Ernst Johann Eitel: A Nineteenth-Century Appraisal of Feng Shui.**	133
	Feng Shui and Siting Practice.	134
	Chinese Protoscience and Experiment	137
	Science and Metaphysics	140
	Romanticism in Chinese Science	142
	Chinese Astrology.	145
	Astronomical Problems for Feng Shui	146
	The Five Elements	150
	Eitel's History of Feng Shui	152
	The Spirit World	153
	Eitel on the Educational Task for China	154
	Conclusion	155

8	**Science, Westernization, and Feng Shui in Early Twentieth-Century China**	157
	The Chinese Naturalist Tradition	158
	Nineteenth-Century Reckoning	160
	The Imperial Examinations	163
	Early Twentieth-Century Adjustments	164
	The 'New Thought' and 'May Fourth' Movements	166
	The Philosophy of Life Debate	173
	Conclusion	176
9	**Feng Shui, Science, and Politics in Contemporary China**	177
	Education, the Enlightenment Tradition, and the Modernization of China	177
	Science, Liberalism, and the Modernization of Chinese Politics	179
	Marxism as Official Philosophy	184
	Feng Shui and the Chinese Communist Party	186
	Feng Shui Rehabilitated	189
	Educational Responses	190
	Conclusion	191

Part IV Feng Shui: Considerations from Philosophy of Science

10	**Joseph Needham on Feng Shui and Traditional Chinese Science**	195
	The Needham Question	195
	Chinese Technology	198
	Internal and External Impediments to Modern Science in China	200
	Neo-Confucianism, the Organicism Worldview, and Chinese Science	205
	Appraisal of Needham	211
	Conclusion	213
11	**The Science and Teaching of Energy**	215
	Energy and Metaphysics	215
	History and Conceptual Refinement	216
	Animal Magnetism	220
	Imponderable Fluids	228
	Spiritualist Science	231
	Conservation of Energy	234
	Teaching About Feng Shui in the Science Programme	236
	Energy Literacy	237
	HPS-Informed Teaching	238
	Multicultural Considerations	244
	Conclusion	247

12	**Scientific Testing of Chi (*Qi*) Claims**	249
	Remarkable Qigong Claims	249
	Paucity of Tests	257
	Philosophical Insulation of Feng Shui	258
	Methodological Naturalism	260
	Chi as an Intervening Variable	263
	Chi as Metaphor	266
	Conclusion	267
13	**Feng Shui as Pseudoscience**	269
	Demarcation of Science from Non-science	272
	Rejecting the Demarcation Project	276
	Pseudoscience as a Warranted Category	279
	Ecology of Science and Pseudoscience	282
	Conclusion	288

Part V   Conclusion

14	**Concluding Remarks**	293
References		299
Figure Credits		329
Name Index		331
Subject Index		337

Part I
Feng Shui: Educational Responsibilities and Opportunities

Chapter 1
Introduction

Feng shui (pronounced *fung shway*) has medical, health, architecture, construction, design, and divination components. It is part of a widespread, long-standing, deeply entrenched Asian view of life, nature, and the world. It is an eclectic, yet distinctive, worldview that is at once environmentally focused, naturalistic, dismissive of supernatural entities and interventions, yet wedded to the reality of astrological effects and the quest for accurate divination or fortune telling. The naturalistic and intellectual core of feng shui is commitment to a putative all-encompassing special energy or life force *chi* or *qi* (pronounced similarly to 'chee')[1] that has existed since the beginning of time; indeed it contributed to the beginning that occupies the entire cosmos – the cosmos, solar system, earth, and human body where it moves in defined meridians – and that has a multitude of causal interactions with everything in its environs. Chi belief has been engrained in Chinese culture – Confucian, Daoist, Buddhist, and other variants – for at least 3000 years. The practice of directing and controlling personal chi is called *qigong* (*gong* meaning work/effort). Feng shui belief bears upon most aspects of everyday life: the design of domestic, commercial, and official buildings; the siting and orientation of graves; the utilization of Traditional Chinese Medicine (TCM) practices such as acupuncture, yoga, and qigong; personal fortune telling and divination; and choosing auspicious times for marriage, commencement of building construction, opening a restaurant, launching a public company, and going on holiday.

---

[1] The traditional Chinese character 氣 was translated as *ch'i* in the Wade-Giles authoritative Chinese-English translation system, began by Thomas Wade and completed by Herbert Giles in his *Chinese-English Dictionary* of 1892. This routinely became *chi* in most English publications; the misspelling was normalized. While in Taiwan, along with traditional, not simplified, characters, *ch'i* is retained. When the Hanyu Pinyin system of Chinese-English translation became official in what was then called mainland China in 1958, the simplified character 气 became *qi*. It has many meanings, including 'life force' and 'energy flow'. The Korean and Japanese preference for 'chi' is *ki*, the word which appears in many of their chi-based martial arts and therapies.

In traditional characters, feng shui is 風水; in simplified characters it is 风水.

Feng shui has been, and is, to Asia like Christian ideas and worldview were to the European medieval world. Although lacking both a judgemental deity and a supernatural realm, feng shui likewise impinges on daily personal and social life; it informs people's understanding of their place in the cosmos. One commentator has said:

> Based on ancient Chinese philosophical traditions, feng shui has developed for over two millennia to include knowledge, rituals, aphorisms, and superstitions from throughout China. As such, it is central to any understanding of Chinese cultural history, life, and psychology, as well as that of many other East Asian cultures that also practice Chinese feng shui. (Puro 2002, p. 108)

A study of feng shui in Korea commenced with the statement: 'The importance of geomancy [divination or foretelling the future] in understanding the East Asian cultural landscape and cultural ecology is difficult to overemphasize' (Yoon 2006, p. xiii).

Throughout its history feng shui has been associated with divination practices that foretell the future – *geomancy*. It is a mega-billion-dollar industry with multiple millions of adherents. In the past century, with the Asian diaspora, feng shui has spread from its Chinese and Asian roots to all parts of the globe; additionally, it has been embraced on a significant scale in both Western counterculture and mainstream culture. Its theory and practice are elaborated on hundreds of thousands of websites, visited probably by countless millions every day. Feng shui is now upmarket; it is no longer just a hold-over belief of aged, rural, and traditional people.

Much has been written on the history, philosophy, and practice of feng shui.[2] But despite feng shui's significant historical, cultural, and economic footprint, there has been precious little critical philosophical appraisal of the theory or practice. For example, for every thousand expository works on the feng shui classic, *I Ching* (*Book of Changes,* ca.1000 BC), there are but a handful of critical philosophical commentaries.[3] And there has been no effort to examine how feng shui as a subject matter might be handled as a topic in schools, specifically in science programmes. This is a pity on both counts. Any philosophical examination will lead directly to fundamental issues about the role of metaphysics in science, the function of evidence in theory appraisal, the demarcation of science from non-science, and more specifically the justification for the category of pseudoscience and how it is characterized.

The educational consideration of feng shui will connect to at least three important topics: first, the decades-long debate about the nature of science (NOS) and how this can be best taught (Matthews 2015a, chap. 11);[4] second, the equally long-debated question of multicultural science and whether culturally significant indigenous sciences should be endorsed, evaluated, or criticized in a science programme (Matthews 1994, chap. 8; 2015a, pp. 366–370); and third, it can promote STEAM

---

[2] See, among others, Bruun (2008), Bruun and Kalland (1995), Callicott and Ames (1989), Eitel (1873/1987), Henderson (2010), Parkes (2003), Smith (1991, chap. 4), Yosida (1973), and Zhang (2002).

[3] The book is discussed in Chap. 4.

[4] For a guide to current literature on the impact of Chinese culture on the teaching of NOS in science programmes, see Wan et al. (2018).

(Science-Technology-Engineering-Arts-Mathematics) education, by demonstrating the productive interaction between historical, philosophical, cultural, economic, and scientific studies (Matthews 2015a, pp. 262, 293).[5]

The hope is that this book will to some degree rectify the almost complete neglect of feng shui by both philosophers and educators. It will lay out something of the cultural history of feng shui, it will indicate some of the core philosophical issues presented by feng shui, it will argue that feng shui is a pseudoscience, and it will suggest ways in which feng shui can be productively discussed in science classes. Given the enormity and significance of the subject matter, it barely needs to be said that treatment in book length by a scholar from outside the Asian cultural tradition will be marked by oversights and shortcomings. Nevertheless, although much delayed, a start on critical appraisal and educational attention to feng shui needs to be made. It is in the spirit of a beginning that this book is written.

What Is Feng Shui?

Feng shui is concerned with identifying, manipulating, and utilizing the supposed all-encompassing flow of chi, the putative universal life force (sometimes rendered as 'vital force'), so that people's environments, homes, workplaces, social places, and, indeed, their own bodies can be brought into harmony with it and thus made more natural and healthy.[6] Good and bad chi are connected to locations, building styles, furnishing, and even handbags. According to different authorities, chi can be manipulated by diet, herb infusions, exercise regimes, meditation, and shamans or masters. The once Eastern, now universal, practice of acupuncture is based on intervening in the supposed 600+ internal routes, or meridians, along which chi moves in the body. Stephen Skinner, a proponent of feng shui who has degrees in philosophy and geography from the University of Sydney (this author's alma mater), writes:

> To understand feng-shui it is essential to appreciate ch'i. On a microcosmic level, ch'i is the energy of the body's breath which if concentrated in various parts of the body can enable the practitioner to perform the more amazing feats of the Chinese martial arts schools.
>
> What is true of the microcosm is also true of the macrocosm, and ch'i is naturally accumulated and may be enhanced at certain points in the earth by the application of landscape alterations made in accordance with feng-shui rules. (Skinner 1982, p. 19)

Feng shui informs the choice of burial sites and tomb structures so that the spirits of the departed can reside harmoniously and not be troubled in their afterlife. Contented spirits release their own good chi that lifts the chi of their descendants: a

---

[5] It is a moot point whether STEAM education means anything other than just simple education. For the liberal tradition, and most others, education has always meant the learning of science, technology, mathematics, and humanities.

[6] A recent study of neo-Confucianism notes that other translations of the Chinese character include 'ether', 'material force', 'matter energy', and 'psychophysical stuff' (Angle and Tiwald 2017, p. 230).

win-win situation. Feng shui fits into a cosmology that has been, and still is, a significant feature of Chinese and south-east Asian cultures. Since its beginnings, there has been a strong astrological and divinatory stream within the feng shui tradition. Foretelling the future and identifying auspicious days and times for staging significant events (weddings, business openings, journeying, conference meetings, beginning building construction, and so on) have always been part of the stock-in-trade of feng shui practitioners.

Feng shui has long migrated from Asia and has an increasing international commercial and personal presence. As a writer in the American Institute of Architects newsletter commented: 'Feng Shui is no longer just an ancient Chinese secret. While slow to take root outside of its original heartland, it is now global and transcends culture and politics' (Knoop 2001). The Amazon Kindle site lists over 1000 feng shui books in English alone; there are many times this number on second-hand book sites; there are countless thousands, if not hundreds of thousands, of commercial feng shui websites; and hundreds of thousands, likely millions, of people throughout the world daily visit these sites and to varying degrees regulate or inform their life by what they read.

On one website, King Wen offers his services:

> As a descendant of the Founding Father of Feng Shui – King Wen, is highly sought after by scores of billionaire customers, with 30 years of experience, Master Feng, the world's only Feng Shui Master who has the gift of SEEING Feng Shui energies, will help you get rid of: Bad Luck, Achieve Success, Health, Wealth and Happiness.[7]

He names one of China's richest men, Li Ka-Shing who Forbes lists as worth $34.4 billion, among his satisfied customers, confirming that feng shui consulting is ubiquitous in the ever-expanding class of Chinese billionaires and millionaires.

Not only does King Wen 'see' energy, but he has 'x-ray vision' that detects ulcers, slipped discs, and a whole range of ailments. And he can predict people's futures. He claims his 3-day course (price on application, credit cards accepted) 'eliminates confusion caused from different conflicting Feng Shui ideologies'.

On one sophisticated website 'for the modern professional' – along with feng shui handbag advice – the feng shui qualities of the HSBC Headquarters Building in Hong Kong are given by Jerry King:

> Jerry King received private BaZi (八字) Four Pillars of Destiny and I Ching training from Dr. Lily Chung in San Francisco, a world-renowned expert in Eastern Metaphysics, specializing in Four Pillars and the I Ching. He has also studied Feng Shui and Four Pillars under various masters in Taiwan and Hong Kong. Jerry specializes in Purple Star Astrology (紫微斗數) readings and travels extensively, consulting globally and obtaining research data and verifying theories of cosmic flow in the Four Pillars of Destiny.[8]

Feng shui is an international growth industry; it is a billion-dollar business; it is not merely some leftover set of beliefs shared by uneducated farmers, peasants, or low-level urban workers; and it is embraced by well-paid, educated professionals.

---

[7] See http://www.masterfeng.com/consultations.html

[8] See https://www.fengshuitoday.com/feng-shui-of-the-hsbc-headquarters-building-in-hong-kong/

What Is Feng Shui? 7

Feng shui is but one component of the wide spectrum of chi-based Eastern beliefs, therapies, and practices. Over some 2000–3000 years, under different philosophical and cultural influences, many versions of chi-based practice have evolved. Along with feng shui, others are qigong, falun gong, tai chi, and Jin Shin Jyutsu. They all combine chi (*qi*) with *gong*, meaning work, practice, or exercise. Two philosophers identify chi as the foundation of Chinese cosmology and medicine:

> *Qi* is one of the most important and widely interpreted concepts in Chinese intellectual history. As a shared notion underlying all schools, *qi* is believed to be a dynamic, all-pervasive, and all-transforming force animating everything in the universe. The air one breathes, the force that drives the flow of blood, the food one eats, the strength of one's mind, the flow of one's thoughts, the deepest urges of one's heart—all of these are understood in terms of *qi*. Thus *qi* extends across realms that might otherwise be divided in the spiritual, mental, or physical. (Wang and Ding 2010, p. 42)

Understandably, chi-related belief dominated traditional Chinese understanding of science, medicine, town planning, and much else. And its influence on each of these fields continues. Chi commitments underlie the practice of acupuncture; they inform the design and location of Chinese gardens and much else.

It needs be recognized that in the Confucian and Daoist (Taoist in the earlier Wade-Giles translation) Chinese traditions, philosophy and science were not separate enterprises. It is not that the first influences the second or that the second corrects the first; there is no such dramatic separation. Traditional Chinese science (natural philosophy) was a part of, or interwoven with, religion and philosophy. The standard Western preoccupation of demarcating one from the other, and examining their interaction, is alien in this tradition. When the issue does arise, the answer is even more complex than that given in the West.[9] Ole Bruun writes:

> The concept of *qi*, which may be translated into 'breath' or 'breath of nature', is fundamental to Chinese natural philosophy. It is strongly indicative of an organic predisposition in Chinese thinking in general, as opposed to the mechanistic orientation that became dominant in European natural philosophy after the Middle Ages. (Bruun 2008, p. 108)

The Wikipedia entry for *qigong* (chi kung) provides a beginner's overview of the notion:

> Qigong is traditionally viewed as a practice to cultivate and balance *qi* (chi), translated as 'life-energy' … qigong allows access to higher realms of awareness, awakens one's 'true nature', and helps develop human potential. … Qigong is now practiced throughout China and worldwide.

Simon Brown, author of *The Feng Shui Bible*, gives an account of chi similar to what can be found in thousands of popular books on the subject:

> Chi is the subtle charge of electromagnetic energy that runs through everything, carrying information from one thing to another. The chi flowing through your body predominantly carries your thoughts, beliefs and emotions. At the same time some of your chi is floating off, while you are also drawing in new energy. … Your energy field connects you to everything else, whether you like it or not. The secret to making this energy work is understanding the process and finding out how you can make it help you in life. (Brown 2005, p. 24)

---

[9] On this matter see Chap. 12.

Aversion to Testing

The foregoing are remarkably confident and scientific-sounding assertions; they are made without hesitation and so confident that there is no need for references or citations; and they appear as if coming from a book of established science. But they should be tentative; the claims should be whispered not shouted. The putative universal life force has never featured in the history of modern science and has never been identified in any science laboratory. Despite the chi beliefs and practices having been around for at least 3000 years, there is no instrument that objectively indicates chi's presence and its flow, or measures its intensity. There is no standard chi unit comparable to the standard units for every quantity measured in science: there is nothing equivalent to a standard metre, kilogram, ohm, calorie, watt, or joule. There is no 'chi-counter' equivalent of the Geiger counter; indeed, there is simply no chi meter at all. In the whole constellation of chi talk, theory, and practice, there are no standards. None of the proponents of the above remarkable chi claims have ever subjected the claims to scientific test, much less have had their research published in mainstream science journals.[10]

This 'no-testing' situation is altogether odd as the foundational energy-centric commitment of feng shui should bring it into the orbit of science laboratories. The very first reaction to the above claim – *Chi is the subtle charge of electromagnetic energy that runs through everything* – would be to ask: When? Where? How much? In a world desperate to find alternate energy sources and less expensive healthcare, one might expect everyone would be chasing it down. But this has not happened. This 'no-testing' commitment, or just practice, is a powerful indicator that chi talk, although sounding scientific, is pseudoscientific.

A 'no-testing' regime or mindset has unfortunate consequences: if citizens are raised to simply accept, without evidence and testing, the foundational claims about chi and its yin-yang governed manifestations, then what other such claims will they accept? Crushed deer antler is a staple in Traditional Chinese Medicine (TCM) because it supposedly contains concentrated chi or somehow beneficially impacts chi flow in a patient. The immediate downside of this untested belief is the money required for the medicine that could have been spent on scientifically proven medicine or on other personal and social necessities. This might be regarded as a small price to pay for the maintenance of a cultural tradition. But for almost 2000 years, crushed rhinoceros horn has also been a stable of TCM; supposedly it has even more distilled chi and performs even more herculean interventions on internal chi. Here the dollar price and the species price paid for the maintenance of a cultural tradition is not small.[11]

---

[10] Some putative scientific testing by those outside the tradition is documented in Chap. 12.

[11] In the early 1990s, rhinoceros horn was taken off the official list of TCM pharmacopoeia products. This was a political-ethical decision, not a scientific one based upon testing of chi claims; antler horn is still on the list. Nevertheless, rhino horns currently sell for around USD300,000 each, and 800–1000 are slaughtered each year to serve the TCM market.

If people are routinely accustomed to accept claims without evidence, then the pool of such 'no-evidence' claims is open to further expansion and can take in all manner of objectionable, dangerous, and abhorrent beliefs. The demand for evidence can be problematic when the claims are about supernatural entities, powers, and activities. It might, with some initial plausibility, be argued that believing supernatural claims is independent of or removed from empirical, 'natural' evidence: to ask for the latter is to misunderstand what the claim is about; it is about *super*natural events and processes. So, to ask for empirical evidence for the belief that God's abhorrence of homosexuality caused the devastating 2004 Indonesian tsunami is to misunderstand the belief.[12]

Feng Shui in a Science Programme

But feng shui is not a supernatural belief. It is about the constitution of, and processes in, the world; its core is explicitly energy-centric, and it claims to be a scientific practice. So, it should be empirically testable; and as such should also be examined in science classes as there are quite general philosophical, cultural, and educational lessons to be learnt from an informed and critical testing of its theory and practice. Such examination offers rich opportunities for science teachers to elaborate on important features of science or 'nature of science' (NOS) as these are often labelled (Matthews 2011), features such as the relationship of science to metaphysics, scientific method: the connection of science to worldviews, the demarcation of science from pseudoscience, and the central scientific concept of energy. This could be an important part of the contribution of science education to cultural health;[13] yet feng shui does not appear in any school programme.

The same education argument favours the examination in science classes, of any competing or alternative system of beliefs about the world and its mechanisms. Such examination can promote better understanding of the methods, methodology, and domain of science.[14] There are practical, pragmatic, and pedagogical considerations about how and when to do this. Minimally, any such examination should fit in with the science curriculum. So, for example, when evolution is studied, then contrast Darwinism with creationist and Lamarckian science; when Mendelian genetics is studied, then contrast it with Lysenkoism; when the Copernican system is studied, then contrast it with the Ptolemaic system; and when the geophysics of earthquakes or volcanoes is studied, contrast it with pre-scientific theological accounts of both phenomena.

---

[12] A claim commonly made at the time by conservative Muslim Imams and even ruling bodies. If you believe that God micromanages the world, then what ever happens can be traced back to some attribute or decision of God. And there is no gainsaying the putative attribute.

[13] The argument is developed in Chap. 2.

[14] Often 'methods' and 'methodology' are collapsed. There is value in using the former to pick out the ways and means whereby different sciences generate hypotheses and gather data and the latter to pick out how different sciences or all sciences deal with data once it is confirmed or how data relates to theory confirmation (Nola and Sankey 2000).

In all of these cases, there is an educationally valuable move from current best science, to earlier but now rejected science, to pseudoscience. This progression mirrors the via *negativa* of Christian and other theologies where one learns about God by learning what he or she is not; things can be learnt about science by seeing what it is not. Science is not private; not subjective; not confined by race, ideology, gender, or class; and not dependent on revelation or sacred text; its claims are not judged against feelings, intuitions, hunches, or personal observation;[15] and it cannot grow in a closed society where communication is controlled by party or business. Some or all of these negations of science are manifested clearly in pseudoscientific practice; examining the latter sheds light on the former. With energy being a constant and universal component of all science programmes, there is ample opportunity to study and discuss the startling chi energy claims of feng shui.[16]

The argument of this book is that feng shui can legitimately and with educational benefit be appraised in science classes. Although it is a pseudoscience, it need not be 'outside' of the science curriculum; its classroom examination can illuminate important features of the nature of science and the role of science in society and culture. The careful and sympathetic teaching of false theories and pseudoscientific theories not only illuminates the methodology of science, such teaching can prepare students for their inevitable daily encounters with such theories (Bhakthavatsalam 2019; Martin 1971, 1994). Further, feng shui can be a rich topic for coordinated, cross-disciplinary teaching between different school faculties: science, history, social studies, philosophy, art, economics, and religion. The cross-disciplinary appraisal of feng shui is an excellent case study for much-advocated STEM (Science-Technology-Engineering-Mathematics) education and its Arts-augmented STEAM education (Kim 2014; Rosicka 2016).

This general argument is at its strongest when feng shui is shown to be a pseudoscience, but it also applies in the weaker case when feng shui is held to be merely a poor or mistaken science. Students can learn the difference between a poor and a good science and be curious about why so many people in so many countries over so many centuries can continue to believe in a poor or mistaken science. This opens a door into anthropological and social-psychological investigations which benefit students' historical and social science studies.

The problems with feng shui that are identified in this book are systematic and essential; they are not just matters that can be dismissed as modern, commercial, web-enabled corruptions of a sensible philosophic tradition. This is the same argument used in saying that Christianity should not be judged by the sordid self-serving behaviour of US and other televangelists or of by the shockingly extensive degree of paedophilia in different parts of the Roman Catholic Church; that Marxism should not be judged by the criminal actions Stalin and of the Communist Party

---

[15] For the debates and literature on the role of observation in science, see Matthews (2015a, pp. 250–256).
[16] This is elaborated in Chap. 12.

wherever it came to power in the twentieth century; and that Islam should not be judged by the brutal and discriminatory actions of many modern Islamic states where blasphemy and aposty are capital crimes and the oppression and exploitation of women are simply part of the social fabric. Whatever the arguments can be about not judging the foundations of Christianity, Marxism, and Islam by their contemporary manifestations (or corruptions), the argument of this book is that the foundations of feng shui are all mistaken, they lack scientific and competent philosophic warrant; there are foundational features of the belief that lead to its aversion to testing and to its proness to fraud and corruption.

Surprisingly feng shui is ignored in all philosophical discussion of pseudoscience; nor is it ever used as an example in the demarcation of science from nonscience. In both cases it can be an informative case study. Mario Bunge, the philosopher/physicist, well expressed the motivation for this book:

> Given the intrinsic interest and the cultural importance of pseudoscience and anti-science, it is surprising that they should receive so little attention on the part of philosophers, particularly in our times of crisis of public confidence in science. (Bunge 2001, p. 189)

His chastisement of philosophers applies equally to educators. But the responsibility is greater for the latter, as they are charged with the task of preparing students to take their part as informed citizens in a global society burdened with serious environmental, economic, and social problems and to play a part in their own societies that are awash with particular pseudosciences. It is irresponsible for educators not to suitably prepare students for life in such a world.

This established argument has recently been well re-stated by Sindu Bhakthavatsalam:

> The point here is rather simple. If we want our students not just to be scientifically informed but also responsible citizens with an active interest in policy and advocacy, we need to educate them about opposition to scientific ideas in as much detail as possible. […] The more in detail they learn about these views, the better equipped they are to challenge them. It could be greatly beneficial to educate students on how these opposing ideas come about, arguments and evidence given for them, and why they are problematic. (Bhakthavatsalam 2019, p. 19)

Conclusion

The claims of this book are not novel. Just over 100 years ago, in 1915 following the collapse of the Manchu dynasty and establishment of the new Republic of China, in the bubbling economic, cultural, philosophical, and warring times of the period, a hugely popular and influential periodical, *New Youth* (*La Jeunesse Nouvelle*), was launched. It was edited by Ch'en Duxiu [*Ch'en* Tu-hsiu] (1879–1942), a noted champion of science, critic of traditional Confucian philosophy and culture, and founder of the Chinese Communist Party (CCP). Its opening editorial opined:

> Our men of learning do not understand science; thus they make use of *yin-yang* signs and beliefs in the five elements to confuse the world and delude the people and engage in speculations on geomancy … The height of their wondrous illusions is the theory of *Ch'I* [primal force] …We will never comprehend this *Ch'I* even if we were to search everywhere in the universe. All of these fanciful notions and irrational beliefs can be corrected at their roots by science, because to explain truth by science we must prove everything with fact. Although this is slower than imagination and arbitrary judgment, every progressive step is taken on firm ground. It is different from those flights of fancy which in the end cannot advance one bit. (*New Youth* 1915, vol. 1, p. 1. In Kwok 1965, p. 65)

Ch'en's lapidary opinion ranges over many of the issues elaborated in this book: the place of science in popular culture and belief; the separation of science from pseudoscience; the role of science in the making of government policy; proper identification of scientific method; the scientific testing of claims about the operation of chi; the possible separation of defendable core metaphysical claims about chi from its historical and cultural corruption; and the separation of rational from irrational belief and the responsibility of schools for promotion of the former and minimizing the latter. Ch'en spoke for, and to, the hundreds of thousands of citizens, students, teachers, and scholars who were addressing the pressing problems of modernization of China; appraisal of the Chinese, specifically Confucian, cultural tradition; and the organization of schooling and curricula so as to advance the modernization project. This task and associated philosophical and cultural appraisals continue. Subsequent chapters of this book indicate some of the steps that have been take and point to some that remain to be taken.

Chapter 2
The Cultural Contribution of Science Education

Countries put effort and resources into education for a variety of reasons: to form citizens that have an appropriate knowledge of how the natural and social worlds work; to give citizens the competence and tools for utilizing such knowledge in more immediate, domestic, and personal matters; to promote a well-functioning economy; and to contribute to the better and more flourishing personal and cultural lives of citizens. All school subjects – history, mathematics, social studies, literature, art, music, drama, physical education, religion, and philosophy – make their own specific contribution to this overall programme of social and personal betterment; they contribute to a flourishing life for all citizens, a life where numeracy, literature, art, music, and systems of ideas are all better known and appreciated and consequently life can be better lived.

Science education has distinctive contributions to make to the overall educational goal. The obvious areas are citizens having a better knowledge and appreciation of the natural world and its processes, an appreciation of anthropogenic climate change and its causal mechanisms, employment-related knowledge as fostered in STEM programmes, the rudimentary knowledge of health and basic medicine, and critical and logical thinking promoted by mastery of scientific method. Further, science education can contribute to the health of societies by criticizing various unhealthy enthusiasms and fads that take off in societies, such as anti-vaccination scares, mistaken racial and gender biases, and follies promoted by commercial interests, such as vitamin C cure-all campaigns.

Beyond such immediate pragmatic contributions, worldviews can be appropriately clarified and even appraised in science classes. Where there are inconsistencies between worldviews and the world as revealed by science, then options for the resolution of the inconsistency can be laid out, and students can be encouraged to explore resolutions and their consequences. To what degree does the worldview adjust? To what degree does science adjust? Is it viable and sensible to embrace a

© Springer Nature Switzerland AG 2019
M. R. Matthews, *Feng Shui: Teaching About Science and Pseudoscience*,
Science: Philosophy, History and Education,
https://doi.org/10.1007/978-3-030-18822-1_2

conflict between science and worldview? Student awareness of and engagement in these questions is a valuable contribution of science education to cultural health.[1]

On the opening page of a recent book on the history of science education in England, Edgar Jenkins writes:

> … even when the importance of science to society is readily conceded, the contribution that science can and should make to the education of pupils is rather more open to debate. The underlying question can be stated baldly. What is the educational function of school science, what is it *for*? (Jenkins 2019, p. 1)

The entirety of this book deals with Jenkins's bald question, a question that so obviously depends upon positions adopted in philosophy of education, a discipline for which, regrettably, teachers are ill-prepared (Matthews 2019b).

Science, Worldviews, and Education

Science literacy is useful in identifying worldviews that are inconsistent with science and especially identifying those purporting to be scientific but which are not. Jon Miller, who for decades has researched science literacy and public understanding of science, well stated this point:

> In addition to understanding basic scientific constructs, it is important for citizens to recognize pseudoscientific constructs that seek to be recognized as scientific. Astrology is a good example. (Miller 2004, p. 278)

Feng shui is an equally good example.

Not all worldviews are scientific, nor should they be. But if they explicitly make claims about the constitution and processes of the natural world including causal influences on people's physical and mental health that are in opposition to established scientific knowledge of that constitution and processes, then this needs to be recognized and adjustments need to be made to the worldview or to science. Adjusting science to accommodate a conflicting worldview is an option oft taken, but the history of such adjustments is not encouraging.

For many people, religions, and cultures, where there is inconsistency between worldviews and specific empirical claims of science, the science is changed, or science is re-interpreted in an instrumental non-realist manner so that contradictions cannot occur. This often amounts to embracing science-enabled technology without the science. This was the option taken in the sixteenth-century debates over the Copernican heliocentric[2] versus biblical geocentric solar system, where Cardinal Bellarmine famously said that the Copernican model could be used for astronomical

---

[1] The complex matter of education and worldviews is discussed in Matthews (2009b, 2015a, chap. 10) and in contributions to Matthews (2009a).

[2] The Copernican system is routinely called 'heliocentric', but Copernicus put the sun near the centre, not at the centre. More correctly it should be labelled 'heliostatic': that the sun was immobile was the important difference.

calculations though it was not true. Changing science was the option taken by many Christians and Muslims in the nineteenth-century debates over Darwinian versus creationist accounts of the origin of species. That option is still taken by many. It is taken in most traditional societies where the Darwinian account is in stark conflict to deep-seated cultural beliefs about the special creation of particular species, especially of course *homo sapiens*. Changing science was the option taken in twentieth-century debates in the former Soviet Union about Mendelian versus Lysenkoist accounts of the inheritance of plant characteristics. Changing science is the formal position of proponents of Islamic Science for whom when there is a conflict with Koranic teaching, science must adjust.

Claude Tresmontant (1925–1997), a French Catholic philosopher, states a comparable, religion-first or primacy-of-faith, view about metaphysics:

> The thesis which I submit to the critical examination of the reader is that there is one Christian philosophy and one only. I maintain, in other words, that Christianity calls for a metaphysical structure which is not any structure, that Christianity is an original metaphysics. I maintain that Christian theology and Christian dogma contain in themselves a metaphysical substructure, a body of very precise and very well-defined theses which are properly metaphysical. (Tresmontant 1965, pp. 19–20)

Where Catholic metaphysics conflicts with that given in science, the latter has to change. He recognizes, as against most fundamentalists, that this Christian metaphysics is not immediately given in scripture but emerges with historical development. Tresmontant proceeds to the claim that:

> But as Christian thought progresses, it rules out certain doctrines and certain theses as incompatible with its internal exigencies. (Tresmontant 1965, p. 33)

It is easy to appreciate that if one's faith is committed to the salvation of souls, then souls must be part of the furniture of the world; one's metaphysical ontology has to provide for them. Likewise, if one's faith involves the notion of sin and divine retribution for committing sin, then one's metaphysics has to allow for free will; and this puts limits on acceptable theories of mind.[3] If God's creation of the world is part of one's faith, then there are immediate implications for cosmogony; certain cosmogenic speculations are just ruled out. Thomas Aquinas thought there was no philosophical reason for the world to have a temporal beginning, but Christian theology dictated that it must have. If one's sacred scripture or cultural belief requires a 'young earth', then certain epistemological views about evidence and theory need to be adopted. And so on.[4]

Exactly parallel arguments are mounted by most Muslim philosophers as certain philosophical positions are 'articles of faith' in Islam (Dallal 2010; Edis 2007); and by Communist Party theoreticians in the former Soviet Union (L. R. Graham 1973)

---

[3] Some Protestant theologians have argued that free will is illusory, as sinning or not sinning, and thus salvation, is predestined and known by God. So not only does theology not require free will; it prohibits it (Alston 1985).

[4] For Roman Catholic accounts of this philosophy/theology interaction, see contributions to Baum (1967).

and in contemporary China where dialectical materialism, based upon Engels' *Dialectics of Nature*, has been ordained as the supreme arbiter in metaphysical disputes (Kelly 1985).

Such arguments are countered by 'pure' metaphysicians who say that accounts of ontology, cosmology, and anthropology are beholden to no one or no institution and that metaphysics has its own internal corrective procedures. The arguments are also countered by 'science-based' metaphysicians who say that it is science, and only science, that gives reliable guidance to ontology (Is the world all matter? Is there free-floating energy? Is there a life force?), cosmology (Did the world have a beginning and if so, how did it happen?), and anthropology (Do humans have souls? Is there free will? Was there a special creation for *homo sapiens*?). The dissident astrophysicist, Fang Lizhi, is an example of the latter group of metaphysicians:

> Most answers given by Marxism with regard to the natural sciences are obsolete, some are even downright wrong. What Marxism has to say about the natural sciences stems from Engels's book *The Dialectics of Nature*. On nearly every page of this book one can detect something either outdated or completely wrong. (Fang 1987/1992, p. 208)

Science is intimately and inescapably entangled with metaphysics; neither can be advanced without the other;[5] and students completing both school and university science programmes should have some inkling and appreciation of the interconnections. Students also need to learn something of the method and methodology of science and to learn how relevant information is gathered and hypotheses are generated (method) and how competing hypotheses, theories, and claims are adjudicated (methodology).

Scientific Attitudes

In addition to these widely accepted content and method goals, it is expected that students completing science programmes will have some appreciation of the features of science or 'NOS' in shorthand terminology (Hodson 2014; Lederman et al. 2014; Matthews 2011). Beyond these three standard components of science literacy, there is a more contentious expectation: namely, students should develop and exercise a scientific 'habit of mind'. In the words of the American Association of the Advancement of Science:

> Taken together, these [scientific] values, attitudes, and skills can be thought of as habits of mind because they all relate directly to a person's outlook on knowledge and learning and ways of thinking and acting. (AAAS 1989, p. 133)

This was later elaborated on in an editorial in the US National Science Teachers Association magazine:

---

[5] Some good accounts of the interrelationship of philosophy and metaphysics are Buchdahl (1969), Bunge (2006), Burtt (1932), Butterfield and Pagonis (1999), d'Espagnat (2006), Gjertsen (1989), Trusted (1991), and Weinert (2005).

> Science process skills and content knowledge are not enough to produce the scientists and scientifically literate citizens we need in the 21st century. Shared values and dispositions within the science community such as curiosity, honesty, openness, and skepticism must also be nurtured, modeled, and practiced continuously in science classrooms at all levels until they become deeply entrenched and respected "habits of mind". (Liftig 2009, p. 1)

In India this expectation was called the 'scientific temper'.[6] Pandit Jawaharlal Nehru (1889–1964) had its promotion enshrined in the new nation's first constitution as a legal responsibility of the State (Bhargava and Chakrabarti 1995, 2010). Nehru characterized such temper or outlook as:

> The adventurous and yet critical temper of science, the search for truth and new knowledge, the refusal to accept anything without testing and trial, the capacity to change previous conclusions in the face of new evidence, the reliance on observed fact and not on preconceived theory, the hard discipline of the mind. (Nehru 1946/1981, p. 36)

He consciously linked this expectation of science education to the European Enlightenment tradition that he embraced during his Cambridge years (1907–10).

Such expectations, as expressed institutionally by the AAAS and personally by Nehru and others, have a long heritage. Since the eighteenth-century Enlightenment, science and science-informed technology have been seen as fundamental for the material and cultural improvement of society or for 'progress' as it used be called.[7] This conviction energized Joseph Priestley and the 'applied' natural philosophers of the UK (Uglow 2002), the *philosophes* and encyclopaedists of Europe (Hankins 1985, chap. 6; Outram 2005, chap. 7), and the political founders of the fledgling USA – Franklin, Jefferson, and others (Cohen 1995; Commager 1977). It was widely held by the Enlightenment thinkers that with the expansion of education, citizens would reject unhealthy, dangerous, bigoted, prejudiced, or superstitious cultural beliefs and practices; these were, and should be, seen as detrimental to social and personal welfare and so to the goals of education.

John Dewey stressed this Enlightenment conviction in all his educational and social policy writing. For example:

> In short, the scientific attitude as here conceived is a quality that is manifested in any walk of life. What, then, is it? On its negative side, it is freedom from control by routine, prejudice, dogma, unexamined tradition, sheer self-interest. Positively, it is the will to inquire, to examine, to discriminate, to draw conclusions only on the basis of evidence after taking pains to gather all available evidence. It is the intention to reach beliefs, and to test those that are entertained, on the basis of observed fact, recognizing also that facts are without meaning save as they point to ideas. (Dewey 1938, p. 31)

In science education, the expectation has also long been held that 'scientific attitudes', the 'spirit of science', or 'habits of mind' should be promoted as much as knowledge of the content and method of science. And not just as an optional extra

---

[6] On scientific temper, see at least Bhargava and Chakrabarti (1995, 2010) and Nanda (2003, chap. 8).
[7] See at least Baillie (1951), Bury (1920), and Passmore (1970).

but as a very part of the purpose of science education.[8] In the USA this aim has been a component of curriculum documents since the 1930s. It was explicitly stated in the 1966 Educational Policies Commission document, *Education and the Spirit of Science* (EPC 1966):

> The spread of science promotes respect for the role of reason in human affairs by demonstrating the power of the mind when used in accordance with the spirit of science. There is a tendency to be suspicious of absolutes, a respect for tentativeness, a kind of working skepticism. Science poses a clear challenge to pretensions of absolute certainty. It promotes respect for intellectual flexibility and creativity, for the ability to revise or discard old hypotheses and to form and substantiate new ones. (EPC, p. 5)

The conviction has been that a good science education would make citizens more thoughtful and concerned about evidence and hence less susceptible to harmful and discredited practices – for example, objecting to child vaccination, rejecting blood transfusion, denying the carcinogenic effects of smoking, racial discrimination based on supposed genetic differences, gender discrimination based on supposed brain differences, advocacy of unhealthy foods, and so on. The expectation is that developing a scientific habit of mind, or scientific attitudes, as distinct from just having scientific knowledge or a store of scientific information, would lead to holding informed and considered views on the foregoing, and other, socio-scientific and cultural issues. Being able to identify good science, bad science, and pseudoscience is integral to developing a scientific habit of mind.

Lee McIntyre, in a focused and detailed work, has argued that a scientific attitude entails the seeking of, and respect for, empirical evidence and a willingness to change belief and opinion in the light of that evidence. Further, it is the presence, individually and communally, of the scientific attitude that distinguishes science from pseudoscience; and it does this in a way that putative methodological criteria have been unable to do (McIntyre 2019).

A Role for History and Philosophy in Science Education

The history and philosophy of science (HPS) can illuminate pedagogical, curricular, and theoretical issues in science education (Matthews 2014a, 2015a). Consider, for instance, the above quotation from John Dewey. His claim is important and supportable, but clearly it needs amplification and clarification:

- What are facts?
- Are facts neutral as between proponents of different theories and ideologies?
- What amount of evidence is required to overturn established opinion or theory?
- Is empirical evidence the only relevant evidence in theory appraisal?

---

[8] Colin Gauld in a series of articles has well documented this tradition and has pointed to the conflict between habits of mind as itemized in education documents and the actual practice of scientists. See Gauld (1982, 2005).

- Can congruence with established theory be part of the evidential base for a claim?
- What happens when empirical facts clash with entrenched metaphysics?
- When is it reasonable to retain a theory when facts contradict it?

All of these questions require a degree of HPS knowledge for their answer; and this should be part and parcel of the preparation of science teachers (Matthews 2015a, chap. 12). With such HPS-informed teachers, students might gradually become more sophisticated in their appreciation of science-related issues, including the appraisal of feng shui.

This book seeks to demonstrate the contribution of HPS to a theoretical issue in science education, namely, the responsibility that science education has for the cultural 'health' of a society. It will support the Enlightenment contention that science education needs to promote a scientific habit of mind, or scientific temper, and that this is the best way for it to contribute to cultural well-being. The book advances this argument by a critical appraisal of just one of the many chi-based worldviews and associated practices, namely, feng shui. Feng shui can reasonably be termed an ideology because of its systematic integration of belief, theory, putative science, worldview, and personal practice.

Prevalence of Unscientific Beliefs

The modern world is high-tech, science-dependent, and globalized. This is commonplace observation and is evident daily to the two-plus billion users of Facebook, WeChat, and the web more generally; tech dependence is obvious to all users of TV, modern appliances, medicine, transport, and occupiers of homes and offices. Yet there is persistent evidence that unscientific, antiscientific, and pseudoscientific beliefs and worldviews are as common as they ever were. Notwithstanding the global reach of high-tech communication and mass education, including science education, the hopes of the eighteenth-century Enlightenment thinkers for a flowering of evidence-based, rational, science-informed thinking, debate, and social policy – the 'Age of Reason' – as many anticipated (Paine 1797/2004) are taking longer to come to fruition than they anticipated. Many doubt whether it ever will; many say it never should.

Indeed, in many contemporary quarters, there is a flight *from* science, rather than the flight *to* science that the Enlightenment anticipated (Gross et al. 1996; Mooney and Kirshenbaum 2009; Otto 2016). This flight has been called Denialism (Specter 2009). The situation is a little complicated because often enough the flight *from* is presented not as from science but to alternative sciences, ethnic sciences, traditional science, or multicultural science. Some universities allow completion of an 'ethnoscience' unit to satisfy the science requirement of their Humanities or Education degrees. Their faculty councils clearly answered affirmatively to the question of whether ethnoscience was science.

The rising popularity of non- or antiscientific beliefs presents science teachers with both responsibilities and opportunities. Science teachers are reluctant to bother themselves and students with scrutiny of esoteric and plain silly beliefs about the world. Learning just the content and methods of orthodox science is time-consuming enough, and there is such a multitude of irrational belief systems many of which have millions of adherents, that selecting any one for classroom critical examination might be regarded as prejudicial.

One ground put forward for disregarding common antiscientific beliefs is Thomas Jefferson's liberal assertion of an individual's right to an opinion, even a discredited one. In Query 17 of his 1787 *Notes on the State of Virginia*, he wrote:

> The legitimate powers of government extend to such acts only as are injurious to others. But it does me no injury for my neighbour to say there are twenty gods, or no god. It neither picks my pocket nor breaks my leg. (Jefferson 1787/1982, Query 17)

This can be advanced to evade any responsibility for examining feng shui or other such beliefs. But it will be argued here that feng shui does 'pick pockets' and it does social damage by acclimatizing citizens to believe in things for which there is no evidence. It puts citizens on a slippery slope toward dangerous commitments.

Michael Shermer lists among the popular but unfounded and scientifically rejected ideas: 'dowsing, psychokinesis, astrology, ghosts, faith healing, antigravity locations, Noah's flood, graphology, clairvoyance, mediums, and Big Foot' (Shermer 1997, p. 27).[9] Shermer's list can be augmented by hosts of other such beliefs and belief systems that are inconsistent with or disproved by science: demon possession, clairvoyance, spirit communication, and so on. These beliefs are not just non-science in the way that music, art, literature, and theology are non-science; the listed beliefs are antithetical to science or disproved by science. Different studies have shown that a quarter of Americans believe that vaccines cause autism, two-thirds of Americans do not believe that carbon dioxide emissions are the main cause of Earth's warming, one-third do not believe that humans evolved from earlier *homo*-like species, and two-thirds believe in telepathy and ghosts.

In 2002 the National Science Foundation in its report on 'Science and Engineering Indicators' showed that 40% of US citizens believe in devils,[10] ghosts, and spiritual healing (NSF 2002). A 2005 US Gallup poll of 1002 citizens (with a sampling error of 3%) showed that 40% of US citizens believed in ESP, 32% believed in ghosts, and 30% believed in telepathy, while 21% believed in witches. In that poll, 75% of respondents believed in at least one of the ten paranormal/pseudoscientific beliefs being presented (Musella 2005).

---

[9] On these topics, see at least Campion (2016), Gallup and Newport (1991), Goode (2011), Radner and Radner (1982), Rice (2003), Specter (2009), Walker et al. (2002), and Zimmerman (1995, chap. 2).

[10] The case of devils is special, as belief in their existence and powers is an integral, and hence required, part of at least the Judaeo-Christian-Islamic religious traditions. Whatever the empirical evidence might or might not be for their existence, believers can say such belief is warranted by the wider umbrellas of Scripture and/or Tradition. On this see Matthews (2015a, pp. 361–363).

A 2008 National Science Foundation (NSF) study found that 34% of the sample thought that astrology was either 'very' scientific or 'sort of' scientific. In the same year, 31% of respondents in a General Social Survey said that their deceased ancestors had 'supernatural powers' and excised them in the world (Losh and Nzekwe 2011, p. 474). A study at the University of Texas revealed 60% of students believed that communication with the dead was possible, while 80% thought that psychics might be able to predict the future (Zimmerman 1995, p. 15).

Over the past decades, this high proportion of US citizens believing in paranormal and pseudoscientific beliefs has remained constant (Alcock 2018; Orenstein 2002; NSF 2002; Stark 2012; Walker et al. 2002; Wilson 2018). All these beliefs can be evaluated and shown to be scientifically unwarranted.

There is wide agreement that at least one in three American adults reject evolution; believing instead that humans were created some several thousand years ago in our present form (Specter 2009, p. 17). Astonishingly a 2008 study found that 13% of US biology teachers reject evolution in favour of creationism (Berkman et al. 2008); an earlier study of Ohio science teachers found 40% favoured teaching creationism in biology as a genuine alternative to Darwinian theory (Zimmerman 1995, p. 64). Surveys in the UK have reported 63% of people believing in the paranormal and 67% in astrology (Preece and Baxter 2000, p. 1147). In one survey of 2000 students, 31% of PGCE would-be science teachers believed in ghosts, and 25% believed in the efficacy of crystals to keep people healthy; the gullibility figures for sixth form students (17–18 years) were, predictably, higher for each category – 60% and 33%, respectively (Preece and Baxter 2000, pp. 1150–1153). These survey results led the researchers to say: 'we live in the most superstitious age ever' (Preece and Baxter 2000, p. 1147). The simple challenge for educators is: What to do about widespread gullibility and the commonplace basing of life on superstition? Looking the other way, closing eyes, or whistling is hardly a responsible or professional option.

In the early 1970s, it is estimated that there were 200,000 practising astrologers in the USA and reportedly there were 5 million Americans spending 200 million dollars annually on astrological consultations (Kelly 1982, p. 47). More recently, Gallup polls indicate that 25% of adults in the USA, the UK, and France give credence to astrology and variously indulge it (Campion 2016). In 2003, the science correspondent of the UK *The Sunday Telegraph* wrote:

> Astrology has grown to be a huge worldwide business … It seems that no sector of society is immune to its attraction. A recent survey found that a third of science students subscribed to some aspects of astrology, while some supposedly hard-headed businessmen now support a thriving market in 'financial astrology'…Astrology supplements have been known to increase newspaper circulation figures and papers are prepared to pay huge sums to the most popular stargazers who can earn £600,000 or more a year. (Campion 2016, p. 4)

One among the hundreds of thousands of astro-medicine books is *Astrology: Key to Holistic Health* (Starck 1982) where the author reassuringly says that: 'In fact, none of us ever need manifest any physiological symptoms of disease if we maintain that constant flow of energy between our spiritual, emotional, and physical selves'

(Starck 1982, p. 3). Optimistically she asserts that utilizing the horoscope is the key to maintaining the required energy balance:

> The astrological horoscope serves as a road map though which each of us individually may attune to the cosmic energies and be guided through our transformations with as much insight as possible. (Starck 1982, p. 6)

In support of this claim, she says:

> One homeopathic physician sees a timetable, given him by a Hindu astrologer, which allocates the systems in the body to particular hours of certain days. He recommends that the remedies be taken at these times. In this manner the vibrational qualities of the remedies are enhanced. (Starck 1982, p. 7)

Her readers are reassured that cardiovascular disease and heart attack can both be averted with suitable astro-counselling:

> Astrologically the heart is ruled by Leo and the Sun. The horoscopes of those with cardiovascular disorders exhibit a strong fixed sign emphasis. … With a blockage of the arteries or arteriosclerosis, Jupiter is often found in Leo or Aquarius, or fifth and eleventh houses with hard aspects from Saturn. (Starck 1982, pp. 100–101)

There are the significant enough problems with 'private' belief and attachments to such manifestly ill-founded (to put it gently) claims; these are magnified when it comes to state and medical insurance support for such discredited and dangerous practices. Everyone wants state-supported medicine and private health insurance: this brings into immediate focus the question of whether such practices as astro-medicine are scientific and deserving of state support or pseudoscientific and undeserving of state funds. Having made that decision, it would be appropriate to move on to considering state support for acupuncture, something experienced by three million US citizens each year: or for homeopathy, on which it is estimated US citizens spend 3 billion dollars per year (McIntyre 2019, p. 174).

In 2017, the Italian National Consumer Association reported that there are about 155,000 astrology/occult practitioners in the country and around 13 million Italians – about a quarter of the adult population – regularly visit astrologers, fortune-tellers, and tarot card readers.[11] The sector is worth an estimated 8 billion pounds a year. Valter Cascioli, a psychologist and supposedly scientific consultant to the International Association of Exorcists, said that:

> The number of people who take part in occult and satanic practices which lead to serious physical, psychological and spiritual damages, is constantly rising. … The lack of exorcists is a real emergency as a result of a significant increase in the number of diabolical possessions that exorcist priests are confronting'. (Squires 2017)

The foregoing population proportions are repeated for belief in special creation, ghosts, communication with the deceased, and numerous other such discredited entities and systems (Hines 2003). It is estimated that three or four billion people worldwide believe in parapsychology in one form or another: that is, the ability to

---

[11] See https://codacons.it/italians-turn-to-fortune-tellers-and-occult-economic-slump/

'read minds', to 'see' through solid objects, to 'know' of distant events happening, to 'mentally' converse with isolated others, to 'levitate', and so on.

A 2004 study of 175 college students by philosopher Matthew Johnson and biologist Massimo Pigliucci fairly represents a tranch of such studies on the impact of enrolment in science and non-science majors on pseudoscientific and magical beliefs (Johnson and Pigliucci 2004). They found little, if any, difference between the biology and philosophy groups but did find a staggeringly high degree of acceptance of pseudoscientific practices and unwarranted beliefs among both groups of US college students. Just on 40% of both groups believed that magnets can heal illnesses; 32% of the first and 38% of the second believe that telepathy and clairvoyance are real.

As with most others, the Johnson and Pigliucci study showed that merely learning science is no predictor of scientific or even rational thinking. There is no guaranteed flow-over of scientific thinking from the laboratory to the market place; science teaching is not a vaccine against pseudoscientific and unsubstantiated beliefs (Fasce and Picó 2019). Those piloting the 2001 Twin Towers planes were engineers, as were many of the 2019 Sri Lankan bombers. Merely teaching scientific content (facts, theories, measurement) has little impact on pseudoscience; teaching scientific method has more impact but still falls short of an effective vaccine; teaching the HPS dimensions of science has better outcomes; but most studies point to the need to explicitly engage with particular problematic beliefs and allow students to see their evidential problems (Bhakthavatsalam 2019; Johnson and Pigliucci 2004; Pigliucci 2007; Walker et al. 2002; Wilson 2018).

James Wilson, the teacher of one mixed-majors course where heterodox beliefs and practices were explicitly addressed, wrote:

> at the heart of the development of this course was an attempt to provide students with scientific evidence that would directly contradict many paranormal and pseudoscientific subjects and the worldviews associated with those beliefs. It is not the goal of the course in this study to change someone's worldview, but rather to provide a view that, if something is real and actually works, there will be empirical evidence to support that subject. (Wilson 2018, p. 205)

The Spectrum of Unwarranted Beliefs

It is educationally and culturally irresponsible to leave *all* antiscientific or unwarranted beliefs unexamined. Not all have to be, or could be, examined in class time, but if the belief system is significant and widespread; if it makes either unwarranted or false claims about the constitution and mechanisms in the world, that is, about the domain of science; if the beliefs, actions, and policies based upon the belief system are detrimental to social and personal well-being; and especially if it connects with curriculum topics, then it is irresponsible for educators, philosophers, and science teachers to turn their backs on it. Indeed schools do not turn their back. Racial, religious, and sex discrimination is not tolerated in most countries, and where 'science' is appealed to justify the discrimination, students are told that the science is bogus. The same argument is applicable to feng shui belief and practice.

The above-mentioned superstitions, irrational beliefs, and pseudosciences range along a spectrum from small-scale, idiosyncratic, silly, but harmless (e.g. do not walk under ladders, avoid unlucky numbers) to large-scale, systematic, and harmful (e.g. spirit possession turns women into witches; lung cancer is unrelated to smoking). The former do not warrant precious classroom time, but, given an appropriate context, the latter do warrant such time. Societies invest in education to have an informed, knowledgeable, and hopefully critical citizenry. It is irresponsible for schools to just ignore large-scale systematic beliefs that spread ignorance and promote credulity in society. Both consequences – ignorance and credulity – are antithetical to the purpose of education and the welfare of society.

These are big claims that have been contested in the history of education. Where 'welfare of society' has been interpreted as 'stability of society' – as it is in many countries, for instance China, – then governments have argued that ignorance and state censorship is a precondition for this stable state of affairs. And if education is interpreted simply in terms of social engineering, then there is no reason why growth of knowledge for all should be regarded as an aim of education. Growth of knowledge for some is a requirement for social engineering, but not for all. The liberal education tradition takes a more intrinsic view of the purpose of education and a wider view of the welfare of society and culture.

For instance, in 2013, the Papua New Guinea government released a report on the AIDS epidemic in the country detailing the prevalence, and uselessness, of traditional treatments such as having sufferers sit atop huts inside of which is burnt 'special fires' in expectation that the rising smoke would carry off the evil spirits inhabiting the person and causing the sickness. In the same year, in just one PNG province, there were 150 sorcery-related, medicine-man-directed attacks on supposed witches, with many unfortunate women being beheaded and burnt alive (Elliot 2013). Belatedly the government passed legislation making the practice a criminal offense and disallowed the defence of 'she was a witch' on the philosophically pertinent ground that there are no such things. The argument to be advanced here is that PNG science teachers have a responsibility to appraise the 'ontology' and 'epistemology' of traditional AIDS medicine and of sorcery: to appraise the efficacy of the former and the claimed powers and mechanisms of the latter. In doing so, science can be learnt and additionally the nature of science better appreciated.

In 2000 in South Africa, President Thabo Mbeki rejected the scientific consensus that AIDS was caused by the human immunodeficiency virus (HIV) infection and further that it was preventable by condom use and alleviated by medication. He then instituted completely ineffective government health and education regimes, based on supposed 'local knowledges'. Further, he restricted the pharmaceutical company Boehringer Ingelheim from distributing its drug nevirapine which prevented mother-to-foetus transmission and prescribing various herbal cures for treatment. Over a 2-year period, this science-denialist policy resulted in an estimated 330,000 deaths and the birth of 25,000 infected babies (Mbali 2004). One estimate is that 10% of the South African population of 50 million are infected with the HIV virus (Pigliucci

2010, p. 60). South African science teachers were ideally placed to appraise the 'epistemology' of the government's AIDS programme; further they had a responsibility to do so; to remain mute would be to forfeit their role as educators.

Such a case offers a clear distinction between teachers and educators; the former are not obliged to concern themselves with personal and social betterment; the latter are so obliged. It is no contradiction to say that someone is an excellent and effective teacher of a false and obnoxious subject or doctrine nor that they have well taught someone to do bad things. Teaching is indifferent to moral outcomes; indeed, it is indifferent to cognitive outcomes. People can be well taught to be stupid and to reason incorrectly. This is the standard pattern of indoctrination. There students are exposed to the most limited subject matter[12] and are fed a diet of illogical reasoning. In contrast, education is conceptually tied to cognitive growth, meaning better and more justified reasoning, and moral betterment. Indoctrinators can be wonderful, engaging, and effective teachers; they are just not educators.

Along with specific cases such as sorcery, there are numerous other commitments, beliefs, and ideologies that make claims about the constituents and mechanisms of the world and hence occupy the same field as science. As such they are able to have their credentials compared to those of science.[13] There may be local circumstances, as with the above New Guinea case, that warrant their specific examination in science classes.

The 2016 US Presidential campaign was marked by millions of people being prepared to believe all manner of outrageous claims made by candidate Trump without giving any consideration to evidence for the claims, much less thinking through what their implementation would involve and what their likely consequences would be for themselves, much less for others. This suggests that for a significant portion of the US population, the practice of seeking evidence and questioning assertions – the beginning of a scientific habit of mind – is not a routine outcome of US education. This should be a matter of national concern (Mooney and Kirshenbaum 2009; Specter 2009). This same argument can be made for societies where there is widespread feng shui belief and practice. This latter argument depends, of course, on showing that feng shui belief and practice is ill-founded, unwarranted, and inconsistent with science and constitutes a pseudoscience. This book hopes to establish all of these assertions.

Against those saying that education authorities and science teachers should leave feng shui alone and just look the other way, it will be argued that the ideology is not just idle or harmless hand-waving; it opens the cultural door to mystification, obfuscation, and easy manipulation by charlatans. If people get accustomed to believing

---

[12] As a minor, but telling, example, the English philosopher Anthony Kenny describes how, when he was a graduate philosophy student at the Pontifical Gregorian University in Rome in the 1960s, copies of David Hume's *On Religion* could only be read in the library if a supervisor signed a note saying it was absolutely necessary to do so (Kenny 1985). Such restriction pales alongside that imposed in lesser Catholic and Protestant seminaries, Orthodox Jewish yeshivas, Islamic universities, or Communist Party training schools. In each case, the curriculum and materials are limited, and ignorance is lauded as faithfulness.

[13] A landmark anthology on this subject is Gross et al. (1996).

the fantastic, evidence-free, or evidence-neutral, chi narrative, then what other evidence-free or neutral beliefs might they be prepared to accept? There is lots of supposed evidence for chi, but none of it stands up to critical scrutiny. European history is full of shocking and bloody episodes of mass hysteria and pogroms sustained by such gullibility; so too is US history (Andersen 2017; Park 2000, 2008), as is the history of all nations. 'Witch craze' can be shorthand for such abominable hysteria; the shorthand name should remind all of the cultural responsibility of schools.

The Cultural Responsibility of School Science Programmes

The US biologist Michael Zimmerman wrote on the baneful state of affairs in the USA concerning deep-seated and widespread acceptance of discredited, to put it mildly, beliefs:

> The problem is less that most Americans share no solid grasp of a body of scientific 'knowledge' (although many surely do not) than that they have a complete misunderstanding of the nature, processes, and purposes of science. Americans lack the critical capacity to distinguish real science from pseudoscience. (Zimmerman 1995, p. 14)

For Zimmerman, among other unfortunate consequences of this situation is that: 'What we have is a public largely anxious to jump to supernatural and so-called alternative conclusions' (ibid. p. 15). And not just 'the public', universities now teach pseudosciences, hospitals provide alternative medicine, and insurance firms pay for a variety of 'holistic' treatments.

Alarmingly, this capacity to distinguish real science from pseudoscience is little affected by simply having scientific knowledge, by having completed a school science programme. This depressing fact has been repeatedly confirmed (Otto 2016; Shermer 1997; Smith 2010). In the USA a study across three universities that comprehensively measured and scored students' scientific knowledge then did the same with, and scored, their belief in 14 paranormal/pseudoscientific beliefs such as:

\# A person's astrological sign can predict a person's personality and their future.
\# The body can be healed by placing magnets on to the skin near injured areas.
\# Houses can be haunted by the spirits of people who have died there in tragic ways.
\# Water can be accurately detected by people using 'Y'-shaped tree branches.

When correlations were done between the first and second scores, the authors concluded that:

> … there was no relationship between the level of science knowledge and scepticism regarding paranormal claims. (Walker et al. 2002, p. 26)

Susan Losh and Brandon Nzekwe provide a useful table of 15 national USA polls (Pew, NSF, GSS) taken between 2001 and 2011 documenting the extent of pseudoscientific belief among college-educated adults and college students.

Numerous different polls returned alarming proportions of belief in astrology (26%), communication with the dead (51%), flying saucers from other planets (11%), psychics who can predict the future (28%), ghosts (15%), and so on down a depressing list (Losh and Nzekwe 2011, p. 481). It is of interest for the argument of this book that feng shui has not yet made it on to the depressing list of pseudoscientific beliefs being documented in these large national studies. In Asia, any such comparable study would include feng shui; given the details provided in Chap. 4 of international, including US, practice of feng shui, doubtless it will soon take its rightful place in such national surveys.

Forty years ago, Martin Gardner documented the tsunami of 'Fads and Fallacies' believed by so many millions in the USA and pointed to the hundreds of best-selling books catering to and promoting those beliefs. In a less than optimistic tone, he remarked:

> It will be a long time until the average citizen is well enough informed about science to make the promotion of popularly written pseudoscientific books unprofitable. And as long as they are profitable, you can be sure they will be written and printed. (Gardner 1981, p. 57)

Although these observations relate to USA education and culture, assuredly they apply to other countries and cultures.

One large Finnish study of 3141 students from 20 higher education institutions used the Tobacyk 26-item Revised Paranormal Belief Scale (Tobacyk 2004) to tease out relations between paranormal belief and length of study, discipline studied, and style of thinking (analytic vs intuitive) promoted in the discipline (Aarnio and Lindeman 2005). Predictably, mere length of study did not have much impact on incidence of paranormal beliefs. Education and Theology students had highest mean scores for paranormal beliefs, while Medicine and Psychology students had lowest, with Natural Science students falling in-between. Overall, while Finnish higher education students' belief in paranormal and religious matters was lower than the USA, the authors point out:

> In this study, belief in the paranormal was more strongly related with intuitive thinking than with analytical thinking. The results are in line with the basic tenet of dual-process theories in that beliefs which are most resistant to change arise from the intuitive, not from a malfunctioning analytical system. (Aarnio and Lindeman 2005, p. 1234)

This suggests that although analytical, evidence-sensitive thinking needs promotion in order for students to be proficient in science, it is not by itself going to result in diminishing of paranormal, superstitious, magical, or no-evidence beliefs. It seems that cognitive compartmentalizing occurs: analytic thinking develops and enhances scientific understanding, but an adjacent 'intuitive' compartment remains unaffected, and this supports superstitious and paranormal conviction.

The results of this study suggest that there is a need to make students more explicitly aware of the existence of different ways of arriving at a conclusion or of holding a particular position. These include an analytical, evidence-based approach and an intuitive one. The former should be highlighted as being verifiable or at least being given an assessment of the likelihood that the conclusion is consistent with reality, whereas the putative conclusions based on the latter approach can only at best lead to a random match with the events in the real world. Given the apparent underlying strength and acceptance of intuitive thinking which was highlighted by the Finnish study, it may not be sufficient to merely promote, in a science education curriculum, the more analytical approach of science, without also demonstrating the weakness of positions held when they are based on paranormal, superstitious, magical, or no-evidence beliefs. In the interests of educational balance, it might also be necessary to admit that some intuitive thinking does occur and even has a place in the institution of science. However, where and when that arises (often in the context of suggesting a new line of research), it is still subject to the analytical tests of evidence for verifiability.

All societies have their share of unfounded, unwarranted, and irrational (to put it mildly) paranormal, superstitious, and magical beliefs.[14] All societies contain a goodly portion of charlatans, fraudsters, and 'snake-oil' merchants whose success depends on the credulity of the population; large commercial enterprises and their associated advertising campaigns depend upon just such gullibility and credulity of consumers. Mega-millions of advertising dollars and political campaign dollars are spent upon that very assumption being true. Schools have a responsibility for preparing students to appraise such belief systems and to give students the competence to identify con artists, fraudsters, programmes, and activities that are detrimental to their immediate health and flourishing and to the overall welfare of their society. In recognition of this responsibility, many countries and school systems have introduced 'Critical Thinking' programmes and subjects into their curriculum.[15] Science courses can make a distinctive contribution to this educational aim by routine appraisal of appropriate pseudoscientific and superstitious beliefs.

Religion and Superstition

It can be a productive exercise with classes to examine what constitutes superstition. The answer is more complex than might appear. For instance, two-thirds of Americans believe that angels and devils are active in the world, while a quarter believe in witches that have power to intervene in nature.[16] How to separate belief in angels, devils, and witches? In the thirteenth century, Thomas Aquinas, one of the

---

[14] The warrant, or otherwise, for purely religious belief can be put aside. That is a matter for sophisticated theology. But when claims about the world and its mechanisms are made by religions, then those claims can and should be scientifically scrutinized like any others (Boudry 2017).

[15] See Bailin and Siegel (2003), Davies and Barnett (2015), Lipman (1991), and Sprod (2014).

[16] For these statistics and more, see Andersen (2017, pp. 5–9). He collected the statistics from survey data collected between 2000 and 2017 by the Pew Research Centre, Gallup, Scripps, Harris, and other sources. It is noteworthy that feng shui belief has not yet made its way onto these standard, routine polls for paranormal and pseudoscientific belief.

most imposing intellects of all ages, put serious effort into determining the properties and numbers of angels and 'Whether the Angels Differ in Species?' (*Summa Theologica*, Part 1, Question 50, Article 4). Henry Gill, a catholic priest, philosopher, and physics lecturer, gave succinct expression to the kind of worldview held by many religious believers:

> It will be useful to recall briefly the Catholic teaching as to the existence of spirits. The Scripture is full of references to both good and bad spirits. There are good and bad angels. Each of us has a Guardian Angel, whose presence, alas, we often forget. Angels, as the Catechism tells us, have been sent as messengers from God to man. (Gill 1943, pp. 127–28)

Belief in Jinn is a formal requirement for the world's 1.5 billion Muslims:

> The Jinn are beings created with free will, living on earth in a world parallel to mankind. The Arabic word Jinn is from the verb 'Janna' which means to hide or conceal. Thus, they are physically invisible from man as their description suggests. This invisibility is one of the reasons why some people have denied their existence. However, (as will be seen) the affect which the world of the Jinn has upon our world, is enough to refute this modern denial of one of Allah's creations.[17]

There are certain parallels between angels, jinn, and the chi beliefs of feng shui: none of the principals are immediately visible; and all have to be inferred from their effects. There is no in-principle problem with invisibility, and this is routine for theoretical entities in science; but there is a problem when the invisible cannot be directly and lawfully linked to occurrences in the visible world and when there is no agreed meter or measuring instrument to indicate their existence and efficacy.

As is apparent in the foregoing quotations, a difference between chi belief and that of Christian and Muslim belief in angels and jinn is that the latter is warranted by Sacred Scripture or Revelation. Chi beliefs are not warranted in the same manner. Their origin in Dao classics is acknowledged, but the beliefs are supposedly warranted by empirical evidence and by experience. Chi is supposedly a naturalistic, impersonal part of nature and the cosmos. For believers, devils, jinn, and angels are individuated, there are guardian angels and individual devils; individuals have chi, but there is not a personalized guardian chi. As will be shown in Chap. 12, there are significant problems with maintaining this naturalist account of chi.

On the other hand, none apart from the most ill-informed fundamentalists can maintain that 'if it is in scripture, it cannot be superstition'. The scriptures of all religions are replete with the superstitions of the age of their composition. All serious religious believers acknowledge this and are committed to separating superstitious chaff from theological wheat. So, the exercise of identifying superstition is not wasted on either religious or chi believers. As will be shown in Chap. 11, this is done best when there is some curricular coordination between science, religion, and philosophy programmes or at least conversation and planning between teachers in the programmes.

---

[17] See https://www.missionislam.com/knowledge/worldjinn.html

Superstition in Asia

As with all parts of the world, Asia has its share of shamans, astrologers, diviners, numerologists, fortune-tellers, sorcerers, and practitioners of other 'dark arts'. Perhaps they are more visible and public in Asia, where their influence reaches up to the highest strata. This is seen in the 2016 travails of South Korea's then President Park Geun-hye who for some years had been under the influence of the shamanesque Choi Soon-sil. Chinese Communist Party elites have personal qigong instructors and advisors, though these are kept out of public view, so their impact on policy is not known. *The Economist* magazine, in a 2016 report on the soothsaying and fortune-telling in Asia, reported that a feng shui master visited its Hong Kong office and left behind, in one corner, old coins for prosperity, a hidden mirror to ward off evil spirits, and a picture of a dragon to enjoy the view of the harbour and to invite in good fortune. It commented that 'In Asia the occult is baked into daily life'.[18]

Two years later *The Economist* published another item on South Korea titled 'Prophets and Profits' that details much of the $3.7 billion per annum fortune-telling and prophecy business in the country.[19] Paik Woon-san, head of the Association of Korean Prophets, estimates that there are over 300,000 fortune-tellers in the country and 150,000 shamans, many of whom provide clairvoyance. More than two-thirds of those surveyed by Trend Monitor, a Korean market-research firm, said they see a fortune-teller at least once a year. At Kyobo, South Korea's biggest bookstore chain, as many shelves are devoted to deciphering destiny as to understanding Korea's modern history, with primers including 'Your Winning Lotto Number is in Your Dreams'. Handasoft, a software developer, has launched 13 apps in the past 5 years. Its most popular, Jeomsin, introduced two years ago, has been downloaded over three million times. Every morning it sends users their personalized fortune for the day. It is easy to imagine the unsavoury and deleterious purposes to which such penetration can be put: 'Your good fortune is enhanced if you vote for presidential candidate X', or '… if you buy medicine Y', or '… if you send money to association Z'. And all of this in a country that routinely wins medals in international science achievement tests.

The cosmological-theoretical foundation of Korean divination practice is *Saju*, an ancient feng shui-related system deriving from the 3000-year-old Chinese classic *I Ching* (the *Book of Changes*). *Saju* analyses the cosmic energy at the hour, day, month, and year of a person's birth from Chinese astrological records and texts and, by whatever means, extrapolates to present life circumstances and decisions. It is recognized as an academic subject in Korea and is widely patronized. Janet Shin, a *saju* master and newspaper columnist, who also lectures at universities, says that her clients include doctors, professors, religious clergy, and church administrators. Tarot-card divination is also omnipresent through Korea. Tarot readers are a fixture in all shopping centres; careers fairs at Hankuk University of Foreign Studies, in Seoul, reserve places for tarot readers as, presumably, do many others.

---

[18] *The Economist*, November 12–18, 2016, p. 26
[19] *The Economist*, February 24–March 2, 2018, pp. 22–24

Since the beginning of the Programme of International Student Assessment (PISA) testing in the, now 69, OECD countries, South Korea has always ranked in the top 10 for science attainment. The juxtaposition of elevated and laudable science performance with high recourse to divination, clairvoyance, and occult beliefs brings into sharp focus issues about the contribution of science programmes to cultural health.

The social psychologist James Alcock, in his comprehensive study of *Belief: What It Means to Believe and Why Our Convictions Are So Compelling*, advances a sober but realistic claim:

> No matter how intelligent we are or how educated we might be, we can all be fooled, although perhaps some more readily than others We are all vulnerable to the development of faulty beliefs based on deceptions perpetrated by others and on our own self-deceptions. Yet we are poor at detecting deceit in others, just as we often fail to recognize that we have deceived ourselves. Learning to apply critical thinking both to our own perceptions, memories, and contemplations and to the attempts of others to influence us is our best, and perhaps only, defense. (Alcock 2018, p. 248)

One of the arguments that thread through this book is that a richer understanding of science can be achieved when students are presented with a contrast between good science and pretend science or pseudoscience. In common parlance, this is 'killing two birds with one stone': core features of science can be better understood and appreciated, while the limitations and deleterious cultural effects of pseudoscience are revealed.[20]

Concerning feng shui, this book will argue:

- That feng shui constitutes a worldview.
- That feng shui is a pseudoscience.
- That feng shui supports detrimental social practices.
- That examination of feng shui enables basic philosophy of science to be appreciated.
- That examination of feng shui can contribute to better understanding of the philosophy, methodology, and ontology of science or, as commonly stated in education, the nature of science (NOS).
- That critical appraisal of feng shui is indicative of a scientific habit of mind.
- That the historical and philosophical examination of feng shui in classrooms can contribute to the cultural health of society.
- That feng shui is not an integral part of the Chinese cultural tradition; the latter can be affirmed without the former.

---

[20] Others have advanced this argument. See especially Bhakthavatsalam (2019), Eve and Dunn (1990), Martin (1971, 1994), Mugaloglu (2014), Turgut (2011), and Zimmerman (1995).

Academic Neglect of Feng Shui

While there have been historical (de Groot 1892–1910; Paton 2013), sociological (Freedman 1974; Yoon 2006), and anthropological studies (Bruun 2003; Feuchtwang 2003; Yang 1970) of feng shui and some philosophical expositions, there have been few critical philosophical appraisals of its theory and practice; it has slipped under the critical academic radar. With few exceptions, science popularizers, educators, and philosophers have ignored feng shui; it has been given a 'free pass' into the Asian and Western cultural realms. Its examination has fallen between stools: it is not in the science curriculum which indicates educators do not regard it as science; yet it is not listed among the multitude of practices that routinely are criticized by philosophers as pseudoscience.

The notable exception to the critical appraisal of feng shui is the thread of articles in the journal/magazine *Skeptical Inquirer*.[21] Likewise there are few substantial criticisms of the theory of qigong, the enormously popular sibling of feng shui, practiced daily by millions throughout the world. Both feng shui and qigong have chi (qi) as their ontological bedrock. The noteworthy exception to this particular neglect is the book co-authored by six Chinese medical researchers *Qigong: Chinese Medicine or Pseudoscience?* (Lin et al. 2000). The authors, who in chapter-after-chapter expose the pseudoscientific claims so commonly stated by qigongists, write:

> In compiling this book we aim not to deny our culture, but to allow the people of the world to have a bona fide comprehension of what Qigong is, to better enable them to distinguish the true from the false, and to join together with them in a scientific attitude to illuminate this treasure of scientific culture that belongs to all humankind. (Lin et al. 2000, pp. 11–12)

Given the multiple millions of serious and not-so-serious feng shui believers in Asia and increasingly beyond Asia, and feng shui's immense social, commercial, and personal impact, it is noteworthy that there has been so little appraisal, critical or otherwise, of feng shui in philosophical or educational literature. This neglect of feng shui by educators and philosophers is an enigma as it is a worldview that makes claims about the constitution of the world and the ways in which humans can most harmoniously live in the world. It makes detailed empirical claims about processes in the world, about the efficacy of acupuncture treatments, about the health and wellness consequences of building sites, and about the impacts of design and living arrangements.

Unlike many comparable worldviews, feng shui explicitly claims to be scientific. The whole theory and practice of feng shui revolves around the identification and utilization of universal energy, or chi, that circulates in a defined manner through nature and through specific meridians in people's bodies where purportedly there are 600+ nodes (acupoints) that can be manipulated by acupuncture. This supposed energy has remarkable properties that are mapped, harnessed, and utilized by

---

[21] See, for instance, Beyerstein and Sampson (1996a, b), Hall (2006), Hill (2013), Huston (1995), and the entry in the *Skeptic Encyclopedia of Pseudoscience* (Puro 2002).

masters; and this knowledge and technique can be appropriately taught and learnt and, of course, paid for.

A raft of popular and highly regarded books devoted to pseudoscientific belief systems simply fail to mention feng shui.[22] Nor is it mentioned in the long review article 'Science, Pseudo-science, and Science Falsely So-called' (Thurs and Numbers 2013) or the edited 23-chapter *Philosophy of Pseudoscience* (Pigliucci and Boudry 2013). And, revealingly, feng shui does not appear in the 35-chapter, 472-page *Chinese Studies in the History and Philosophy of Science and Technology* (Dainian and Cohen 1996).

The foregoing books, journal special issues, and research papers contain expositions and criticisms of practices such as alchemy, astrology, allopathy, alternative medicine, anthroposophy, astral projection, aural photography, dowsing, chiropractic, Christian Science, clairvoyance, cold fusion, creationism, dowsing, ESP, Gaia, graphology, homeopathy, Kirlian aura photography, magic, mesmerism, N-rays, occultism, parapsychology, past-life regression, phrenology, poltergeistism, polywater, psychokinesis, psychoanalysis, scientology, séance communication, spiritualism, telepathy, UFOlogy, vitalism, witchcraft, astro-therapy, and more obscure and doubtful practices. Yet not one of the books has 'feng shui' as an index entry. This is noteworthy that feng shui has so completely slipped under the critical radar. This book is a belated attempt to subject it to the level of examination to which all other such comprehensive and consequential practices have been subjected.

It is common to find education articles whose titles contain the words 'astrology' (Turgut 2011), 'paranormal' (Eder et al. 2010), 'psychic' (Eve and Dunn 1990), 'magic' (Losh and Nzekwe 2011), 'creationism' (Pennock 2002), and 'pseudoscience' (Martin 1994; Mugaloglu 2014; Pennock 2002), but 'feng shui' never appears.[23] Many years of science education journals can be searched without finding any mention of feng shui; nor does it appear in the Subject Index of school science textbooks even ones promoted in the Asian market. Especially for Asian education, but increasingly beyond Asia, feng shui is the elephant in the room; curriculum writers and teachers know it is there, but it is not mentioned.

---

[22] Feng shui is not mentioned in any of the following best-selling books on pseudoscience: Martin Gardner, *Science: Good, Bad and Bogus: A Skeptical Look at Extraordinary Claims* (Gardner 1981); Michael Friedlander, *At the Fringes of Science* (Friedlander 1995); Patrick Grim, *Philosophy of Science and the Occult* (Grim 1990); Terence Hines, *Pseudoscience and the Paranormal* (Hines 2003); Wendy Kaminer, *Sleeping with Extra-Terrestrials: The Rise of Irrationalism and the Perils of Piety* (Kaminer 1999); Robert Park, *Voodoo Science: The Road from Foolishness to Fraud* (Park 2000), or his *Superstition: Belief in the Age of Science* (Park 2008); Massimo Pigliucci, *Nonsense on Stilts* (Pigliucci 2010); Michael Shermer, *Why People Believe Weird Things: Pseudoscience, Superstition, and other Confusions of our Time* (Shermer 1997); Carl Sagan, *The Demon-Haunted World* (Sagan 1996); and Victor Stenger, *Physics and Psychics: The Search for a World Beyond the Senses* (Stenger 1990). It is a productive exercise to explain why, for these major books, feng shui is 'missing in action'.

[23] Good reviews of science education research on 'contentious' belief systems can be seen in Eder et al. (2010), Fishman (2009), Fishman and Boudry (2013), Lindeman and Aarnio (2007), and Martin (1994).

The Scientific Habit of Mind

Feng shui warrants the attention of educators because of its deep cultural roots, its significant personal and social impact, and its propensity to accustom people to believe in things about the world for which there is no compelling evidence. Or perhaps more accurately, feng shui belief accustoms people to believing in things for which there is, at best, only 'positive' evidence, and it inculcates a disregard for negative evidence and for alternative interpretations of the positive evidence. This more informed and critical approach to evidence should be something fundamental in the outcomes of good science education. Fortunately, feng shui's seemingly 'scientific' appearance, its constant appearance in 'scientific' clothes while making 'scientific' claims about energy flow and reservoirs, allows meaningful comparison with science and makes it a legitimate subject for investigation in the classroom.

This neglect is noteworthy as Asian countries famously top the world in school science achievement: Singapore, South Korea, Hong Kong, Japan, Shanghai, and other Chinese centres define the gold standard in the field. These countries are also the gold medallists in feng shui belief. The concurrence of high science achievement along with widespread feng shui belief does raise the question: To what extent does science education promote a scientific habit of mind? Which invites a further question: Can belief in feng shui theory and practice be compatible with a scientific habit of mind? Is something missing from 'gold standard' science education if it seemingly has no impact on the cultural breadth and depth of feng shui belief and practice? Is development and exercise of a scientific habit of mind part of the meaning of 'gold standard' science education? Can a science education be gold standard if it does not promote a scientific habit of mind? For the Enlightenment tradition, formation of a scientific habit of mind is the core responsibility of science education.

This disjunct between educational results and commitment to non- and antiscientific beliefs is not confined to Asia and feng shui. In the USA the same pairing of advanced gullibility and high-level education is common. In the 2016 presidential election, 63 million Americans voted for a climate-change-denying candidate proposing a host of economic policies, most of which were fanciful and contrary to the self-interest of the majority of those voting for him. The majority of voters were well educated, yet credulity and gullibility were rampant.[24]

Sedona (AZ), with a population of 10,000, is a standout case of alternative medicine, earth-energy tourism combined with elevated education levels. Every possible kind of complementary and alternative medicine or health regime is available. These include Reiki Drumming Classes that enable, for a mere USD475, clients to 'harmonize their energy with the heartbeat of the Earth'.[25]

---

[24] The sorry 500-year history of tsunami-scale gullibility in the USA is depressingly documented in Kurt Andersen's *Fantasyland* which ranges across deep-seated religious, political, economic, cultural, and educational delusions (Andersen 2017).

[25] https://reikiclasses.com/product/reiki-healing-drumming-class-april-27-28-2019/

One of the hundreds of what in other towns would be called a counter-culture business but in Sedona is just a 'normal' business says that Sedona is an 'interdimensional portal', and the company takes paying customers to specific energy sites (vortexes), claiming that:

> The energy inside a vortex is similar to that inside every human being and when these two energies meet they create a resonance, described by people as a faint vibration. Those who experience this strange phenomenon, feel tranquillity or rejuvenation.[26]

Another business operator says:

> There are four main energy vortexes in Sedona. The subtle energy that exists at these locations interacts with who a person is inside. The energy resonates with and strengthens the Inner Being of each person that comes within about a quarter to a half mile of it. This resonance happens because the vortex energy is very similar to the subtle energy operating in the energy centers inside each person. If you are at all a sensitive person, it is easy to feel the energy at these vortexes.[27]

Other Sedona businesses speak of Mother Earth's Natural Electromagnetic Fields (NEMFs) and oversee proper sitting and mind-opening routines so as to let the NEMF flow into the body. With multiple thousands of tourists each year, one might have expected that long ago there would have been reliable scientific documentation of these vortexes. There has not been. There are countless confirmatory testaments in which individuals say how they 'felt the energy', how they were 'cured of bad thoughts', how their 'persistent headaches have gone', how their 'destructive relationships have been changed', how 'Reiki drumming eased my pain', and so on. Predictably, there are no testaments from people who paid their money and sat in the vortex circle or played their drums but nevertheless still have their headache, relationship problem, and low self-esteem or remain unemployed. This focus on confirmation is a standard ploy of pseudoscience and would be seen as such in any classroom appraisal of pseudosciences.

Scientific Literacy

The foregoing questions can be recast in terms of scientific literacy, a routine subject for educational deliberation and research.[28] The American Association for the Advancement of Science, in its landmark 1989 publication *Science for All Americans*, opened with the statement that:

> This book is about scientific literacy. *Science for All Americans* consists of a set of recommendations on what understandings and ways of thinking are essential for all citizens in a world shaped by science and technology. (AAAS 1989, p. v)

---

[26] http://www.messagetoeagle.com/is-mysterious-sedona-a-portal-to-another-dimension/

[27] http://www.lovesedona.com/01.htm

[28] Issues around the definition and educational value of scientific literacy are well discussed in Arons (1983), DeBoer (2000), Hodson (2008), Kolstø (2001), Laugksch (2000), Roberts (2007), Shamos (1995), and Trefil (1996).

And later:

> Becoming aware of the impact of scientific and technological developments on human beliefs and feelings should be part of everyone's science education. (AAAS 1989, p. 173)

Table 2.1 outlines standard categories of scientific literacy. At the first level, there can be non-committal literacy where folk know about science and are able to perform adequately on international TIMMS or PISA tests or on whatever local tests are relevant, but they do not believe what they learn or guide life choices by it. This is the sort of 'spectator' or 'academic' knowledge that anthropologists or sociologists have when they study other cultures, religions, or political systems. They might well know, for example, Roman Catholicism, Islam, or Stalinism, and be able to competently answer, from the inside, difficult questions within the subject – 'Is there salvation outside the Church?', 'Are contemporary interpretations of the Koran legitimate?', 'Can there be other worker's parties apart from the Communist party?' – but do not share the beliefs or the practices. They are knowledgeable or literate in the subjects but have no commitment to them and need not have any such commitment. A question that then arises is: Does a society or culture warrant being called 'scientifically literate' where spectator literacy is the norm?

On the other hand, most liberal educators, especially science educators, want scientific literacy tightened to include belief, commitment, and trust in science, saying that without these, a person is not really scientifically literate. This could be called 'participant' or 'commitment' literacy. It does not mean that students or citizens participate in science or are mini-scientists; it just means that they support science, are committed to its goals, and utilize scientific knowledge in making life choices. Such commitment is a hallmark of the Enlightenment tradition in science education (Matthews 2015a, chap. 2). A scientifically literate person must act scientifically; and a scientifically literate society must organize itself and develop social, economic, and cultural policies that are informed by science.

For example, on all criteria, President Trump's Border Wall proposal is not informed by science; it can only find support in a scientifically illiterate society, a society that does not ask scientific questions about policies or bring science into policy appraisal, and a society too easily persuaded by rhetoric and theatrics. Ditto for the anti-vaxxing movement; it can only be supported if science and scientific thinking is put aside and marginalized (Mnookin 2011, Offit 2013).

But there is a third level of literacy concerning the scope of science. To be committed to science and scientific method within the laboratory is one thing; it might be called 'limited' or 'technical' science. For the Enlightenment tradition, scientific literacy means extending science and its methodology outside of the laboratory and

Table 2.1 Scientific literacy

Scientific literacy			
Spectator/academic	Participant/committed		
	Limited/technical	Expansive/social	
		Uncritical/scientism	Informed/critical

into the analysis and appraisal of social and cultural matters and problems. This might be called 'expansive' or 'social' science.

Enlightenment thinkers believed that the Galilean-Newtonian method, the 'New Science', had brought unheralded knowledge of the natural world. Further they thought that the same method would bring comparable knowledge of the social, historical, and cultural realms. This conviction was led from the top. Isaac Newton (1643–1727) in his *Opticks* said: 'If natural philosophy in all its Parts, by pursuing this Method, shall at length be perfected, the Bounds of Moral Philosophy will be also enlarged' (Newton 1730/1979, p. 405). David Hume (1711–1776) echoed this expectation with the subtitle of his famous *Treatise on Human Nature* which reads, *Being an Attempt to Introduce the Experimental Method of Reasoning into Moral Subjects*.[29] In the preface he says he is following the philosophers of England who have 'began to put the science of man on a new footing' (Hume 1739/1888, p. xxi). The Marquis de Condorcet (1743–1794), a leading *philosophe* of the French Enlightenment, said in his 1782 acceptance speech at the French Academy that: 'the moral [social] sciences' would eventually 'follow the same methods, acquire an equally exact and precise language, attain the same degree of certainty' as the natural sciences (Condorcet 1976, p. 6). Science was to be pursued outside of the laboratory and utilized beyond the natural world to produce comparable historical, social, and cultural knowledge.

In this final subclass of literacy, the philosophical and educational problem is to distinguish informed from ill-informed commitment to science. The latter is frequently dismissed as 'scientism', the view that science knows everything and is needed to solve all problems (Haack 2016; Sorell 1991). But this is a caricature of scientism; it is pseudoscientism. The dismissal does not touch or engage with the philosophically informed application of science to social and cultural domains and issues.[30] Such examples should be provided in good science programmes; the examples strengthen a society's commitment to science. Science should be criticized, philosophically ill-founded positions should be called out, and naïve reductionisms and positivisms should be exposed. But this does not amount to the rejection of science; it means support for robust science that is philosophically and historically informed, science that is confident in its truth-seeking capacity, but not overconfident (Bussmann and Kötter 2018). Bringing the rudiments of epistemology and ontology into science programmes prepares students for this sort of supportive yet critical embrace of science.

---

[29] At the time, 'moral subjects and philosophy' included present-day history, social sciences, politics, economics and ethics.
[30] See, for example, Bunge (1986, 2010, chap. 13; 2014, chap. 2), McIntyre (2019), and Ladyman (2011, 2018) and discussion in Chap. 14.

Francis Bacon and Critical Evidential Support

The foregoing Sedona problem, the problem for all feng shui adherents, and the problem besetting most belief systems has long been known and was explicitly formulated by Francis Bacon in his famed 1620 *Novum Organum* where he discusses the four classes of idols that beset people's minds leading them to error and keeping them there (Bacon 1620/1939). He writes:

> The human understanding when it has once adopted an opinion (either as being the received opinion or as being agreeable to itself) draws all things else to support and agree with it. And though there be a greater number and weight of instances to be found on the other side, yet these it either neglects and despises, or else by some distinction sets aside and rejects; in order that by this great and pernicious predetermination the authority of its former conclusions many remain inviolate. (Bacon 1620/1939, p. 36)

Bacon is identifying a most general feature of human cognition, namely, the seeking of support for beliefs, the favouring of positive evidence, and the neglect and ignoring of contrary evidence. In a much-cited comment about a traveller entering a foreign port city, Bacon goes on to say:

> And therefore it was a good answer that was made by one who when they showed him hanging in a temple a picture of those who had paid their vows as having escaped shipwreck, and would have him say whether he did not now acknowledge the power of the gods, - 'Aye', asked he again, 'but where are they painted that were drowned after their vows?' (ibid)

Bacon might well have been addressing feng shui, or indeed the whole host of cognate chi-beliefs and practices, when he proceeds to say:

> And such is the way of all superstition, whether in astrology, dreams, omens, divine judgement, or the like; wherein men, having a delight in such vanities, mark the events where they are fulfilled, but where they fail, though this happen much oftener, neglect and pass them by. (ibid)

Bacon's early seventeenth-century discussion of the idols of the mind has been foundational in all subsequent discussion of the psychology of belief and the processes of scientific thinking and practice, especially experimental practice (Agassi 2013; Gaukroger 2001). A great deal of Bacon is echoed in Karl Popper's writings on scientific method where he relentlessly criticized Marxism and Freudianism for amassing confirmations yet ignoring disconfirmations (Urbach 1987, chap. 4). Bacon is also echoed in the Nobel Prize winning research of Daniel Kahneman, Amos Tversky, and Paul Solvic on confirmation bias in judgement, such bias going all the way through to the subjective attribution of values for prior likelihoods of probability in sophisticated Bayesian analysis in science and social science (Dawes 2001; Kahneman 2013; Kahneman et al. 1982; Sutherland 1992).

Conclusion

Appraisal of feng shui (or astrology, alternative medicine, or any other pseudoscience) in classrooms, either in science alone or coordinated across subjects, requires explication of a number of core issues in philosophy and in science education:

- What are the central features of science?
- What constitutes a scientific attitude or habit of mind?
- How can we separate science from proto-science, pseudoscience, and just bad science?
- In what ways is pseudoscience a legitimate and useful category?
- What was and is involved in transition from proto-science to science?
- How do we evaluate competing accounts of phenomena?
- What personal and social factors support the near-universal embrace of non-scientific explanations of natural phenomena and the denial of established scientific facts and theories?
- How can the Asian cultural tradition be respected yet feng shui belief be critically appraised in schools?

For teachers to discuss and elaborate any of these issues with students requires that they are interested and competent in the history and philosophy of science as well as philosophy of education. This in turn means that HPS needs to be included in programmes of science teacher education, along with, hopefully, the rudiments of philosophy of education (Matthews 2015a, chap. 12, 2019b).

Part II
Feng Shui: Its Theory and Practice

Chapter 3
Feng Shui and Chi

Feng shui had its origin, but not its name, 3000–4000 years ago in China as part of a family of cosmological worldviews in which an all-pervasive cosmic energy, material force, or life force known as chi (or *qi*) bound together nature, man, and the universe.[1] Stephen Skinner writes:

> Ch'i acts at every level – on the human level it is the energy flowing through the acupuncture meridians of the body, at the agricultural level it is the force which, if not stagnant, brings fertile crops, and at the climatic level it is the energy carried on the winds and by the waters. The various forms of ch'i include cheng ch'i or vital ch'i, and ssu ch'i or torpid ch'i. (Skinner 1982, p. 14)

While another theorist writes:

> Chi is the subtle charge of electromagnetic energy that runs through everything, carrying information from one thing to another. The chi flowing through your body predominately carries your thoughts, beliefs and emotions.
>
> All the times some of your chi is floating off, while you are also drawing in new energy. The fresh chi that you draw into your own energy field brings with it something of the world around you. This includes the energy of the weather, the chi of other people, the atmosphere of your home and the living energy of the food you eat. As this chi enters your energy field, it alters your own chi, resulting in you feeling different and having new thoughts. (Brown 2005, p. 24)

Chi has an annual seasonal waxing and waning which underlies or is associated with the same seasonal variation of yin and yang. Spring and summer were yang seasons; their opposites, autumn and winter, were yin seasons. Between each season there were transition periods (yang-yang, yang-yin, yin-yin and yin-yang). A traditional representation is (Fig. 3.1).

---

[1] Many works discuss the notion of chi and its place in Chinese philosophy and protoscience. See at least, Chan (1963), Fung (1947, 1949), Liu (2010, 2015), Skinner (1982), and Zhang (2002).

Fig. 3.1 Yin-Yang seasonal variation

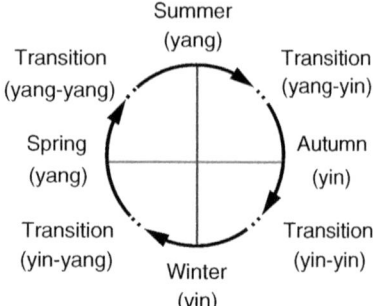

For Confucians, heaven-man-nature – the Confucian Trinity – were melded, not as a naturalistic and materialistic whole but as an organismic whole. Confucian thinkers did not have the pre-Socratic and later European mechanical worldview but had an organismic worldview in which all levels of the cosmos were interconnected; they had a holist, verging on monist, worldview, thus, the common Chinese saying: *Tian Ren He Yi* (Man and Heaven combined into One). What in the West became laws of nature were rather laws of custom, and natural law reflected custom, and in turn social custom had cosmological warrant, hence the inherent conservatism of Chinese Confucian thought. According to Hsun Tzu (Xun Zi, 298–238 BCE), a leading Confucian of the third century BC:

> Custom [*li*] is that whereby Heaven and Earth unite, whereby the sun and moon are brilliant, whereby the four seasons are ordered, whereby the stars move in their courses, whereby rivers flow, whereby all things prosper, whereby love and hatred are tempered, whereby joy and anger keep their proper place. It causes the lower orders to obey, and the upper classes to be industrious; through a myriad of changes it prevents going astray. If one departs from it, one will be destroyed. Is not custom [*li*] the greatest of all principles? (Dubs 1928, p. 223)

Daoist Origins

Wing-Tsit Chan (1901–1994), the Harvard-educated Chinese philosopher and translator, writes that 'Chinese civilization and the Chinese character would have been utterly different if the book *Lau Tzu* had never been written' (Chan 1963, p. 136). The short book, also known as *Tao-te Ching* (Wade-Giles) or *Dao De Jing* (pinyin) (*The Classic of the Way and Its Virtues*), which is the foundation of Daoism

Daoist Origins

(Taoism),[2] was written by Lao Tzu in the fifth century BC and currently sustains 300+ commentaries.[3] In Chapter (stanza) 42, an account of foundational chi cosmology is given:

> Tao produced the One.
> The One produced the two.
> The two produced the three.
> And the three produced the ten thousand things.
> The ten thousand things carry the yin and embrace the yang, and through the blending of the material force (*ch'i*) they achieve harmony. (Chan 1963, p. 160)

The passage is hardly pellucid, but Chan comments:

> It is often understood that the One is the original material force [*ch'i*] or the Great Ultimate, the two are yin and yang, the three are their blending with the original material force, and the ten thousand things are things carrying yini and embracing yang. However there is no need to be specific. The important point is the natural evolution from the simple to the complex without any act of creation. (Chan 1963, p. 161)

And then adds 'This theory is common to practically all Chinese philosophical schools' (Chan 1963, p. 161).

What needs to be appreciated is that Lao Tzu's accounts of Dao are basically accounts of Chi; they are one and the same. Tzu writes in chapter 25:

> There was something undifferentiated and yet complete,
> Which existed before heaven and earth.
> Soundless and formless, it depends on nothing and does not change.
> It operates everywhere and is free from danger.
> It may be considered the mother of the universe.
> I do not know its name, I call it Tao. (Chan 1963, p. 152)

And in chapter 14:

> We look at it and do not see it;
> Its name is The Invisible.
> We listen to it and do not hear it;
> Its name is The Inaudible.
> We touch it and do not find it;
> Its name is The Subtle (formless).
> These three cannot be further inquired into,
> And hence merge into one.

All of this becomes one notch less mysterious if it is understood that the claims are about the properties or attributes of all-pervasive and primeval chi. The book is a testament of chi cosmology.

---

[2] The convention in this book will be to use 'Dao' instead of 'Tao'. The former is the modern Pinyin rendering of the ancient Chinese character for 'way' or 'path'; the latter is the older Wade-Giles Romanization of the character. 'Dao' is also the formal state-required rendering in mainland China.

[3] The entire book is translated, with commentary, in Chan (1963). The book, its author, and period are discussed in Wong (1997, Chap. 2).

A mid-second century BC Daoist text gives a common account of ancient Chinese cosmological understanding of origins and chi (qi):

> The Dao began in the Nebulous Void. The Nebulous Void produced spacetime; Spacetime produced the primordial qi.
> A shoreline (divided) the primordial qi.
> That which was pure and bright spread out to form Heaven; The heavy and turbid congealed to form Earth.
> The conjoined essences of Heaven and Earth produced yin and yang. The supercessive essences of yin and yang caused the four seasons. The scattered essences of the four seasons created all things. (in Parkes 2003, p. 192)

Daoist commentators conjectured that chi was first transformed into wind or breath in heaven and into water on earth. Chi animated the earth and all things in it, including human bodies. Recognizing this, and living accordingly, was the mark of wisdom. This whole picture had some parallels in the Western, Aristotle-inspired, tradition of natural law anthropology and moral theory: things unfolded according to their natures, moral action, and law were identified as something supporting this natural unfolding (Jacobs 2012).

This point marks a fundamental difference between the Chinese traditions of Confucianism and Daoism. The former is primarily an ethical and social system preoccupied with correct personal, social and political conduct; it is detached from considerations of nature; and understanding how the world is constituted. The latter asserts that good living has to be in accordance with nature and with how the world is; consequently, Daoism values and requires knowledge of nature, that is, science. This was just how it was for the medieval scholastics in Europe. But for Daoists, the fundamental knowledge was the science of chi; life had to be regulated by, or lived in accordance with, chi patterns and movement. So, in *Dao De Jing* we have:

> There are four great things in the universe, and the king is one of them.
> Man models himself after Earth.
> Earth models itself after Heaven.
> Heaven models itself after Tao.
> And Tao models itself after Nature. (Chan 1963, pp. 152–153)

This sense of living in an alive yet ordered world, or being one with the world, is well captured in an ancient chronicle of the death at the end of the third century BC of General Meng Tien, the Ch'in dynasty builder of the Great Wall of China. On being informed of the Emperor's sentence of death against him, he supposedly:

> Sighed bitterly and said, 'What is my guilt before Heaven, that I must die for no crime? Then after a pause he said gravely, 'I am guilty, and assuredly should die. From Lin-t'ao all the way to Liao-tung, a moated wall of more than 10,000 li [about 6,700 kms]; in the course of this work I cannot have avoided cutting through the earth's veins [or ridges; *ti mo*], this is my guilt'. And he swallowed poison, killing himself. (March 1968, p. 260)

The excavations and building of the Wall did cause widespread resentment and complaint about the disturbance of local spirit veins. This consciousness predated by nearly 1000 years the first formalization of feng shui as an ideology or 'energy theory'. And the same consciousness has been a feature of all subsequent Chinese life. This was noted by Edwin Joshua Dukes who observed in his 1885 *Everyday Life in China* that:

> The making of a path or building of a house is not, therefore, a matter in which the workman or his employer alone is concerned. Everyone who lives within sight of it, and every spirit whose bones repose in a grave near it, is intensely interested in the questions where and in what style that house or road is going to be made. (Dukes 1885, in March 1968, p. 254)

Feng shui (風水) translates literally as 'wind water'. It is not surprising that this became the name for the all-encompassing cosmological system, as wind and water were the primeval forces that formed landscapes and governed much of life. And they were not just physical processes as now understood; the wind and water were somehow embodiments of the basic all-pervading chi energy. This was the conceptual milieu from which Taoism emerged and provided the metaphysical and proto-scientific component of systematic feng shui. Chi (氣) is fundamental to the worldview, metaphysics and 'science' that underpinned ancient and contemporary feng shui practices.

The name *feng shui* was introduced to Chinese vocabulary and culture by Guo Pu (or Kuo P'u in Wade-Giles) (276–324 AD) in his early fourth century AD *Book of Burial* (Guo 2001, 2004).[4] Guo was a prominent scholar in the Jin dynasty (265–420 AD). He was also a diviner, omen interpreter and, as with all Confucian scholars, a commentator on ancient texts. In Guo's account, feng shui described the properties and behaviour of chi which was the basic all-pervasive constituent, or energy, of the world. For Guo, as with all commentators, chi is borne by wind and diverted or absorbed by water. The qi lines were orientated north-south, a property utilized in the early Chinese geomantic compasses, which then became the technological foundation for the lodestone-based magnetic compasses that were later used in the West for navigation and in Asia for both purposes.[5]

Chi is more ontologically fundamental than the four elements of Empedocles – air, water, fire, earth – and the five elements of Aristotle who added the ether to the list; it is more fundamental than the atoms and void of the pre-Socratic atomists. Guo in his *Book of Burial* writes:

> The Classic says that if qi rides the wind it is scattered; if it is bounded by water it is held. Ancient men gathered it, causing it not to be scattered and curtailed its area of circulation. Hence this is referred to as *fengshui*. The method of *fengshui* is, first of all, to obtain water and secondly to store from the wind. (Paton 2007, p. 427)

Guo also elaborates on the mechanism of chi, saying:

> *Qi* circulates through the earth according to the geodetic force of the earth. It gathers where the geodetic force stops. The qi follows the trunk of a hill and branches along its ridges. (Paton 2007, p. 427)

---

[4] There is some contention as to how much of the standard *Book of Burial* can be ascribed to the historical Guo Pu (Bennett 1978, p. 9).

[5] On the original Chinese geomantic interpretation and use of the compass, see Aczel (2001, Chap. 7) and Skinner (1982, 2008). On the technologically much more sophisticated south-pointing chariot, with its differential gearing, see Lu (1996).

And adds:

> The Classic says that when *qi* circulates through landforms, entities are thereby given life. The geodetic forces of the earth are the basic veins. The geodetic forces of the mountains are the basic bones. They snake either west to east or north to south, curling back on themselves as if crouching and waiting, as if with something in their grasp. *Qi* desires to proceed but it is cut off. It desires to halt and becomes deep. Where it approaches and accumulates, stops and gathers, there will be a clashing of *yang* with a harmonising of *yin*, the earth will be rich and the water deep, the grasses lush and the forests luxuriant. (Paton 2007, p. 427)

A contemporary astrologer and feng shui consultant, Elliot Jay Tanzer, elaborates on the connection of chi with the totemic yin-yang monistic, dualism of Chinese thinking, and philosophy. Yin energy is yielding, soft and traditionally identified as female; yang is hard, penetrating and traditionally identified as male. Tanzer writes:

> *Qi* is the Chinese word for "energy." Everything animate and inanimate, real or conceptual, has *qi*. Different people have different *qi*. Each kind of animal has its own kind of *qi*. A nation has its *qi* and a religion has its *qi*. There is roadway *qi*, rock *qi*, locational *qi*, and vocational *qi*. There is soft-yin *qi* and hard-yang *qi*. There is children *qi*, male and female *qi*. Each item of food has its unique *qi*. To identify the *qi* of anything animate or inanimate, real or conceptual, is to understand its essential nature. *Qi* is the Isness [sic] of whatever is – the essence of the thing or situation. If your goal is good health and success in all areas of your life, there is no other concept more important than the study and understanding of *qi*, and how *qi* flows.[6]

Traditionally, feng shui practitioners have distinguished good or vital chi (*sheng chi*) from bad or torpid chi (*ssu chi*) according to the function the chi is performing, and this function varies in daily and seasonal cycles. So, *sheng chi* refers to the effects of chi during the yang hours of the rising sun; it invigorates processes. *Ssu chi* refers to the effect of chi during the yin hours of the sinking sun (Bennett 1978, p. 7). It is noteworthy that Guo's account of the properties and dimensions of *qi* has characterized all subsequent traditional Chinese understanding of nature, cosmology, and science (Callicott and Ames 1989; Henderson 2010; Porkert 1974; Yosida 1973). In contrast to Western science of energy, there has been no development of the basic theoretical idea, no experimentally informed development of the concept, and no productive engagement with parallel investigative or theoretical systems. As will be elaborated in Chap. 13, these are not intellectual strengths, but weaknesses; they indicate closed and pseudoscientific thinking.

A proper, historically informed understanding of chi is clearly a major challenge, yet Stephen Field, a translator of Guo's classic text says that '*Qi* is the *sine qua non* for any discussion of feng shui' (Guo 2001). So, this challenge cannot be avoided, even if it needs to be done here at an introductory level.

---

[6] At http://abodetao.com/feng-shui-guidelines-to-energy-flow-analysis-what-is-qi-and-how-qi-flows/

Historical Development

The outstanding eleventh-century, Song dynasty scholar Zhang Zai [Chang Tsai] (1020–1077) warrants attention because he wrote extensively on the core feng shui notion of chi, with one commentator writing: 'The most original contribution of Zhang Zai to Chinese philosophy is his concept of *qi* as the basis of his onto-cosmology' (Huang 1999, p. 60).[7] Early in his career, Zhang put aside Confucian thought and studied Buddhist writings which had been introduced into China from India sometime in the third century BC and also Daoist literature and learning. But he returned to Confucianism with a greater insight into the classic *I Ching* [*Book of Changes*]. His combination of Confucianism with Daoism constituted the first neo-Confucian school of Chinese philosophy.[8] Wing-Tsit Chan comments that:

> Chan [Zhang] replaces the traditional theory of spiritual beings or spirits of deceased persons and things with a completely rationalistic and naturalistic interpretation, and establishes a doctrine from which later Neo-Confucianists have never deviated. (Chan 1963, p. 495)

Further, his metaphysical system was, in the context, materialist and naturalistic, so it did not of itself impede a possible Chinese science in the modern Western form; indeed it provided much of the intellectual prerequisite required for just such a development.[9] Whether it can *now* be regarded as naturalistic is more than a moot point; the identification of chi-cosmogony as pseudoscience hinges on this question.

Chi, or vital force, was for Zhang, as it was for Mencius (385–303 BC) the early Confucian philosopher, and for the Daoists, the fundamental 'stuff' of the universe. Chi was a material force, at least as a material as Newton's force of attraction was material. He writes:

> Material force [chi] moves and flows in all directions and in all manners. Its two elements unite and give rise to the concrete. Thus the multiplicity of things and human beings is produced. In their ceaseless successions the two elements of yin and yang constitute the great principles of the universe. (Chan 1963, p. 505)

For philosophers developing his system, chi was the ultimate 'something' by which all processes can be explained.[10] Zhang took into his cosmology the Daoist idea, cited above, of an original Great Void but thought that:

---

[7] See Angle and Tiwald (2017, Chap. 2), Chan (1963, Chap. 30), Huang (1968, 1999, Chap. 4), Kasoff (1984), Kim (2015), and Wang and Ding (2010). Kai-wing Chow (1993) provides an elaboration of his moral theory and its relation to the practice of ritual, something so embedded in Confucian, and more generally Chinese, culture.

[8] The term 'neo-Confucian' is loosely used to refer to the renaissance of Confucian writing that emerged in the Song dynasty (960–1270) and continued to the end of the Qing dynasty (1912). On the terminology, and the differences with modern 'New Confucianism', see the Introductions to Angle and Tiwald (2017) and Makeham (2010).

[9] Western science was developed by devout Christian believers – Copernicus, Galileo, Boyle, Kepler, Newton, Faraday, and Wallace – who had a Christian worldview but whose ontology and epistemology was naturalistic, and in Priestley's case was materialistic. And leading contemporary scientists include religious believers of all kinds.

[10] Angle and Tiwald comment that 'For subsequent Neo-Confucians, Zhang's use of vital stuff [chi] as a central category becomes common property' (Angle and Tiwald 2017, p. 29).

> The Great Void cannot lack the *ch'i*; the *ch'i* cannot but condense to become the ten thousand things; the ten thousand things cannot but disperse to become the Great Void … the Great Void has no form and is the Primordial Substance of *ch'i*. (Chap. 2, p. 22 of *Correct Discipline for Youth*, cited in Huang 1968, p. 249)

So, as opposed to much Buddhist, some Christian, and Newtonian understanding, there was no original void into which 'the world' was placed by a Creator.[11] There was never a void; the primeval 'space' or 'Great Vacuity' of earlier Daoists was filled by chi.[12] Further, this vital force is forever changing: 'The ch'i pervades the Great Void; in its ascending and descending, soaring and moving about, it is perpetually ceaseless' (Chap. 2, p. 22 of *Correct Discipline for Youth*, cited in Huang 1968, p. 249).

There is a certain naturalism in Zhang's cosmology and metaphysics: change and events in the world are to be explained by procedures occurring *within* the world, not by intervention from outside, not by 'non-natural' causes. In Zhang's system, there was no scope for miracles, for change coming from outside the system:

> In spite of the wide variety of the ten thousand things we may know that their actuality is one single thing alone, and that there is nothing in them which does not pertain to the *yin* and the *yang*. Consequently, we understand that the changes of heaven and earth depend solely on these dual principles. (Chap. 2, p. 26 of *Correct Discipline for Youth*, cited in Huang 1968, p. 250)

To contemporary Western ears, yin and yang might not seem very naturalistic, but for Zhang they were principles that governed all change, pertained to matters that were within, and a part of, nature. To this degree Zhang's system is not only naturalistic, but it is deterministic: change is not arbitrary and chaotic; it is in accord with 'principles'. Contrary to expectation, this naturalistic and determinist cosmology did provide a basis for geomantic thinking and practice: all is somehow ordained, and it is a matter of finding signifiers of the future. In his system, and more generally the neo-Confucian system, there is a connection between cosmos, cosmogony and individual. The Chinese tradition is seriously holistic; heavens, nature, and man are all causally interconnected. Zhang wrote:

> There is a chronological order in birth; this is what constitutes the temporal order of Heaven (t'ien-hsu). Small and big, high and low, all appear in juxtaposition; this is what is called the hierarchical order of Heaven (t'ien-chih). Heaven creates things in an orderly manner, and, after taking shape, things come to have a hierarchical order. (in Chow 1993, p. 202)

Not only is Zhang's system naturalistic and deterministic, it is realist, conservative and stable; things in existence stay in existence; they may disappear but are not destroyed in their entirety; there is no passage from being to non-being, only from one form to another, from visible (liquid) to invisible (air). Bodies undergo changes

---

[11] There is a sense of Deism about Zhang's system, but Deists have a role for Creation that Zhang does not countenance. For an account of ten dominant and popular cosmologies, see Bunge (2001, Chap. 2).

[12] The metaphysical question of whether there was an original void in which chi materialized, or whether chi was a part of the original void, is a lively question among some Chinese metaphysicians (Liu and Berger 2014).

of state, but not elimination or annihilation. Zhang illustrates the claim by reference to melting ice:

> The *ch'i's* condensation from and dispersion into the Great Void is similar to the freezing and thawing of ice in water. He who understands the identity of the Great Void and the Vital Force understands that there is no such thing as non-being. (Chap. 2, p. 22 of *Correct Discipline for Youth*, cited in Huang 1968, p. 250)

And the never-ending changing of seasons: what there is in summer is still there in winter but in a different form.

Zhang keeps returning to the fundamental concept of chi. He elaborates on the idea as found in Chinese scholars of the Ch'in (221–206 BC) and Han (206 BC–220 AD) dynasties and uses his own formulation to address the challenge of Buddhist idealist philosophy that had gained considerable traction in China. One modern historian has commented:

> In fact, Chang [Zhang] and the other neo-Confucians were forced to face a tremendous problem. This problem, unknown to the early Confucians and even more difficult to deal with than that of Taoism, was how to prove that the universe, as perceived, is real and not illusory. Hence, Chang Tsai's concept of *ch'i* serves for him the double purpose of attacking the Buddhist [idealist] theory of reality on the one hand and of constructing a sounder one of his own on the other. (Huang 1968, p. 255)

Two contemporary philosophers write of the breath-taking integrative role that chi plays in Zhang's system:

> Zhang Zai presents a critical element, *qi*, for constructing and sustaining a system of Confucian thinking. His theory of *qi* supplies the resources for conjoining ontological and cosmological concerns with Confucian moral cultivation and justification. (Wang and Ding 2010, p. 54)

Another, Liu JeeLoo, writes:

> Zhang Zai's *qi*-monism cannot be categorized into the materialist camp, because *qi* is not purely matter. The tradition of Chinese philosophy does not subscibe to the dichotomy between matter and spirit or the material and the immaterial. *Qi* is both the constitutent of material things and the essence of immaterial things. Under Zhang Zai's development, the realm of qi covers the mechanistic, the organic, and the spiritual dimensions of existence. (Liu 2018, p. 77)

Zhang's critique of Buddhism was forceful. He accuses the Buddhist masters of identifying the real world with the perceived world and consequently declaring the real to be illusory. He writes in *Correcting Ignorance:*

> Buddhists are preposterous and arrogant to discuss only the nature of heaven but do not know the vast field of heaven's functions. ... so they falsely charge that heaven, earth, sun and moon are all delusions. (in Wang and Ding 2010, p. 48)

What Zhang identifies is a well-worn idealist path from empiricist epistemology (we know only what can be sensed) to idealist ontology (only sensations are real).[13]

---

[13] Hence Bertolt Brecht's retelling in his 'Turandot' of the story about the Buddhist congress of scholars called together in the Mi Sang monastery on the banks of the Yellow River to settle the question of whether the world was real. Unfortunately, a huge flood occurred and drowned all the

This path was most notably trodden in the West by George Berkeley (1685–1753) but also by countless phenomenalists who followed in Berkeley's footsteps, right up to, and including Ernst von Glasersfeld (1917–2010), the influential educational constructivist (Matthews 2000, 2015a, Chap. 8). Zhang's metaphysics was realist and monist; he was a monist realist and so distinct from dualist realists. It was the life force or *ch'i* that pervaded everything and gave unity to the whole.

The coherence and comprehensiveness of Zhang's chi-cosmogony warrants it as being called chi-naturalism. For Zhang, and for the whole Chinese tradition, Chi is natural; it is a component, an everlasting one, of the world; it has its own rhythms and ordered motion. In these respects it meets the requirements of any naturalistic philosophy. The moot question is whether chi-naturalism can be understood as a version of scientific naturalism.

Moreover, and importantly, Zhang had what might be called a 'critical' or 'sceptical' attitude; he warned against being beholden to authority, texts, or revelations. His advice for his followers was 'If one can doubt what seems to others not to be doubtful, he is making progress' (Chap. 6, p. 108 of *Correct Discipline for Youth*, cited in Huang 1968, p. 256). Such a sceptical attitude is one component of a modern scientific habit of mind; it was a pervasive element of early European Enlightenment thinking. Nevertheless, despite possessing so many of the components that together led to modern science in the seventeenth-century Europe, such science did not emerge in China. Just why this was the case is the much-discussed 'Needham Question' to be discussed later.

Zhang's system or worldview had flow-on effects for ethics and politics: his monism meant there was not much 'space' to fit in some realm where an afterlife could be lived and where rewards or punishments were awarded for personal behaviour in this life. Consequently, he rejected the whole Hindu and Buddhist idea of the Wheel of Karma, and doubtless would have rejected the Christian ideas of Heaven and Hell had he been familiar with them. It is hardly surprising that the Chinese Communist Party was quick to recognize Zhang as, if not a precursor, then at least as someone in the Chinese philosophical lineage that their Maoist version of Marxism might embrace for cultural legitimation.[14]

Five hundred years after Zhang, Luo Qinshun, also known as Lo Cheng-an (1465–1547), advanced the same chi-naturalism, writing in his 1528 *Knowledge Acquired through Adversity*:

> Penetrating heaven and earth, persisting from the ancient past to the present day, *qi* is the only thing there is. *Qi* is inherently one, and yet it has motion and rest, it goes to and fro, it opens and closes, rises and falls. All developments of *qi* continue in endless cycles. What was obscure becomes salient; what was salient would later return to obscurity. It constitutes the temperature change of the four seasons; it enables the germination, growth, storage and

---

scholars before they could agree on an answer and so philosophers lamented that 'The proof that things exist externally to us, self-sufficiently, independently of us was not furnished' (Suchting 1986, p. 53).

[14] Mainland publications on Zhang up to the Cultural Revolution are listed in Chan (1967, pp. 38, 190–193).

dormancy of all things; it sustains the daily human ethics and it substantiates the success or failure in human affairs. (Liu 2015)

One century later, Wang Fuzhi (1619–1692), the most prolific philosopher in Chinese history, embraced and developed Zhang's theory of chi. Like Zhang, Wang based his ontology and metaphysics on the ninth century BC *I Ching* or *Book of Changes* (discussed in Chap. 4). Wang wrote:

> [*The Book of Change* says,] 'Yi has the Supreme Ultimate (*taiji*), whence generates the Two Modes (*lianyi*).' The Two Modes are nothing but *qi*, only when it is good (*shan*) can it become the mode. Therefore, the six yang in Qian and the six yin in Kun all contain the four virtues of greatness, endurance, benefit, and stability (*yuan heng li zhen*). (Liu 2010, p. 356)[15]

And he restates the chi base of Traditional Chinese Medicine:

> What heaven endows in men is uninterrupted *qi*. If *qi* is uninterrupted, then *li* must also be uninterrupted. Therefore, as long as life continues, one's nature gets daily renewal. (Liu 2010, p. 365)

Cosmology and Science

It is noteworthy that Zhang was writing in China at the very time of the Scientific Revolution in Europe. Galileo's *Chief World Systems* were published in 1633 and his *New Sciences* in 1638, Harvey's *Circulation of Blood* in 1628, Descartes' *Principles of Philosophy* in 1644, Boyle's *Skeptical Chymist* in 1661, Newton's *Principia* in 1687 and Huygens' *Treatise on Light* in 1690. Beginning in the seventeenth century, all major European philosophers (Descartes, Locke, Hume, Berkeley, Leibniz, Spinoza, Kant, d'Alembert, Diderot, and others) developed their metaphysics, ontology and epistemology – their philosophical 'systems' – through engagement with the New Science. Nowhere else, and at no other time, has there been such intense and productive engagement of philosophy with science (Matthews 1989); this was unique to the European Scientific Revolution and philosophical tradition.

All eighteenth-century European Enlightenment thinking was a response to and an elaboration of the implications of the new science for the study of society, law, history, education, and religion (Matthews 2015a, b, Chap. 2). Philosophers of the time who did not engage with science left little trace in the history of European, or wider, ideas and are now known only in specialist history of philosophy courses. Chinese philosophers did not have this modern scientific tradition available to them. So, Wang's metaphysics derives from the reading and interpretation of equivalent Chinese science as found in the *I Chin*, a text even older than Aristotle's that the now 'forgotten' philosophers of Europe were interpreting at the same time as Wang wrote. But Aristotle did of course engage with the science of his time and indeed contributed to it.

---

[15] The 'six yang' and 'six yin' are the lines in the Qian and Kun hexagrams of the *I Chin*.

Wang proposed something akin to a conservation of matter doctrine prompted by the burning of a cartload of timber. He conjectured that the amount of 'substance' before burning equalled the total substance after burning (ash, smoke, other). This was a consequence of the larger understanding of change being one thing passing into another. As there was, typically, no quantification and no precision, there could be no controlled experiment; it remained metaphysical and provided no direction for any scientific development. This is in contrast to, for instance, the detailed measurements and repeated experiments of Jan Baptist van Helmont (1580–1644) whose life overlapped with Wang's.

Van Helmont conjectured that mass was conserved in all change, and in his famous 5-year-long willow tree experiment, he carefully measured the weight of water added to a sapling in a pot, the change in soil weight which was close to zero, and concluded that the 164-pound of wood, bark, roots, and leaves had come from the added water. He missed the contribution of air (carbon dioxide), but this would come later (Hershey 1991).

This conjecture and measurement tradition was firmly established and demonstrated in Antoine Lavoisier's (1743–1794) work. Newton's attraction at a distance was metaphysical, as was incessantly pointed out by George Berkeley (1685–1753) and by his continental opponents, but with its 'inverse square law' formulation, it became testable and scientific (Shank 2008).

The central Dao and neo-Confucian philosophical notions are *Li* and *Qi* (chi). The former refers to the principle of organization that is manifested everywhere in the world, from plants, animals to the solar system; the latter, as has been shown, refers to an all-pervasive special 'energy' in the world. Both ideas present insuperable problems for testing and appraisal.

Concerning *Li*, there is already a division within Chinese thought on whether the principle of organization is inherent in the world which was the position of Chinese Aristotelian-like naturalism and realism as seen in Zhu Xi's (1130–1200) orthodox neo-Confucian writings in the Song dynasty or whether the principle of organization is just in the mind of individuals which was the position of Chinese idealism and quasi-Kantianism, as seen in Wang Yangmin's (1472–1529) unorthodox neo-Confucian writings in the Ming dynasty.

Concerning *Qi*, there is no quantification, no standards for measurement, and no established objective instrumentation despite the optimistic claim of some practitioners that the long-awaited 'Meridian Energy Analysis Device [MEAD] has finally arrived and is making valuable contribution to the diagnosis of Qi-blood dysfunction' (Tsai et al. p. 254).[16] Despite much talk of calibration, ohms, volts, amps, skin moisture, and much else, the proponents of the MEAD device admit:

> … the reliability of most electrical devices for the detection of meridian energy may appear low, or even unsatisfactory, for even the better-performing indices often double or halve the values from test to retest. (Tsai et al. 2017, p. 254)

---

[16] They say their device is 'Basically an amperometer employing a DC voltage of 12 V with output current of 0–200 uA' (Tsai et al. 2017, p. 254).

What is the difference between no instrument and a chronically unreliable one? Here the unreliability is not the instrument makers' fault; it is the supposed 'thing' being measured.

Chi-fusion

Each of the thousands of contemporary exponents of feng shui and the hundreds of feng shui schools give their own account of chi. This conceptual confusion, or chi-fusion, to coin a term, is indicative of the pseudoscientific status of feng shui. Genuine sciences can commence with confused, overlapping or competing fundamental concepts – the differentiation of velocity from speed and of force from energy are examples – but it is a mark of science that over time, with experiment, conceptual analysis and public checking, confusion gives way to agreed community-wide concepts, agreed standards and common measuring instruments. This does not happen with chi belief.

Sarah Rossbach, a best-selling author of numerous feng shui books, writes:

> Chi is the vital force that breathes life into the animals and vegetation, inflates the earth to form mountains, and carries water through the earth's ducts ... Without chi, trees will not blossom, rivers will not flow, man will not be. (Rossbach 1984, p. 21)

Another author and practitioner, who sells chi machines on the web, elaborates on chi as being:

> Chi is the energy of life itself, recognized as the balance of Yin and Yang (female and male, positive and negative, electromagnetic energy), which flows through everything in creation.[17]

Some feng shui exponents maintain that chi changes daily in a 12-day cycle; this acknowledgement opens up the whole business (so as to speak) of ascertaining auspicious days. Other interpreters deny the 12-day cycle. Grand Master Stephen Skinner claims that:

> ... there is a change in direction and quality of ch'i flows every two hours of the day and these ch'i flows are not exactly repeated for the next 60 years. Consequently exact determination of the best starting time for any venture which might involve ch'i – and this extends to activities other than building – is quite an exacting science. (Skinner 1982, p. 18)

He is just one of hundreds of thousands who caters to, or perhaps cashes in on, this oft-overlooked temporal property of chi:

> The dynamic movement of energy changes with the passing of time — in the year, month, day, and hour. Flying Star Feng Shui shows you how these changes and movements of energy can affect space and people. That's why Flying Star is considered one of the most dynamic and effective techniques in Classical Feng Shui. It is used not only to assess the

---

[17] At http://www.chimachine4u.com/chi.html

current condition (i.e., the energetic quality) of the home, but it can also be used as a valuable forecasting tool.[18]

Skinner sells his *Flying Star* book for $72 which is cheap because:

> Every home has its own Flying Star chart based on the year the home was built, the facing direction of the property, and the composite make-up of natural landforms (external) and the layout of the floor plans (internal). In other words, every home has an energy map, just like a person has his/her own natal birth chart.[19]

He proceeds:

> This Flying Star chart will consist of many numbers (1 to 9) and will occupy as a Water Star position (energy governing money prosperity) or a Mountain Star position (energy governing health and human harmony). These Stars will be activated or suppressed depending on the Feng Shui of the home. Over the course of time, events will unfold in the home and affect its residents in a positive or negative way. That's why most people, especially in Asia, work out the yearly influence before the current year's end so that they can make plans and adjustments for the placement of Feng Shui remedies and enhancers for the upcoming new year.[20]

What Skinner is doing might be exacting, but it is hardly exact – this being a notion minimally requiring some measurement and empirical test. Further, as will be argued in Chap. 13, Skinner and all the other feng shui exponents are not even engaging in a proto- or immature science; they are engaged in a pseudoscience.

As the saying is of Skinner's foregoing claims: 'if you believe them, you would believe anything'. And that is the dangerous consequence of leaving feng shui unexamined. Students and citizens get accustomed to believing things with no evidence and no warrant; this prepares them for believing far worse and more disturbing things from politicians, demagogues, and fraudsters – as has been evidenced more than enough times in witch crazes, pogroms, purges, and mass delusions.

One consultant turns from the Grand Master Dong and other methods because 'These methods require an in-depth knowledge of Chinese Metaphysics' and settles on a simpler method for identifying good times:

> The 12 Day Officers refer to the 12 types of ch'I or energies that prevail in a given day and these are common to everybody for that particular day. However, once the dynamics of your Four Pillars is known this information can be used to fine tune the selection of auspicious dates to maximize your success.[21]

Zhang Danian's authoritative modern text on Chinese philosophy devotes a chapter to Chi (Zhang 2002). The book provides an informative listing of scores of representative and influential Chinese chi beliefs from ancient times to the present. Some examples from authors cited in the book are:

---

[18] At https://www.fengshuitoday.com/ and http://www.fengshuitoday.com/shop/advanced-flying-star-feng-shui-gm-dr-stephen-skinner/

[19] At https://www.fengshuitoday.com/ and http://www.fengshuitoday.com/shop/advanced-flying-star-feng-shui-gm-dr-stephen-skinner/

[20] At http://www.fengshuitoday.com/shop/advanced-flying-star-feng-shui-gm-dr-stephen-skinner/

[21] At http://www.shenchi.com.au/services/personal-analysis/personal-services#Annual_Review

Qin dynasty (200 BC)

> This *qi* of heaven and earth does not lose its order. If it goes beyond the order, the people are in confusion. *Yang* bends over and cannot go out; *yin* rushes and cannot distill away, so there is an earthquake. (Zhang 2002, p. 46)

Han dynasty (200 BC–200 AD)

> When the will is concentrated then it moves the *qi*; when the *qi* is concentrated then it moves the will. Now if someone should fall or run this is due to *qi* and will come back and affect his mind. (Zhang 2002, p. 47)

Song dynasty (960–1280)

> Everything with shape exists; everything which exists has a visible form; everything with visible form has *qi*. (Zhang 2002, p. 57)

> Within heaven and earth there is principle and *qi*. Principle is the Way above form and the root that produces things. *Qi* is the vessel that is below form, that produces the frame of things. (Zhang 2002, p. 60)

Ming dynasty (1370–1650)

> The two *qi* [*yin* and *yang*] touch and transform; the mass of shapes appears and is constituted; this is what heaven, earth and the myriad things are produced from. (Zhang 2002, p. 61)

> *Qi* transforms, flows and goes, 'producing and reproducing without ceasing'; hence it is called the Way ... Yin-yang and the five agents are the embodiment of the Way. (Zhang 2002, p. 63)

Chi and Science

Feng shui and its chi beliefs are not meant to be outside of or alien to science. As one exponent confidently asserts:

> [Feng shui] is a science, an ancient science much like the science that once existed in lost cities of Atlantis and other ancient and forgotten civilization that marvel and baffle us today [...] only difference is, this science called feng shui has weathered the test of time.[22]

Zhang summarizes the 3000-year tradition of East Asian chi beliefs as follows:

1. *Qi* is the original material out of which all things are formed by coagulation.
2. *Qi* has breadth and depth and can be spoken of.
3. *Qi* is contrasted with the mind. It exists independent of the mind.
4. *Qi* can move. Indeed, it is normally in a state of flux and transformation. (Zhang 2002, p. 63)

And he adds:

> Hence, what Chinese philosophy says about *qi* is basically the same as what Western philosophy says about matter. The Chinese theory, however, is distinctive on two counts:

---

[22] See http://efengshui.org/articles/basic/what-is-feng-shui%2D%2D1530.html

1. Qi is not impenetrable; rather, it penetrates all things;
2. Qi is intrinsically in a state of motion and is normally in flux. (Zhang 2002, p. 63).

Edmund Ryden, the translator of Zhang's book revealingly says:

> Perhaps the best translation of the Chinese word *qi* is provided by Einstein's equation, e=mc². According to this equation, matter and energy are convertible. In places the material element may be to the fore, in others, what we term energy. *Qi* embraces both. ... Qi is both what really exists and what has the ability to become... *Qi* is the life principle but it is also the stuff of inanimate objects. ... As a philosophical category ... [this] meaning is then expanded to encompass all phenomena. (Ryden Introduction to Zhang 2002, p. 45)

Zhang indirectly and Ryden more directly state the commonly made connection of chi-theorizing with basic theory of modern science. It is unfortunate that Ryden brings scientific error to the discussion. He means to say that for Einstein matter and energy are interconvertible, not convertible. But with this correction, he is still wrong. Mass does not convert to energy; it converts to something else that has energy. Energy is not free floating; it is always a property of some concrete existing thing (Bunge 2000). Xiong Shili is formally more correct when he interprets chi as an alias for mass energy, saying that mass and energy are not two separate things, rather they are two propensities of chi; the movement from one to the other is just a transition. He Zouxiu in 1997 asked 'Has the theory of primordial *qi* really influenced the formation of the concept of 'field' in contemporary physics?' He answered 'yes'. Yi Desheng gave the same answer when in 2003 he detailed 'Some impacts and inspirations Chinese ancient theory of primordial *qi* has on contemporary physics'.[23] These are optimistic answers deserving close scrutiny.

Newton's idea of an ether that he entertains in the *Opticks* is closer to traditional chi belief. He thought, as did Descartes, that there could be some sort of ultra-fine material filling empty space and seeping into bodies, perhaps serving to 'cement' the corpuscles of bodies together and perhaps being involved in the transmission of magnetism, gravitation, and electrical effects. These were the functions of Descartes vortices. He calculated that its density had to be less than 1/700,000 the density of air. Newton's ether came close to being part of the 'sensorium of God', a thoroughly unscientific and empirically untestable concept. But after its postulation, he was prepared to reject the idea, along with Cartesian vortices, as idle, saying such a fluid:

> can be of no use for explaining the phenomena of nature; the Motions of the Planets and Comets being better explain'd without it. It serves only to disturb and retard the Motions of those great Bodies and make the Frame of Nature languish. And in the Pores of Bodies, it serves only to stop the vibrating Motions of their Parts, wherein their Heat and Activity consists. And as it is of no use, and hinders the Operations of Nature, and makes her languish, so there is no evidence for its Existence, and therefore it ought to be rejected. (Newton 1730/1979, p. 368)

Despite Newton's dismissal, the idea of an ether lingered on the fringes of science, until finally shown by Einstein to have no possible function and consequently

---

[23] Details of these papers in Chinese philosophy and physics journals are given in Liu JeeLoo (2015).

deserving of no belief. Nevertheless 'ether' remains one of the standard English renderings of chi and qi; proponents routinely use a century-long discarded scientific term to give 'scientific' clothing to their system.

Chi beliefs are not meant to be removed from, outside of, or alien to science. Chi believers, or at least theoreticians, avowedly embrace and accommodate the most sophisticated foundational statement of modern science. But unlike orthodox science, there is no history of the conceptual, much less experimental, amplification and refinement of the basic idea. Matter theory in science is assuredly metaphysical, but this metaphysics has always been elaborated and refined in close conjunction with scientific and experimental developments.[24] This has not been the case with chi as either a matter or an energy theory; it has for thousands of years remained a metaphysical 'theory' divorced from scientific practice.

Feng shui's 'changelessness' is an important consideration in determining its scientificity or, more pointedly, its non-scientificity. There is something to be said for the everyday advice 'when you are on a good thing, stick to it', but sticking to something unchanged, unclarified, and undeveloped for 2500 and more years suggests what you are sticking to is not science; the latter is marked by examination, criticism, development, and change. Fundamentalist religions, of all persuasions, might value things and commitments that are timeless and unchanging; but science, and science-informed philosophy, does not.

Thales of Miletus, in the sixth century BC, made the metaphysical claim that water was the origin of all things; it was understood as a foundational and unchanging commitment. One century earlier, the same claim was made by Kuan Chung in China (Fang 1982/1992, p. 15). Such commitments fitted in with Western (four) and Eastern (five) theory of basic elements. For two and a half thousand years, this was a comfortable and unassailable belief. Then in 1777, Antoine Lavoisier (1743–1794) combined oxygen with hydrogen and created water and, more importantly, dissociated water into its constituents of oxygen and hydrogen. Water was no longer an element; the tradition of elemental cosmology just collapsed.

For instance, Zhang's description of 'coagulation' of elements might parallel a kitchen chefs' use of the term or maybe an artist's when mixing paint, but it has no connection or comparison with the long history in chemistry of investigating what elements combine with what others to form compounds, the conditions under which the combinations can occur, the detailed theory of valency which governs such combinations, and attention to the quantification of all claims.

Identifying with science and seeking the public esteem and legitimation that it brings are a two-edged sword for feng shui: the more explicitly its advocates tout its scientificity, the more pressing is the need for scientific-philosophical appraisal of the worldview and associated geomantic practices. The enigma is that this has not happened in either education or philosophy communities.

The contrast with the long history of atomism (Chalmers 2009), or even the shorter history of relativity theory (Ray 1987), is dramatic. In the latter fields, there is detailed, precise testing and elaboration of theory in the light of exact experimen-

---

[24] See Brooke (1995), Bunge (1973, 2000), McMullin (1963), and Siegfried (2002).

tal testing, all done in a public, published tradition. Once feng shui claims to be on the same field as science, then this contrast is invited, and the deficiency of the former is obvious. An educational and philosophical challenge is to make and appraise the contrast, drawing out implications for the nature of healthy science.

Most of the foregoing statements about the theory and practice of feng shui are almost identical to those made almost 2000 years earlier by Pu who was himself codifying a living tradition which at the time was itself 2–3000 years old. So, in 5000 years not much changed. There has been little conceptual or metaphysical development in the foundations of the feng shui tradition; it is a stable rather than changing tradition. This is of course in dramatic contrast to the history of science which is characterized by change, clarification, growth, and development both in its empirical claims about the world and in its metaphysical commitments.

If the feng shui tradition has correctly identified fundamental cosmological and ontological realities, then as Tanzer says, 'there is no other task more important than the study and understanding of *qi*, and how *qi* flows'. But then if the tradition has not correctly grasped fundamentals about the world, then such study is a scandalous waste of everyone's time and money. It is a pursuit of a Holy Grail that is not and never was there.

In as much as feng shui practice depends on the understanding, interpretation and utilization of chi, then from the foregoing affirmations, it is clear that feng shui and orthodox science are in the same pond; feng shui and science make comparable claims about the natural world and fundamental mechanisms within it. So, as before, two questions can be reasonably asked. First, a philosophical question of whether feng shui is a good science, poor science, pseudoscience, or non-science. Second, an educational question of how, when, where, and why might discussion of feng shui occur in science programmes.

Chi-Based Worldviews

The widespread East Asian chi (*qi*), and Japanese *ki*, beliefs and practices have affinity with versions of Hindu yogic understandings. They are components of a chi-based worldview. In the Hindu tradition, the vital energy of the cosmos is *Prāṇa*. At the end of the nineteenth century, Swami Vivekananda (1863–1902), in a well-publicized address to the 1893 World Parliament of Religions held in Chicago, provided a 'scientific' rendering of *Prāṇa* that was enthusiastically embraced by many in the West, initially by those already committed to some form of spiritualism (mesmerism, theosophy, Christian science) and then by much wider and more secular and countercultural circles right up to contemporary Hari Krishna devotees.[25]

---

[25] On Vivekananda's efforts to give a scientific rendering of ancient Hindu practices, see Nanda (2016, Chap. 3).

A challenge for most religions in the late nineteenth century was to reconcile their theology and practice with Western science which was so dramatically expanding knowledge and control of the natural, biological, and social worlds. One oft-taken option was to retreat and adopt a pietistic or 'other worldly' stance. Another option was to make religious claims scientific, or to reinterpret them in contemporary scientific language, a 'get with the strength' option.[26] This was famously taken by Mary Baker Eddy (1821–1910) who in 1875 published her 600-page *Science and Health with Key to the Scriptures* (Eddy 1875/1990) of which, by 1990, ten million copies had been sold. A few years later, in 1879, she founded the Church of Christ, Scientist. The title – *The Pure Science of Christian Science* (Doorly 1946) – of one primer for *Science and Health* leaves no doubt about this religion's affirmation of science. And the author is confident in saying that *Science and Health* is 'the most metaphysically scientific book in the world' (Doorly 1946, p. 6), and further:

> The demand of to-day is that men and women of all ages and clases be trained to think scientifically and accurately concerning the Science of being ... Progressive thinkers and searchers for truth are not interested in anything which does not enlarge understanding and result in demonstration. (Doorly 1946, p. 10)

It was not just Western religions that were embracing science. Vivekananda did the same for Hinduism. In the opening page of his eight volumes of *Collected Works*, he writes:

> All our knowledge is based upon experience. What we call inferential knowledge, in which we go from the less to the more general, or from the general to the particular, has experience as its basis. In what are called the exact sciences, people easily find the truth, because it appeals to the particular experiences of every human being. The scientist does not tell you to believe in anything, but he has certain results which come from his own experiences, and reasoning on them when he asks us to believe in his conclusions, he appeals to some universal experience of humanity. In every exact science there is a basis which is common to all humanity, so that we can at once see the truth or the fallacy of the conclusions drawn therefrom. Now, the question is: Has religion any such basis or not? I shall have to answer the question both in the affirmative and in the negative. (Vivekananda 2016, vol. 1, Chap. 1, p. 1)

This was reassuring for his earliest followers. He then proceeded to elaborate the scientific foundation of Hinduism saying:

> Just as Akasha is the infinite, omnipresent material of this universe, so is this Prana the infinite, omnipresent manifesting power of this universe. At the beginning and at the end of a cycle everything becomes Akasha, and all the forces that are in the universe resolve back into the Prana; in the next cycle, out of this Prana is evolved everything that we call energy, everything that we call force. It is the Prana that is manifesting as motion; it is the Prana that is manifesting as gravitation, as magnetism. It is the Prana that is manifesting as the actions of the body, as the nerve currents, as thought-force. From thought down to the lowest force, everything is but the manifestation of Prana. (Vivekananda 2016, vol. 1, Chap. 3, p. 1)

---

[26] For examples and discussion of these options, see at least Dupree (1986).

This clearly is a very 'loose' scientific translation of basic Hindu metaphysics, but it suffices for many to establish the scientific credentials and compatibility of the latter. This underpins the powerful and institutionalized Vedic science movement that has done such damage to Indian universities and schools (Nanda 1998, 2003, Chap. 4).

Prāṇa was supposedly a divine emanation, yet was a form of energy, indeed the ultimate energy, and so was within the realm of scientific adjudication. The ultimate goal of Vivekanandra's particular Raja yoga was to 'see' this energy and control it; thus he makes claims such as:

> Suppose, for instance, a man understood the Prana perfectly, and could control it, what power on earth would not be his? He would be able to move the sun and stars out of their places, to control everything in the universe, from the atoms to the biggest suns, because he would control the Prana. This is the end and aim of Pranayama. When the Yogi becomes perfect, there will be nothing in nature not under his control. If he orders the gods or the souls of the departed to come, they will come at his bidding. All the forces of nature will obey him as slaves. (Vivekananda 2016, vol. 1, Chap. 3, p. 1)

Such hubris makes Francis Bacon's hope that the new natural philosophy appearing around him in the late sixteenth century would lead to ever better control of nature and hence to more comfortable and bounteous lives seem positively restrained and modest. Yet the followers and publishers of Vivekananda's views see nothing preposterous in his pronouncements:

> What Hinduism needed, amidst the general disintegration of the modern era, was a rock where she could lie at anchor, an authoritative utterance in which she might recognise herself. And this was given to her, in these words and writings of the Swami Vivekananda.[27]

Conclusion

The mutual interaction of science with cultural worldviews has been a feature of world history. The world's major religions have had an ongoing engagement with science, investigating how their own ontological, epistemological, and ethical commitments – their worldviews – are to be reconciled with both scientific findings and putative scientific worldviews.

Philosophical systems have likewise engaged with science. The Kantian programme in metaphysics and epistemology was erected in response to Newton's science. The positivist programme whose foundations were laid down by Ernst Mach was a philosophical reflection upon the achievements of 200 years of Newtonian science. The engagement of philosophical systems with science has been especially urgent when the systems have had political and institutional embodiment – such as Marxism within the former Soviet Union and in contemporary China, and Thomism within the Roman Catholic Church. In these cases there are educational imperatives

---

[27] See www.advaitaashrama.org

Conclusion

for addressing the question of the relationship of science and cultural worldviews, though in closed societies such educational imperatives are discounted.

Questions that naturally arise and that can, with whatever moulding might be appropriate to local educational realities, be raised in teacher training programmes and in science classrooms are:

- What constitutes a worldview?
- How do worldviews impinge upon and in turn be modified by ontological, epistemological, ethical and religious commitments?
- What worldview commitments, if any, are presupposed in the practice of science?
- What is the overlap between learning about the nature of science (NOS) and learning about worldviews associated with science?
- What is the legitimate domain of the scientific method? Should scientific method be applied to historical questions, especially to historical questions concerning scriptures and sacred texts?
- To what extent should learning about the scientific worldview be a part of science instruction?
- Should science instruction inform student worldviews or leave them untouched?
- What judgement do we make of science education programmes where the scientific view of the world is not affirmed or internalized, but only learnt for instrumental or examination purposes, where learning science is akin to an anthropological study?
- What judgement do we make of proposals that students should become just 'border crossers' moving from their own culture with its particular worldviews to the science classroom in order to 'pick up' instrumental or technical knowledge and then back to their 'native' culture without being affected by the worldviews and outlooks of science? This is the anti-Enlightenment idea that science should leave culture untouched.

Chapter 4
Feng Shui Practice

Feng shui was never just a speculative or metaphysical worldview; from the beginning, it was connected to practice; it impinged on all features of life. 'Qigong' means 'working with qi'. Masters and practitioners ascertained the good and bad chi quality of an area to determine the location and orientation of tombs, the siting, configuration, and fit-out of domestic and government buildings, to advise on lifestyles, to determine where in the body to place acupuncture needles, to identify auspicious times for significant state and personal occasions, and to foretell futures.

The Domain of Feng Shui

In China, for at least 3000 years, feng shui in one form or another has dictated major commercial and domestic siting and construction decisions as well as the proper internal arrangement of offices, homes, kitchens, gardens, furniture, and decorations (Mak and So 2015). One expositor of Daoism has described traditional feng shui siting procedure as:

> The expert would try to find a site that sloped to the south while it was protected from the north, with a hill on the east (wood) larger than one on the west (metal), so that the Green Dragon of Spring might prevail against the White Tiger of Autumn. Valleys were *yin*, but so were rounded hills – in contrast to the *yang* of precipitous heights. The west and north were *yin*. The east and south were *yang*. The perfect site was three-fifths *yang* and two-fifths yin. (Welch 1957, p. 133)

For the same period, countless millions have relied on feng shui astrological guides to make business decisions and for the timing for significant personal and family events (Lim 2003; Lip 2008) and to guide their decisions in romantic and personal life (Hsu 2003; Leung 2010). Feng shui, for example, has an impact in the Asian hospitality industry where decisions about the location, layout, furnishings, decorations, and marketing of hotels, resorts, and restaurants are all dependent on advice about good feng shui. If an institution's bad feng shui is known or advertised by rivals, poor patronage and economic loss follow.[1]

This commitment to the reality of universal chi is well stated by Lillian Too:

> Feng shui prescribes auspicious orientations for harnessing mysterious metaphysical forces that float in the air and the space that surround us. Practitioners describe these forces as chi and colourfully describe it as the dragon's cosmic breath. (Too 1998, p. 6)

She goes on to say:

> The influence of yin and yang cosmology in the practice of feng shui is a universal reflection of the way the Chinese view the Earth's energies. …Yin and yang are viewed as primordial forces that possess completely opposing attributes. (Too 1998, p. 48)

Too, in one book that had 11 printings between 1994 and 1998, offers advice on the chi-enhanced shape of dwellings:

> Avoid triangular shaped houses, or houses that have too many corners. The angles created give off unfortunate *sha chi* and are not conducive to the attraction of good chi flows. … Another unlucky configuration is the U-shaped house. Residents living in such homes suffer from unhappy marriages, and will be plagued with frequent quarrels. (Too 1994, p. 45)

And in the same book:

> A single tree facing your front door could cause havoc in our family life by blocking good chi from entering your home. If it is on your own land, cut it down. If it is outside … you may have to re-orientate your main door. … A steep angled roofline facing your front door brings severe bad luck, causing illness amongst children and family members, and even worse, creating problems for your career and business …. because *sha chi* created is powerful. (Too 1994, p. 50)

The International Feng Shui Society (UK) also affirms the omnipresent reality and therapeutic property of chi, stating:

> Chi, or *qi*, is the oriental word for the vital intangible natural energy that emanates from everything in our universe, a combination of both real and abstract forces: energy from the earth's magnetic field, sunlight, cosmic influences, colour vibrations, the nature of our thoughts and emotions, the form of objects, the quality of the air around us. Depending on whether it flows harmoniously or not, chi influences how a place feels and how we feel in it.[2]

Chinese architecture and construction styles have long been informed by feng shui (Chen and Wu 2009). The styles were largely unchanged over thousands of years, and they were widely admired. The individual palaces, homes, offices, gardens, tombs, bridges, and so on were not conceived or understood in isolation or just by reference to architectural norms or principles. They were understood as part of the overall cosmos inhabited by man with this understanding explicitly given by feng shui theory. The architectural norms were to reflect this underlying, stable cosmic reality; they were not expressions of changing fashion blown hither and thither by commercial or other interests. Nigel Pennick writes:

---

[1] In the marketing of Hilton International Hotels, prominence is given to their high feng shui ratings (Perry-Hobson 1994).

[2] http://www.fengshuisociety.org.uk/

> Chinese practitioners of Feng-Shui, believing that spirits travelled in straight lines, combated them by making winding paths to prevent demons from approaching temple doors. For additional protection, they constructed a spirit wall in front of the entrance, creating a corner unnegotiable by straight-line fliers. (Pennick 1979, p. 66)

The situation was comparable to that informing the construction of churches and mosques. In these cases, it was not just architectural principles or norms that governed design but Christian and Islamic theology. The clear difference with China was that feng shui guidance went all the way down the construction ladder from palaces to city gardens, to humble homes. This is nicely seen in a report of Ernst Boerschmann who went to China in 1906 at the bequest of the German government, who long had Treaty Ports in China, to report on Chinese culture and construction. He travelled widely and reported on feng shui and its impact on the built environment:

> One imposing conception of the universe is the mainspring of all Chinamen, a conception so comprehensive that it is the key defining all expressions of life . . . especially fine arts and architecture. They exhibit in nearly every work of art the universe and its idea. The visible forms are the reflex of the divine . . . In the microcosm is recognized and revealed the macrocosm. (Boerschmann 1912, p. 542; in Parkes 2003, p. 190)

And he was impressed with the harmonizing of construction and landscape:

> The large cities and almost all others are located in the most clever concord with the natural conditions to combine most advantageously the industrial interests with the most beautiful environment possible. The manner in which the Chinese artistically build their structures to harmonize with the natural environment is astonishing. (Parkes 2003, p. 190)

In a following publication, Boerschmann comments:

> That feeling of restful comfort and harmony of our soul [that] arises at the sight of Chinese buildings. For we not only enjoy the unity of the extensive edifices and grounds with the immediate surroundings and nature, with which we feel ourselves a part in the picture of the buildings and the landscape. We also feel that the buildings themselves, nay, even their ornaments must somehow be imbued with nature's living spirit for them to evoke this mood of consummate peace. (Boerschmann 1924)

The Chinese, of course, had no monopoly on building design, garden layout, and landscaping that evoked 'feeling of restful comfort and harmony'. Capability Brown (1716–1783), the eighteenth-century English landscape designer, had a renowned reputation for doing just that; so did Peter Lenné (1789–1866), the Prussian landscape designer, and many more all over the world. Restful comfort and harmony can be had without feng shui.

Along with guidance in construction and outfitting of dwellings came various forms of geomancy, divination, or fortune telling (Han 2001). Traditionally feng shui was linked to geomancy; it was an Eastern version of astrology that was so routine across the entire ancient and medieval worlds and that still lingers in the modern world, both East and West. William Spear the author of *Feng Shui Made Easy* writes: 'feng shui tells you more than how to arrange your furniture, it tells you how to change your life' (Spear 1995, p. 8). Another advocate stresses:

> The aim of geomancy [feng shui] has always been the reestablishment of balance, the restitution of the cosmic order by modifying human activity according to complementary rather than contrary deeds. The dual forces of construction and destruction have been harmonized as far as possible. (Pennick 1979, p. 161)

Form and Compass Schools

The first formalization, and naming of feng shui, occurred in Guo's [Kuo P'o] fourth-century AD *Book of Burial* (Zhang 2004). Guo made more systematic what had been a long pre-existing, heterogeneous, cosmological, and divination tradition. By the beginning of the Tang Dynasty (618–906 AD), feng shui had divided more sharply into two schools: the Form or Landscape School, which was the orthodoxy of the time, and the later Compass or Directions School which was the emergent or innovative stream. Both presuppose the existence and influence of chi; they digress about its origins, indicators, influences, and means of utilization.[3] As with the Islamic schism in the seventh-century Middle East, the Christian schism in sixteenth-century Europe, and the Communist schism in twentieth century Russia, these schools continued, predictably with a degree of internecine feuding and mutual charges of fraud and heterodoxy, up to the present time.

The Form School attended to gross and dominating features of locations – valleys, ridges, mountains, rivers, lakes, and winds – as indicators of chi lines and chi movement. As Guo, who has been quoted above, stated:

> The Classic says that when *qi* circulates through landforms, entities are thereby given life. The geodetic forces of the earth are the basic veins. The geodetic forces of the mountains are the basic bones. (Paton 2007, p. 427)

This proto-geophysics is overlain by a descriptive 'metaphysics' wherein the ridges are called dragons, with eastly hills being home to the green dragons, westly hills being the domain of white tigers, northern hills are black turtles, while the southern features are vermillion phoenixes. These are all metaphors for something in a metaphysical geography. Thus sculptures, drawings, and paintings of dragons are omnipresent in China. Using the theories and devices available, the feng shui masters identified beneficial sites and made recommendations about how best to master the site's good chi.

Traditionally they attended to the seven natural features of any site: mountains, water, soil, direction, wind, time, and land-shape. Surface and visible water were important for the chi-rating of a site but so also was subterranean water, the location of which was ascertained by dowsing methods – either a dowsing rod or a pendulum (Birdsall 1995, pp. 193–204). In modern times feng shui masters consider high-rise buildings as well as mountains and fish ponds as well as streams. It is important for

---

[3] For the origins and history of the two major feng shui streams, see at least March (1968) and Paton (2007).

them to refer to municipal drainage plans so as to know the location of major sewer pipes and their relation to the site. Building over a sewer pipe is a feng shui no-no.

A fourteenth-century Form School classic is illustrative of feng shui core ideas and also of the schismatic tensions in the tradition at the time:

> The whole of this work discusses the form, force, feeling and nature of water. It is never ignorant of the important principles as are the practitioners of the theories of direction [compass school] who absurdly match longevity, the receiving of favours, becoming an official and imperial prosperity with good and evil spirits and good and ill-fortune, consequently causing the lucky not to be buried and those buried not to have good fortune. In deluding the world and misleading the people, nothing is worse than this. (Paton 2007, p. 429)

In the seventeenth century, Shen Hao, another promoter of Form School orthodoxy, took serious exception to the Compass School heterodoxy of his time, saying the practitioners were lazy charlatans who preyed on the credulity of the people. In a work published in 1652, 20 years after Galileo's *Two Chief World Systems,* and 30 years before Newton's *Principia*, the 'mid-point' of the European Scientific Revolution, he wrote:

> There have since appeared in the world eyeless ones, who climb and gaze out but cannot tell departing from approaching [dragons, or waters]; go up and go down but do not know the fronts [of landforms] from their backs. They see only the confused motions of the compass needle, think it is a marvellous object and deify it. ... They say: I remember clearly, there is no need to climb mountains, we can sit down and discuss geomancy. These are called the Directionists [*Fang wei chia*]. Each makes his own theories. From the T'ang dynasty to the present there has always been a great ruck of them. (March 1968, p. 262)

And he warns against taking the easy route to feng shui knowledge on account of it being so well trode by imposters and charlatans:

> Those who first made up the Planets and Trigrams theories, and talked of Directions, were deluded men; others after them perpetuated the theories. As the then masters [of geomancy] refuted them with might and main, they dissembled by giving themselves the name *Li ch'I chia*; and those who heard their theories supposed that what was being imparted was the *li* [principle] and *ch'I* [breath, material force] [of the Neo-Confucians] – which, however, it is not. I know of no 'principle' if one departs from shapes [*hsing*] nor any 'breath' if one discards terrain [*shih*]. But climbing is laborious, and you cannot let someone else go and look in your stead; whereas sitting and talking is easy, and the compass is convenient to handle comfortable for the consultant, comfortable for the client. This is why those who talk of directions and *li ch'I* have multiplied and have written book after book. (ibid.)

He offers some explanation for the constantly regenerating large pool of believers in the fanciful:

> But if you must denounce heterodoxy without its rationale, it can argue against you, so you must know all twenty-four authorities' Yin-Yangs and Five Elements before your refutation will succeed. Strong-minded and impatient people usually just decide anyhow, knowing nothing, and heedless people also dislike the trouble of climbing mountains and enjoy the ease of sitting and talking. This is why the words of the Directionists or *Li ch'I chia* take fresh life each day and proliferate more each month. Those struck by the poison are like smoke dissipated and fire extinguished, and never have a chance to reason out the source of their misfortunes. (ibid.)

Shen Hao, of course, does not reject feng shui or its principles as developed in his own Form School tradition. It is only the beliefs of the *rival* feng shui school that are foolish and that cater for the 'lazy minded':

> I grieve that the gentry will not interest themselves in (geomantic) lore, but put their trust in rustic matters and so imperil their parents and prepare disaster for their prosperity. (ibid.)

Modern-day Form School adherents frequently speak of it as the 'ecological' or 'environmental' school of feng shui. One Australian Form School maintains that:

> Form School Feng Shui follows natural principles and works with chi energy. The knowledge of form school Feng Shui regarding chi energy is based on the concept that form defines energy. Although this understanding of how the surrounding forms of the natural world affect human beings has been the basis of traditional wisdom all over the world throughout history, it was in China that it was honed to a fine science, due to the long period of continuous civilization there. Thousands of years of accumulated and synthesized knowledge are contained in the universal principles of Form School. … It is the true traditional Chinese Feng Shui.[4]

As its older Chinese precursors did, this Australian operation is scathing of the competing Compass School, saying that these schools are mostly faith-based, they are not commonsensical, and they do not have a solid theoretical foundation. Rather they have:

> Imported and bundled Asian superstition, folklore, and regional beliefs together confusing many on the true practice of Feng Shui. They also use many so-called Feng Shui gadgets such as miniature fountains, hanging crystals, and Bagua mirrors as remedies in a house to attract prosperity or expel evil. As many people always like a silver bullet approach to solve their personal problems and in hopes to bring love and fortune, this practice became popular in a rather short time. But, this new age hype also declined quickly.[5]

Lillian Too, a Singapore-based feng shui consultant to 'the stars' and big business, author of numerous feng shui books, and international feng shui entrepreneur,[6] recommends:

> Look for land with compact, reddish loam. Such soil is full of the celestial breath of life or CHI. Avoid hard rocky soil. …If the grass on your land is especially green and lush, it is a place of good feng shui. Fertile land is good feng shui. Dry, arid, rocky land is not. (Too 1994, p. 23)

She advises not to buy and build on land where hills are being cut immediately above because 'Injured dragons hovering above, create bad feng shui' (ibid., p. 23). She does say that 'Buildings with a view of water in front is excellent feng shui' (ibid., p. 25) but warns that:

> Dirty water creates *sha chi* [bad chi] which brings unlucky vibrations, especially ill health for residents. It is worse if the water smells of decaying materials, or is muddy. It requires the presence of clean, free flowing water before wealth feng shui exists. (ibid., p. 25)

---

[4] See http://www.bluemountainfengshui.com/about/form-school/, accessed 4.12.2017).

[5] Ibid.

[6] Her website says she is the author of '80 best-selling books [on feng shui] that have been translated into 30 languages'; after a corporate career, she learnt feng shui from a master who had made 'many businessmen into truly prosperous billionaires and multi-millionaires' (Too 1998, p. 8).

And further suggests:

> Avoid locations on hilltops; in cul-de-sacs; facing a T-road junction, or a straight-line structure like transmission lines and railway tracks. These areas are most susceptible to being influenced by poisonous CHI which bring ill health and bad luck. (ibid., p. 27)

Maurice Freedman, in a study of feng shui site selection in the New Territories of Hong Kong, found that for acknowledged good sites:

> A site is protected from high winds by its hills. Places from which streams and rivers flow too directly and too fast are avoided. An ideal site is one which nestles in the embrace of hills standing to its rear and on its flanks; it is then like an armchair, comfortable and protecting. (Bennett 1978, p. 122)

Simple common sense and experience provide the same guidance for many of the foregoing decisions as does expensive feng shui consultation – don't build on windy hilltops or alongside polluted streams, etc. – so in these cases feng shui is harmless, though expensive. But in many cases, the expensive advice though novel is useless and detrimental to self-interest. Consider the advice: 'do not build along or near straight-line structures'. If this means do not build beside a busy highway or railway, then there might be common-sense grounds for not doing so (provided of course that you can afford to build in a more expensive, less-trafficked location). But if your business depends on easy access to road and rail, then the advice is silly. To build a home alongside an abandoned rail line or little-used road might be perfectly sensible. To bring into the already complex decision-making (location, cost, mortgage repayments, distance to work, etc.), feng shui considerations about straight lines encouraging bad chi are just a distraction and a useless one.

Dismissing Form School criticism, Too says: 'Compass school feng shui takes you deeper and allows you to discover powerful methods of seriously enhancing your luck' (Too 1998, p. 16). Another adherent says:

> The 'Compass School sect', is more technical, since a compass, protractor and meticulous graphs, as well as auspicious days determine a configuration. Feng shui practitioners of this discipline are mathematically inclined and have the minds of structural engineers. (Rolnick 2004, p. 13)

Though this author does disarmingly add: 'Though it must be said a great deal of "magic" goes along with this science' (ibid.). Stephen Skinner, another contemporary advocate who has degrees in philosophy and geography from University of Sydney, tries to balance the two schools by saying:

> In the Form School the principles are clear but the practice is difficult ... with the Compass the principles are obscure but the practice is easy. (Skinner 1982, p. 13)

As with religious and political schisms and heresies, how is the right and wrong side of this feng shui Form/Compass cleavage to be settled? There is disagreement in science and prolonged competition between theories, paradigms, and research programmes, but a feature of science is that these, Thomas Kuhn not withstanding, can by deliberate and widely accepted methods be settled (Machamer et al. 2000; McMullin 1987).

Rolnick does not ask the philosophically interesting question: How much magic can be mixed into a system and it still remain science? Magic, religion, and astrology were, famously, all pursued by Isaac Newton (Figala 2002; Mamiani 2002). But they were not 'mixed in' with his science; the latter could be, and of course was, separated from the former. The latter thrived and was the foundation of modern science; very few developed Newton's astrology, or his distinctive, heterodox, Arian religious position. In Newton's equations, there are no terms for God, spirit, or magic.

Feng Shui in Hong Kong

Hong Kong is the pulsing centre of the feng shui industry. One scholar writes that: 'Nowhere in the world is feng shui so intensely integrated into every aspect of social, religious and commercial life as in Hong Kong' (Bruun 2008, p. 136). One website announces that:

> Today Hong Kong is the unofficial heart of feng shui practice. In a city where more than 10,000 masters (called geomancers) now ply their craft, you can feel the power of feng shui everywhere.[7]

Feng Shui has dominated construction and related decision-making in Hong Kong for centuries, if not millennia. In 1964 the government was to erect a water reservoir in Sheung Kwai Chung district; residents objected because the construction would cut veins of the green dragon that protected the district. The *Hong Kong South China Morning Post* of 9 June 1964, reported that:

> As a result of these objections, the District Officer, Tsuen Wan, accompanied by PWD (Public Works Department) engineers, visited Kwei Chung and located another site for the proposed service reservoir. This does not interfere with the villager's *feng shui* and the issue was thus settled. (in Bennett 1978, p. 24)

Jack Potter, an anthropologist, wrote in 1970 that 'Most roads in the New Territories have a serpentine quality that is due more to feng shui requirements than to bad engineering' (in Bennett 1978, p. 24). At the same time, the villagers of Ping Shan claimed: 'that numerous people who once dwelt at the rear of their village died off because the Hong Kong government erected a police station on the hill behind their village which cut off their flow of feng shui' (ibid).

In the 1960s feng shui consultants lobbied successfully for a new high-rise, high-cost residential block in Repulse Bay to have a huge gap (eight floors by four floors) constructed in its centre to allow for the uninterrupted passage of *qi* (chi) from the mountains behind to the ocean in front. The *qi* lines traced the movement of the northern dragons that came through the valley to their regular resting place in the bay.

---

[7] http://www.hotelclub.com/blog/the-power-of-feng-shui-in-hong-kong/#full-info

Feng Shui in Hong Kong

Fig. 4.1 Feng shui dragon-hole buildings, Hong Kong

Subsequent buildings in Repulse Bay, and then many other locations, also had to provide a 'dragon hole' for chi to flow from interior hills to ocean (Fig. 4.1).[8]

No real estate, from humblest apartment to grandest residence, is sold without feng shui advice or certification. In 2010 a multimillion-dollar lawsuit rumbled through the court system because a large tower proved unprofitable and the developer sued his feng shui advisor for providing poor advice which it was claimed resulted in low occupancy of the building. The legal case, and philosophical interest, depended on being able to identify poor feng shui advice independently of the occupancy rate. Suing the advisor *before* the building was opened would have crystallized this matter, but that did not, and does not, happen. That would be too much like a controlled scientific test, something requiring more diligence than is commonly found in the feng shui business. The case suggests that the adequacy or otherwise of feng shui advice is not in-principle determinable but only in practice, namely, who believes in it and who does not.

Nearly all Hong Kong businesses, big and small, and government departments consult feng shui experts to determine auspicious dates for deals and programme launches, to create interiors and environments that will bring good fortune for a business, and to guide against creating bad feng shui for others that might result in costly legal battles and compensation claims (Emmons 1992).

---

[8] Dragons have a special place in Chinese thinking and culture; they subsist in a shadow world between the real and imaginary. Some mountain chains and ridges are deemed to be dragons and are conduits for the passage of chi.

In 2010 the *South China Morning Post* forced the Hong Kong government to admit that it regularly pays millions of dollars in feng shui compensation when it realigns roads, digs tunnels, builds bridges, plants trees, or in any other conceivable way disturbs feng shui or the flow of positive energy. The government admitted to payments of £6 million but refused to open its books, so most commentators think the figure is distinctly understated.[9] The government also freely admits to paying millions of pounds in compensation to residents and businesses whose feng shui is adversely affected by private construction that it approves; the government becomes a feng shui insurer. This predictably generates a huge mega-million-yuan feng shui litigation industry.

When the new rail link between Hong Kong and the mainland city of Guangzhou was constructed in 2011, 17 residents whose feng shui was adversely affected were each given half-a-million US dollars to pay for a consultant to perform 'cleansing rituals'. Sceptical observers, and those with a more scientific habit of mind, might say that the 'cleansing rituals' amount to a commercial shakedown, with feng shui masters and local landlords colluding to launch outrageous claims against the government before splitting the proceeds. Doubtless this is indeed what happened. The mafia could take lessons from feng shui consultants.

On the advice of a feng shui master, Hong Kong Disneyland in 2005 realigned its entrance gates 12 degrees, tilted the whole site several degrees to obtain a better alignment with the mountains behind and sea in front, and placed a curve in its entry road so as to better capture for itself the positive *qi* flowing in the vicinity. Additionally, Disney set up 'no fire' zones in kitchens to balance the central five elements of feng shui (fire, water, earth, wood, and metal) and chose an auspicious opening date according to the traditional calendar (Bruun 2008, p. 140). It is indicative of the reach of feng shui belief that one of the world's biggest companies made these mega-million-dollar adjustments to its construction programme just on the advice of, highly-paid, feng shui masters. All of this extra expense would have been added to the admission price and eventually paid for by Hong Kong families patronizing the park. Feng shui practice has impact beyond its principals.

The getting, and acting on, feng shui advice is expensive. Countless Hong Kong feng shui consultants have become millionaires. Tony Chan, one such high-end consultant, extracted, in the years up to 2007, £60 million from one very rich woman developer, while another consultant charges $16,000 per hour for auspicious real estate investment advice. The standard 'tycoon' rate is £2–3 per square inch of building (Moore 2011).[10] These might be towards the upper end of the feng shui fee schedule, but all the way down the status pole, such consultative service is obligatory and not cheap. Even when consultation is done via the telephone, as many web services offer, it is still expensive. Money so spent is not spent on other necessities of life: food, education, leisure, and lodging.

---

[9] http://www.telegraph.co.uk/news/worldnews/asia/hongkong/8206601/Hong-Kong-government-spends-millions-on-feng-shui.html

[10] This is not a typo; the cited price is per square inch of dwelling.

All of these examples, disputes and legal cases, raise the general question of the ability and legitimacy of science to settle such issues. This is a fundamental decision that needs to be made across countless areas of life; and science classes are ideal places for students to receive training in such decision-making. It will be argued in Chap. 11 that, given the massive, consequential, and very public nature of feng shui practice in Hong Kong, Taiwan, and elsewhere, its basic claims about chi, its mechanisms, and its human impacts should be appraised in science classes and, even better, appraised in some coordinated teaching between science, social science, history, and philosophy classes.

Feng Shui in Taiwan

In 2004 the *Taipei Times* (17 October 2004) estimated that there were 30,000 feng shui practitioners in Taiwan. And they are kept busy. After an especially bad train derailment in Taipei, the Transport Minister sought the advice of a feng shui master who divined that the main station had bad feng shui, including a faulty bagua symbol on its premises, and so the ministry realigned at considerable expense the station's south entrance (Bruun 2008, p. 135). The opposition transport spokesman counter-claimed that the accident had nothing to do with feng shui but rather with inadequate technical systems and human error due to poor management practices.

Whether the opposition claim was an in-principle claim that feng shui has nothing to do with anything, and so nothing to do with the derailment; or a particular claim that in the case of this accident, feng shui misalignments could not be blamed, is unclear. The former claim, if made explicit, could be politically damaging; it would be akin to a US politician declaring for atheism or for higher taxes. But the disagreement about the cause of the derailment is one that only a scientific investigation can resolve, and hence the recurrent question is: Can science judge feng shui claims? The affirmative answer is argued in this book.

As with elsewhere in Asia, feng shui matters regularly end up in Taiwanese courts that determine culpability for bad feng shui advice or for engaging in practices that negatively impact the feng shui of both the living and the dead. Planting trees that overshadow a grave is taken as adversely affecting the spirit of the person buried there and is an actionable offense in law. A simple classroom exercise would be to have students be 'expert witnesses' for the prosecution and defence in such trials. They could become familiar with notions of evidence, connections between evidence and hypothesis, the authority of testimony, the demarcation of scientific from other claims, the separation of science from pseudoscience, and much more.

Feng Shui and Western Architecture and Construction

Richard Taylor, the author of a contemporary manual giving advice on utilising feng shui in the modern city, relates that: 'By interpreting the hidden and mysterious forces of the universe, Feng Shui provides a practical approach to environmental planning' (Taylor 2002, p. 9). He goes on and states its foundational principle as:

> The theory of Feng Shui, just like Chinese medicine, is based on the five elements. The five elements control and oversee everything in the universe, and channel and balance the chi of the individual and of his surroundings. Each of the elements – fire, earth, metal, water, and wood – represents a specific energy. These energies are found in a perpetual interrelationship, and their composition, or 'arrangement', creates harmony or disharmony. (Taylor 2002, p. 20)

As with so much, if not all, feng shui advocacy, this has an initial appearance of science; scientific language is in fine display. The manual talks of positive chi lines (curvy) and negative chi lines (straight and angular) and over many pages advise on how, in construction and fit-out, to maximize the former and minimize the latter. For instance, do not build in the shadow of a tall building that blocks sunlight, do not build at the crown of a T-intersection, build near water, and so on. If negative chi cannot be avoided, it lists the eight traditional feng shui remedies for alleviating the situation: mirrors, light, plants, water, crystals, wind chimes, flutes, and colour (Taylor 2002, p. 137). It offers homely, so to speak, advice on creating, among many other things such as toilets (do not have them opening into a living room), bedrooms (located away from the front door), and good house entrances because:

> The entrance to the house is of primary importance. It must be well lit, welcoming, inviting, and pleasant to the eye, and it must be sufficiently broad to permit beneficial chi easy entrance to the house. … A sufficient amount of space in the entrance area is an important characteristic in good Feng Shui. It enables a greater amount of chi to enter the house, broadly and freely. (Taylor 2002, p. 89)

Needless to say, there is no specification of how wide, or of entrance area to volume of house, nor any guidance about obtaining chi measuring instruments in order to determine optimum chi movement. The entire chi domain is a 'measurement-free' zone. There is the appearance of science, maybe even the pretence of science, but there is no science. Measurement, and concern with exactitude, is a requirement of scientific practice.

Multimillionaire (by her own reckoning) consultant Lillian Too's advice on building a good feng shui home has already been given:

> Avoid triangular shaped houses, or houses that have too many corners. The angles created give off unfortunate *sha chi* and are not conducive to the attraction of good CHI flows. … Another unlucky configuration is the U-shaped house. Residents living in such homes suffer from unhappy marriages, and will be plagued with frequent quarrels. (Too 1994, p. 45)

She is clear about the necessity of good planning and siting. In particular:

> Protect your homes and offices from killing *shar Chi* that are created by the presence of malignant poison arrows which point at your place of abode or your place of work. … These poison arrows are symbolic instruments or structures which are pointed, straight, angled, sharp or hostile. (Too 1994, p. 118)

Too is not alone in offering this advice. There was uproar in Hong Kong when the new Hong Kong and Shanghai Banking Corporation headquarters was built with sharp 'poison arrow' lines sending *shar chi* into adjacent buildings. Some of these placed cannons on their roofs to kill the bad chi. Eventually under legal pressure, the HSBC had to alter its arrow lines upwards.

Such manuals are sold by the thousands in both the East and the West, becoming in many cases university textbooks. Consequently, it is increasingly common for architectural design and building construction in San Francisco, Los Angeles, Vancouver, Toronto, London, Sydney, Auckland, and most major cities around the world to be directed by feng shui principles.

Serious Western architectural and construction books[11] and scholarly articles[12] are devoted to feng shui informed location, design, placement, and construction. Indicative of the construction industry embrace of feng shui, and its 'scientific' patina, is the title of a recent book: *Scientific Feng Shui for the Built Environment: Theories and Applications* (Mak and So 2015).

For decades developers in Vancouver have been building 'four-free' apartment blocks (no fourth floor), and in the Toronto suburb, Richmond Hill has banned the numeral 4 in street numbers. Lillian Too informs readers that:

> … the number 4 is regarded as being inauspicious. Anything that ends in this number spells death, loss and problems. Thus, level 14 in a multi-storey building carries bad luck connotations that are even more severe than the western 13 … House addresses that have the dreaded 4 usually cannot get sold, while apartments on level 4 are hugely unpopular. (Too 1998, p. 68)[13]

Indicative of the normalization of feng shui is that universities in China and the USA now teach courses on feng shui and architecture. One scholar has written:

> In recent years feng shui has grown surprisingly popular in Western Europe despite a lack of clear understanding about how and why it is practised. Its proliferation within the architectural profession can be observed at all levels from the selection of building sites to interior design. (Hwangbo 1999, p. 191)

An Australian architect, in the Introduction to his Feng Shui manual, wrote:

> I first became aware of feng shui about 10 years ago when a $20 million project I was building in Sydney's Chinatown district was assessed by a Chinese geomancer and feng shui practitioner. His major recommendations included realignment of the escalators so that the 'wealth did not flow out' and a delay of one week before we could start demolition and construction. We were [only] allowed to start on a particular day at a particular time … (Birdsall 1995, p. 9)

In Southern California, having positive feng shui lines, namely, soft or curvy ones, is a featured consideration in the marketing of certain cars, and there is even a radio station that gives regular updates on the state of regional and local chi energy lines. It is reported that in Los Angeles, feng shui consultants, some earning as much

---

[11] See Alexander (1987), Gallagher (1993), Rossbach (1987), and Taylor (2002).

[12] See Marafa (2003), Chen and Wu (2009), Hwangbo (1999, 2002), and Mak and Ng (2005).

[13] In Pinyin, the number 4 is *sì*, while death is *sǐ*, so by familiarity, number 4 should be avoided.

as $750 an hour, were recruited from Hong Kong and Taiwan to redesign entire buildings and, where necessary, install structures that could ward off negative energy surreptitiously infiltrating a room or edifice. Clearly there can be a feng shui aesthetic in architecture, and many might find it attractive; but the aesthetic can be had without feng shui ontology and cosmology.

Feng Shui on the Web

Advent of the World Wide Web has dramatically expanded feng shui influence and business opportunities. Even in the most isolated places on earth, a feng shui consultation is only one click and one credit card charge away. There are thousands and thousands of feng shui sites on the web that list hundreds of thousands of happy corporate and individual clients. One UK site alone mentions among its satisfied customers: Coca Cola, Orange, British Airways, Hiscox Insurance, Hilton Hotels, Marriott Hotels, BRE (Building Research Establishment), University of Westminster School of Architecture, the National Health Service, and many more. There are feng shui sites for consultants, businesses, retail outlets, teachers, societies, associations, colleges, practitioners, conferences, seminars, publications, diplomas, and much else. The following are examples of feng shui services found in a 15-minute English language, web search:[14]

> (1) Jerry specializes in both residential and commercial Feng Shui utilizing Xuan Kong Feng Shui. In addition, Jerry is an expert in Four Pillars of Destiny, Plum Blossom and *Qi Men Dun Jia* divination. Jerry is currently researching on a higher level of divination known as Da Liu Ren. Jerry has taught Chinese Metaphysics which included topics in Feng Shui and Destiny in London, Hong Kong, Singapore, Australia, Russia, Spain, and Mexico.[15]

> (2) There are many schools of Feng Shui and we utilize as many as we can in order to give our clients optimum results. Some of the schools of Feng Shui we use are: 4 Pillars of Destiny, Flying Stars, Bagua, Form School. In laymen's terms, it is about laying out rooms, furniture, introducing colours, elements and objects in order to create a positive change in one's life, family, home and/or business.[16]

> (3) Feng Shui store, the place where Practitioners, Masters, consultants and enthusiasts from every corner of the world buy their enhancers, cures, crystals, water fountains, Chinese coins, kwan yin, pi yao, wu lou, 2015 Tong Shu Almanac, six coins, salt water cure, 2015 cures kit, six rod wind chime, fu dogs, bagua mirrors, Buddha's, three legged toads, dragons and so much more.[17]

---

[14] Using French, Spanish, Portuguese, Danish, Thai, Japanese, or any of probably 100 other languages and spending not 15 min but 15 h searching, hundreds of thousands of feng shui websites could probably be located. All offering advice, commodities, and taking money.

[15] See http://www.whitedragonhome.com/about/masters/jerry-king

[16] See http://www.globalpalaceoffengshui.com.au/

[17] See http://www.fengshuiweb.co.uk/

(4) For those of you who are previous clients make sure you book in for your Annual Update for 2015 – the Year of the Wood Goat. Remember, for best results you need to have this information PRIOR to the changeover date of 4 February 2015. This service is highly recommended to ensure a healthy, wealthy and positive relationship filled year in 2015.[18]

(5) Time is the single most important difference between the types of feng shui. Black Hat centers around 'placement' whereas classical feng shui understands that place is relative to time – just like we are subject to the changing of time. Time is a very important construct of classical feng shui. In other words, you can't just 'feng shui and forget it'.

Feng shui is a dynamic *living energy* that shifts and changes with the months and year. That's why there is so much emphasis on the annual feng shui, that details about each year – whether it's a rooster, dog, or boar – play a role in the energies that occur during the year and that act differently upon your home, you and everyone in your home.[19]

(6) Good feng shui in your home or office does not bring you wealth if you do not strive for it, but it does give you the necessary support to attract the energy of money. Feng shui helps you create an environment, both at home and in your office, that strengthens you and helps attract the energy of wealth.

The best feng shui money energy foundation is having your home and business free of clutter, displaying symbols that speak to you of wealth, using specific feng shui cures such as crystals, specific images and even essential oils to raise the feng shui energy in your home.[20]

(7) Asia's leading feng shui expert and mentor is coming to Australia to spend 3 hours sharing Real Feng Shui secrets previously kept hidden behind closed boardroom doors. This is the opportunity for you to learn authentic Feng Shui tools and knowledge from a mentor with over $5 Billion Dollars of property consultations under her belt.

Aur's Authentic Feng Shui is entirely different from any other style anyway. Aur has spent the last 25 years conducting property private consultations for many of Asia's wealthiest and most influential individuals and companies, including the First Lady of Malaysia, Property Developer TCC Group (owned by Charoen Sirivadhanabhakdi, net worth: $15.5 Billion), Beauty Gems (jeweller to Asia's Royalty and elite), plus Big Brands such as Makro, Philips, Swarovski and Cisco. Aur also hosted her own TV show on the topic of Feng Shui for nearly 10 years in Thailand, sharing her wisdom and giving live examples of Feng Shui by visiting fan's homes around the country. However, Aur is not like other Feng Shui consultants. She practices a completely unique style of Feng Shui currently unheard of outside of Thailand.[21]

(8) There are many systems of date selection. There is the Xuan Kong Da Gua Date Selection method, the Grandmaster Dong System, the Great Sun Formula, Qi Men Dun Jia etc. These methods require an in-depth knowledge of Chinese Metaphysics.

The 12 Day Officers method is a simple system that is based on the concept that each day is governed or influenced by a certain type of 'qi'. There are altogether 12 types of 'qi' that is repeated every 12 days, hence the 12 Day Officers.

---

[18] See https://www.everythingfengshui.com.au/

[19] See https://redlotusletter.com/classical-feng-shui-and-western-black-feng-shui-the-6-critical-differences-confessions-of-a-former-black-hat-practitioner/

[20] See http://www.knowfengshui.com/feng-shui-money-wealth-tips/

[21] See http://www.meetup.com/vital-life/events/235134045/?rv=ea1

The 12 types of 'qi' have names and are order in a fixed sequence. They are Establish, Divest, Full, Balance, Stable, Initiate, Destruction, Danger, Success, Receive, Open and Close. Establish is the leader of group and it falls on the day where the earthly branch is the same as the earthly branch of the month.

Each of the 12 Day Officers (or types of 'qi') has a function and is suitable for certain type of activities. Of the 12 days, only Success Day and Destruction Day have clear cut positive and negative aspects respectively. The others have both positive and negative aspect and the trick is to match the correct day to the right activities.[22]

(9) Feng shui somehow always seems to offer a solution to both interior design and healthy living conundrums. Truthfully, it's no wonder that people throughout centuries have turned to this ancient Chinese system of special arrangement to bring harmony into their homes—which is why we're not surprised that this practice can also be used to crank up the heat in our bedrooms. For those curious about this Eastern practice's power to boost the energy flow between the sheets, feng shui pro and relationship expert Alison Lessard has some sage advice to impart.

The easiest way to feng shui your way to better sex? According to Lessard, you've been doing it (and by that, we mean organizing your furniture) all wrong if your headboard is under a window or your feet are facing the door. "Both of these placements make your chi (energy) fly out of the room, which could leave you feeling drained and not well-rested," she says. To maximize energy flow, she recommends positioning your bed in a place where you can clearly see the door but aren't directly in line with it. If that's not an option, try hanging a small mirror on the wall that's opposite of your bed in order to reflect the door.[23]

Such examples go on without end. The educational and cultural challenge is how to prepare students to deal with such arrant and fradulent nonsense if they, friends or family, happen to enter its orbit. Comparable psychological and social mechanisms are at work when people are attracted to and then ensnared in any of thousands of quasi-religious cults.

The above-mentioned Lillian Too can be taken as an example from the hundreds of thousands, if not millions, of feng shui consultants and practitioners who offer their services on the web. She has clearly contributed to and benefited from the transcultural migration of feng shui ideology out of its Chinese cultural base. She offers rolling residential courses which move around the globe and around the calendar. For payment of USD 2,200, participants in these courses can attend lectures covering topics such as:

The fundamentals of Feng Shui practice
The anatomy of luck
The different types of Chi in Feng Shui
Understanding the *iching*, the *taichi* and the two *pakuas*
Unlocking secrets from the *yang pakua* compass,
Practicing defensive Feng Shui & disarming 'killing breath'
Feng Shui for interiors and auspicious configurations
Using the *pakua* in the microcosm and macrocosm of dwellings

---

[22] See http://www.absolutelyfengshui.com/dateselection/the-12-day-officers-pt-1/

[23] See https://www.mydomainehome.com.au/feng-shui-for-better-sex

The three methods of demarcating Chi
East west compass method of Feng Shui for individuals
East west compass method of Feng Shui for dwellings
Flying star Feng Shui
Finding and activating your personal 'bag of gold'

It is staggering that this speil fools anyone, but fool millions it does. But, of course, feng shui has no monopoly on hucksterism; the world is awash with it.[24] The current US President has ridden its wave all the way to the White House. Two scholars have drawn a sobering conclusion:

> People just love to believe, and our research shows that eighty percent of them will believe things a gorilla wouldn't. They'll cheerfully empty their wallets to anyone with a twinkle on his tongue and a pseudoscience in his pocket. Astrology, biorhythms, ESP, numerology, astral projection, scientology, UFOlogy, pyramid power, psychic surgeons, Atlantis real estate – they're all good business. (Glymour and Stalker 1982)

Characteristically, feng shui fails to make this pseudoscience team; it mysteriously flies under the critical radar. All of the foregoing examples of contemporary feng shui practice might be seen as the investigation of *external* or *environmental* feng shui and the making of appropriate lifestyle, design, and construction decisions in the light of ascertained local feng shui conditions. But there is also *internal* chi which is the basis of all Traditional Chinese Medicine, qigong belief and the daily practice of millions and millions of people. Once you commit to the existence of an all-pervasive life force, energy, and chi or *qi* that binds together the heavens, nature, and man and governs their harmonization, then the next step is to admit its existence and powers *within* bodies and to then move to control, redirect, and manipulate its flow to enhance personal good effects (wellness) and minimize bad effects (illness). Qigong masters supposedly can externalize their own chi for the good or ill of their audience, patients, or clients. This is a multimillion-dollar business, more properly called a racket.

Feng shui is not always presented so clearly as nonsense. The fundamental, and defining basis, of Traditional Chinese Medicine (TCM) is identification and mastery of chi, hence enabling the balance of the vital yin and yang 'forces'. TCM is enthusiastically endorsed and supported by the Chinese Government; it is a multibillion-dollar export industry. Likewise, chi-manipulating acupuncture is a mega-billion-dollar industry and professional occupation. In the UK, 65% of respondents to a YouGov poll (2017) thought that acupuncture should be included in the National Health Service as a free treatment. This medical/health dimension of feng shui will be discussed in Chap. 5. It is an obvious case of philosophy of science meeting serious social policy and economics. Perhaps more generally, it is a matter for serious social psychology: What leads millions of people to believe feng shui claims and regulate their life by them?

---

[24] There is a large literature by historians, sociologists, social psychologists, and philosophers on the subject large-scale gullibility and credulity. See at least Alcock (2018), Andersen (2017), Pigliucci (2010), Shermer (1997), and Stalker and Glymour (1989).

Divination

Since the beginning, feng shui has been intimately associated with geomancy, one of numerous systematic forms of divination that have a long history in Asia[25] but also in Africa, the Middle East, and Europe. In the Chinese tradition, divination was anchored in cosmology: humans and the cosmos were one, and the energies of the latter were manifested in the movement of heavenly bodies, in the shape and configuration of the environment, and in the constitution of the individual. The diviner's task was to see or intuit and then calculate the local and personal aggregation or trajectory of energy and ascertain how it would unfold in the future. As Joseph Needham relates:

> From the highest antiquity the Chinese had the conviction that it was possible to foretell the future, at least in so far as the affairs of princes and States were concerned, by processes of divination which gave a yes-or-no answer. The oldest technique was no doubt scapulimancy, the heating of tortoise carapaces or ox and deer shoulder-blades with red-hot metal, and the interpretation of the resulting cracks. (Needham and Ling 1956, p. 347)

He details, at length, various other standard divination techniques – physiognomy (from facial characteristics), chiromancy (from finger prints), oneiromancy (from dreams), and glyphomancy (from ideographic name characters) – and mentions their occurrence across Asia, the Middle East, and Europe. In the Dao tradition, divination was more than just foretelling; its cosmological foundations were explicit:

> In Taoist thinking, divination is not simply predicting the future and relying on these predictions to live. Rather, it is a way of appreciating the flux and permanence of the Tao and directly perceiving the interdependency of all things. (Wong 1997, p. 119)

The name 'geomancy' has its roots in two Greek words, *gaia* (earth) and *manteia* (divination); so there is *geōmanteía* which translates literally to 'foresight by earth' or 'earth divining'. Other widely used divination techniques were pyromancy (divination by fire), hydromancy (by water), and aeromancy (by air). One practitioner and author of numerous books on the subject writes:

> Geomancy could be defined as the art of obtaining insight into the present or future by observing the combinations of patterns made in the earth or on paper by a diviner allowing his intuition, or 'the spirits of the earth', to control the movement of his hand or pencil. (Skinner 2011 p. 33)

Clearly there is a gap between 'spirits of the earth' and 'movement of the diviner's hand'. This gap can be, and has been, filled in many ways, most notably by charlatans, the desperate, and the deluded.

The practice of divining good and bad fortune by the spread of blown sand, the fall of nuts, and other geo-related means had older origins in the Arab world (Skinner 1980, p. 14). The mixed Arabic-Sino tradition created, in the second century BC, the geomantic compass, or luopan compass, consisting of a circular, flat, inscribed

---

[25] See March (1968), Needham and Ling (1956, pp. 346–365), Pennick (1979), Skinner (1980, 1982, 2011), and Smith (1991).

Fig. 4.2 Geomantic (Lu-pan) compass

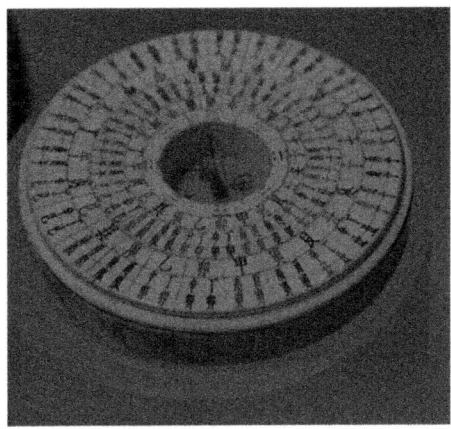

wooden or brass plate, usually with 40 concentric rings and 24 direction lines, and a centred south-pointing lodestone needle (Skinner 2008). Most science museums have exhibits from ancient times, and modern ones can be bought across the counter or online in most 'esoteric' stores. Lillian Too explains the geomantic function of the feng shui compass (Fig. 4.2):

> In feng shui, the attributes of the elements influence each of the eight sectors of the compass. The four cardinal and four secondary directions each have a corresponding element. So the easiest method of creating good feng shui is simply to energize the element of each compass sector. Understanding element attributes is thus essential. (Too 1998, p. 42)

She elaborates the five-element ontology, or Wuxing, of feng shui, saying:

> The theory of the five elements is basic to all branches of Chinese divinitive practice. From astrological fortune-telling to oracle forecasting, understanding the productive and destructive cycles of the five elements – how they interact with each other to create positive auspicious energy or negative inauspicious energy – is what offers potency to the practice. (Too 1998, p. 42)

She provides a standard elaboration of these elements. Fire is 'the ultimate yang element'; water is 'the wealth energy in feng shui'; earth is 'the grounding energy and epitomizes the heart of feng shui'; wood 'brings growth, expansion and advancement, it is the direction of the east and the southeast'; metal 'is the element of the west and the northwest, it symbolizes the strength of heaven and the power of the patriarch' (Too 1998, p. 43).

Feng shui and geomancy had separate but related origins and catered to different concerns – the former for locating local chi currents and hence propitious locations for tombs and dwellings, the latter for foretelling a person's future good or bad fortune. The sixteenth-century Jesuits referred to feng shui as 'geomancy', and subsequently in many circles, the terms became interchangeable (Skinner 2011, p. 36).[26] There was, of course, some connection, as in Chinese cosmology, the heavens,

---

[26] Nigel Pennick in his book on *Geomancy* says that feng shui is simply the Chinese variant of universal geomancy (Pennick 1979, p. 10).

earth, and man are all interconnected; one's location and connection to the earth are not unrelated to one's good and bad fortune; everything in the world was related through chi movement, and so an understanding of the present configuration allowed extrapolation to the future. Nigel Pennick, a student and promoter of international geomantic practices, writes:

> In geomancy, the world was conceived as a continuum in which all acts, natural and supernatural, conscious and unconscious, were linked in a subtle manner, one with the next. In this world view, the incorrect performance of an act, such as misorientating a building, was not merely doomed to fail in achieving its desired objective, but would also bring unforeseen and uncontrollable consequences. Conversely, if the correct manner was applied at the right place and time, the procedures would reflect not only what had gone before, but also what was about to happen. (Pennick 1979, p. 12)

He adds:

> This philosophy underlies the Tarot, the *I Ching*, the Malagasy Sikidy, and numerous other forms of divination involving the creation of patterns to foretell the future. (Pennick 1979, p. 12)

I Ching or *Book of Changes*

The ancient cosmological fount of geomantic dimension of feng shui was codified during the Bronze Age (tenth to fourth centuries BC) in the Confucian classic *Book of Changes* (*I Ching* pronounced 'yee jing'). It is commonly thought to have been first composed at the end of the Yin and beginning of the Zhou Dynasty, about 1000 BC. It is not so much a book in the ordinary literary sense of the term; it is a collection of 64 6-line diagrams, with titles and interpretation. The volume is foundational for both the Confucian and Taoist traditions in China and beyond. It has influenced the entire history of Chinese culture to the present day, being aptly described as:

> The mother of Chinese divination, having fostered both its diversity and persistence; … feng shui is anchored in its perception of reality … it stands out as the single most important book in Chinese civilization … comparable to the sacred scriptures of the other great civilizations. (Bruun 2008, pp. 100–101)

The book's impact has extended over 3000 years and has now spread well beyond China, being embraced in the countercultural, multicultural, and postmodernist West over the past 50 years.[27] The book is based on the metaphysical idea that the universe is founded on two opposing forces, Yin and Yang. Yin is the female, passive, accommodating force; while Yang is the male, hard, decisive force. What is pervasive in this ancient understanding, cosmology, or worldview was the convic-

---

[27] Fritjof Capra (1984) being perhaps its most influential promoter. For the text and commentary, see Huang (2010), Reifler (1974), Rutt (1996), Shaughnessy (1997), Sorrell and Sorrell (1994), and Wilhelm (1950, 1960, 1977). For a 'Plain English' interpretation, see Hulskramer (2004). On the book's Western dissemination and influence, see Smith (1998). The classic critical commentary is Needham and Ling (1956, pp. 304–345).

I Ching or *Book of Changes*

tion that somehow the heavens, the earth, and human life were interconnected, coordinated, and interdependent; they constituted a trinity, and future circumstances could be foretold. Divination, or future telling, is the book's *raison d'être*, though 'fortune telling with guidance' seems to be the common practice. That is: 'this is your fortune unless you stop this' or 'this is your fortune if you do this'. George Hulskramer, an interpreter, translator, and enthusiast says: 'In the *I Ching*, all forms of life and also the functions of the human spirit are associated with Yin and Yang' (Hulskramer 2004, p. 1). This was the 'perception of reality' that underwrote the theory and practice of geomancy or divination; movement in the heavens can be studied so as to inform terrestrial events and social and personal circumstances.

A populariser of *I Ching* wrote:

> As with other divination, like the Tarot, this is a way to obtain guidance about our current situations and possible future events. The true power in any divination practice is to connect with spirits and unseen powers. Spirits and powers that our conscious selves refuse or are unable to acknowledge.
>
> Connecting with these unseen powers with openness allows them to guide us to our unconscious wisdom we hid deep within our minds. The I Ching is a way of communicating and seeing solutions that are otherwise missed by our conscious selves.[28]

Tradition has yin and yang represented by a trigram of three lines where each line is either a solid (yang) or broken (yin) line. There can be eight combinations of the three lines. These are standardly arranged in an eight-sided shape, around a squared yin-yang symbol. This arrangement is called the Pa Kua. According to legend, 4000 years ago a celestial turtle emerged from the Lo river, and an arrangement of numbers on its back held the secret to unlocking the mystery of the trigrams. Subsequently centuries of scholarly debate have gone into these unlocking mysteries (Fig. 4.3).[29]

The *I Ching* is structured by creating an eight-by-eight matrix of these trigrams with one row along the top and one column down the side. The result is 64 hexagrams. Each hexagram, with its top and bottom trigram and a 300-word commentary, constituted one of the 64 chapters in the original *I Ching* book (Fig. 4.4).

Ch'ien	Chên	K'an	Kên	K'un	Sun	Li	Tui
Heaven	Thunder	Water	Mountain	Earth	Wind	Fire	Lake

So in the following table, hexagram #1 is a purely yang, and all its six lines are solid; hexagram #2 is purely yin, and all its six lines are broken. The remaining 62 hexagrams are mixtures of yang and yin.

An early interpretive ten-chapter book, *Annotations on the Changes*, published about 500 years after the *I Ching*, maintained that:

---

[28] See http://www.theartofancientwisdom.com/i-ching/
[29] The central yin-yang symbol is the core of the South Korean flag.

Fig. 4.3 The yin-yang Pa Qua

Hexagram Key

TRIGRAMS UPPER LOWER	Ch'ien	Chên	K'an	Kên	K'un	Sun	Li	Tui
Ch'ien	1	34	5	26	11	9	14	43
Chên	25	51	3	27	24	42	21	17
K'an	6	40	29	4	7	59	64	47
Kên	33	62	39	52	15	53	56	31
K'un	12	16	8	23	2	20	35	45
Sun	44	32	48	18	46	57	50	28
Li	13	55	63	22	36	37	30	49
Tui	10	54	60	41	19	61	38	58

Fig. 4.4 *I Ching* hexagram table

> Therefore, change (*yi*) has a supreme force (*taiji*) which generates two '*yi*', that is, a yang line (-) and a yin line (--); from these four diagrams, 'xiang' are generated, and then eight 'gua. (Dong 1996, p. 125)

The master's skill is to interpret each hexagram and to know its applicability to a person's life and circumstance. As Hulskramer explains:

I Ching or *Book of Changes*

If we consult the *I Ching*, seeking one of these sixty-four hexagrams, what we end up with is not only a Yin-Yang pattern consisting of six lines, but also practical advice based on this pattern. (Hulskramer 2004, p. 3)

Bruun, Hulskrammer, and other commentators repeatedly say that the *I Ching* is comparable to the sacred scriptures of the other great religions. It has been commented upon by countless thousands of scholars over a period of 3000 years, beginning with a philosophical essay, the *Great Commentary*, attached to one of the earliest versions of the book. The Enlightenment philosopher, Gottfried Wilhelm Leibniz (1646–1716), was perhaps the first of the many European scholars to interpret and struggle with the work. His immediate interest was its binary arithmetical, ON/OFF, SHORT/LONG structure (Dong 1996).

But the well-known interpretative problems involved in reading and understanding Biblical, Koranic, Talmudic, Upanishadic, Verdic, Mormonic, and other sacred scriptures pale into insignificance compared to interpreting the *I Chin* text, or more correctly hexagrams. For example, consider hexagram #8 (Fig. 4.5).

Hulskrammer writes:

All Yin lines are grouped around a central leader, the Yang line in the central position. Water and earth are working together harmoniously. Water descends unhindered to earth and the earth absorbs it willingly. This produces the image of brotherhood. ... Leader and followers respect, complement and help each other ... for the wellbeing of the whole. (Hulskramer 2004, p. 24)

While Sam Reifler, who provides 'A New Interpretation for Modern Times' understands the same hexagram ('Seeking Union') as:

You are the creation of a culture, of a society, of a brotherhood of shared experience, and of a family. Even your own individuality is only a concept that – like all others – owes its existence to the community of man. (Reifler 1974, p. 48)

How to decide between Hulskrammer and Reifler or thousands of other diverse interpretations of hexagram 8? It is a moot point how any interpretation could be disputed or at least disputed in a way that had a conceivable end point. Much less of course how any 'true' lesson might be inferred from whatever interpretation might

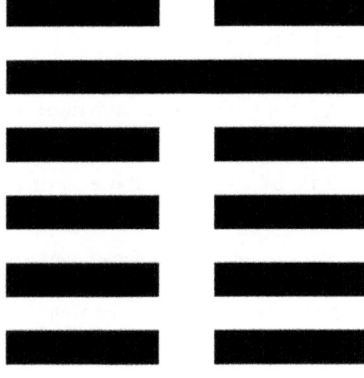

Fig. 4.5 *I Ching* hexagram #8

be agreed upon. The very nature of the *I Ching* book lends itself to interpretation and lesson-drawing without end. It is hardly surprising that one simple listing of the chapter/hexagrams with basic commentary runs to 560 pages (Huang 2010).

As the text is acknowledged as a religious classic, it is instructive to compare and contrast its translation and issues of interpretation with those confronting the sacred texts of other religions – *Torah, Bible, Koran, Ganth, Book of Mormon, Science and Health with Key to the Scriptures*, and so on. In these cases, there is a text to work with, and hermeneutical principles are applicable; nevertheless, each has a multiplicity of interpretations that have given rise to sects, schisms, and religious wars. *Sans* wars, the *I Ching* is no different. Maybe worse, as it is more explicitly Rorschachian than other scriptures.

Whatever the interpretation of any hexagram, the method of its application for any individual is peculiar. Hulskrammer says that to use or consult the *I Ching* for any question or circumstance, you do not begin with a Dictionary or Concordance, rather you begin with a lottery:

> First formulate your question. Take three identical coins. The value is irrelevant. Heads is generally allocated the value of three and tails two, but it works just as well the other way around, as long as you are consistent. Enclose the coins in your cupped hand and shake until you feel the time is ripe to let them fall. … According to tradition, a seven indicates a solid or Yang line and an eight a broken or Yin line. … To end up with a six-line pattern, or hexagram, you need to throw the three coins six times and work from the bottom up. (Hulskramer 2004, pp. 5–6)

Each of the six throws of the three coins draws each of the six lines of the hexagram, one at a time. Having drawn the hexagram, one then looks up its 'meaning' in the *I Ching* table cum chapters.

Sam Reifler, in his 'New Interpretation,' says that the above coin oracle needs to be supplemented by the 'yarrow-stick oracle' in order to fully grasp the message of the *I Ching*. Beginning with a bundle of 50 yarrow stalks cut to the same length and held in the left hand, he guides the reader or initiate through 16 steps:

> Remove one stick with the right hand and set it aside, divide the bundle into two random batches, reduce the batch by counting off bundles of four sticks with the right hand, grasp the right-hand batch between the thumb and forefinger of the left hand, … the total number of sticks in the left hand is now either nine or five, set them aside …using sticks in the discard pile, repeat above steps …the first line, the lowest line of the hexagram, is now determined by the number of sticks in each of the remainder piles … (Reifler 1974, pp. 8–9)

But that is not all:

> To determine each of the remaining lines of the hexagram, going from the lowest to the uppermost, repeat the entire ritual, steps 1-16, using the entire bundle (including the stick first cast off in Step 1. (Reifler 1974, p. 9)

And if, by chance at the end of all of this, you have constructed hexagram # 43 'Breakthrough', then you have learnt that (Fig. 4.6):

> You are threatened by forces opposed to your principles. For the moment these forces have been diverted and the threat is minimal, at an ebb. Take advantage of this weakness both to protect yourself and to exert your own influence on your adversaries. (Reifler 1974, p. 185)

I Ching or *Book of Changes*

Fig. 4.6 *I Ching* hexagram 43

Fig. 4.7 *I Ching* hexagram 64

But if the 16 steps lead you to hexagram #64, then you have learnt that (Fig. 4.7):

> You are trying to make sense out of a strange, unprecedented situation. You and the people involved with you are so disparate, so unsympathetic to each other, so out of touch with each other, that you must exist in a constant state of alertness and readiness. (Reifler 1974, p. 271)

Both the yarrow-stalk and three coins method of being directed to a particular *I Ching* hexagram are both cumbersome. Not surprisingly, a more immediate, low-attention, high-tech method has been developed. Hence:

> The easiest way of casting the I Ching is to use our "Visionary I Ching" app, which preserves the mathematical odds of the yarrow stalk method for each line, as well as preserving an energetic connection, because it depends on the way you shake or the timing of how you click on the app to determine which of the four kinds of line you get.[30]

At this stage, one wonders how different such an outcome might be from just spinning a wheel with 64 sectors, or simply picking one of 64 cards, and going to the selected hexagram and then reading one's future.

---

[30] From https://divination.com/how-to-consult-the-i-ching/

Conclusion

What can be said about these *I Ching* interpretative disputes about 'the single most important book in Chinese civilization', about which hundreds of thousands of books have been published? A sympathetic interpreter writes:

> The *I Ching* is a hodge-podge of Chinese culture from pre-Taoist wisdom to Confucianism and even more decadent, Machiavellian-type philosophies. (Reifler 1974, p. 6)

Joseph Needham discusses, at length, the book, its history, and contested interpretations (Needham and Ling 1956, pp. 304–345). He correctly identifies it as an inhibitor of the development of modern science in China:

> I fear that we shall have to say that while the five-element and two-force theories were favourable rather than inimical to the development of scientific thought in China, the elaborated symbolic system of the *Book of Changes* was almost from the start a mischievous handicap. It tempted those who were interested in Nature to rest in explanations that were no explanations at all. The *Book of Changes* was a system for *pigeon-holing novelty* and then doing nothing about it. Its universal system of symbolism constituted a stupendous *filing-system* (Needham and Ling 1956, p. 336)

He summarizes that it:

> Originating from what was probably a collection of peasant omen texts, and accumulating a mass of material used in the practices of divination, it ended up as an elaborate system of symbols and their explanations (not without a certain inner consistency and aesthetic force), having no close counterpart in the texts of any other civilisation. (Needham and Ling 1956, p. 304)

For Needham, its deleterious effects on general thought, and particularly on incipient scientific thought, arose because:

> These symbols were supposed to mirror in some way all the processes of Nature, and Chinese medieval scientists were therefore continually tempted to rely on pseudo-explanations of natural phenomena obtained by simply referring the latter to the particular symbol to which they might be supposed to 'pertain'. Since each one of the symbols came, in the course of the centuries, to have an abstract signification, such a reference was naturally alluring, and saved all necessity for further thought. (ibid.)

In the West, Needham's negative judgement fell upon unhearing countercultural and counter-scientific orientalist ears that were singing the praises of the classic text. But if nothing else, it needs to be recognized that there is a huge 'opportunity cost' involved in *I Ching* 'studies': the millions of person-hours that have gone into interpretation and exposition of the hexagrams could have been more productively spent on advancing Chinese science, technology, medicine, infrastructure, housing, diet, cooking, or any other socially productive occupation; in other words, on the science-informed modernization of China.

Chapter 5
Feng Shui and Traditional Chinese Medicine

Commitment to chi (*qi*) is a fundamental tenant of Traditional Chinese Medicine (TCM). Without chi, the intellectual backbone of TCM, which is a multibillion-dollar, state-supported business in China, and a much-supported and growing business outside of China, is simply removed. Chi's supposed therapeutic influence spreads well beyond TCM. A whole raft of alternative medicines and practices are based upon the marshalling of life energies, or chi by other names. These include many in the anti-vaxxing movement (Mnookin 2011; Offit 2013). The Chinese State Council Information Office (SCIO) issued a report documenting that in 2015 there were 42 institutions of higher learning in TCM, 3966 TCM hospitals, 42,528 TCM clinics, 452,000 practitioners, about 60,000 TCM medicines have been approved by the government regulator, and 26+ million in-patients were treated. In the same year, the TCM pharmaceutical industry generated 200 billion dollars, of which nearly 4 billion dollars came from exports, mostly to the West. TCM is massive state-capitalist business. President Xi Jinping called TCM a 'gem of the country's scientific heritage'.

For the past 50 years, TCM has been growing in the West and in other Asian countries where it is frequently subsumed under the name 'alternative medicine', 'holistic medicine', or 'complementary medicine'. In numerous countries the practice is both state sanctioned and state supported: public hospitals offer TCM, insurance policies cover its fees, university medical schools offer TCM subjects if not whole degrees, state libraries stock hundreds of TCM books, practitioners require state licences, and so on. In the early 1970s, there were no state-licenced acupuncturists in California, now there are multiple thousands; at the same time, there were no colleges of Chinese medicine in the USA, now there are scores, if not hundreds. One estimate is that three million Americans receive acupuncture treatment each year, meaning there are multiple millions of treatments. Of note for the argument of this book is that the fundamental, and defining basis, of TCM is identification and mastery – by diet, exercise, manipulation, or needles – of internal chi and hence the enabling of the balance of the vital yin and yang 'forces' inside the body. As stated by two American promoters of TCM:

> That which animates life is called *Qi*. The concept of *Qi* is absolutely at the heart of Chinese medicine. Life is defined by *Qi* even though it is impossible to grasp, measure, quantify, see, or isolate. Immaterial yet essential, the material world is formed by it. An invisible force known only by its effects, *Qi* is recognized indirectly by what it fosters, generates, and protects. (Beinfield and Korngold 1991, p. 30)

It is noteworthy that the authors explicitly embrace 'mysteriumism': that is, commitment to the existence of something which is 'impossible to grasp, measure, quantify, see, or isolate'. Such commitment is commonplace in religion where 'seeing through a glass darkly' (1 Corinthians 13:12) is the default position that is illuminated by faith, but the same commitment is also a requirement for chi believers. In science many postulated entities are *difficult* to grasp, measure, quantify, see, or isolate, but the scientific task is to overcome whatever the theoretical, conceptual, or technological difficulties might be. To say in advance that this is *impossible* to do is one indicator of being involved not in science, but in pseudoscience, as will be shown in Chap. 13. This is a challenge for something touted as the 'gem of the country's scientific heritage'.

Many examples of external or environmental feng shui were given in Chap. 4. These were cases where the siting, orientation and fit-out of graves, and of domestic and commercial properties, was determined according to knowledge of local chi lines and movements. So also the layout and design of public spaces such as gardens were governed by chi maps provided by feng shui consultants. But once you commit to the existence of an all-pervasive life force, energy, or chi that binds together the heavens, nature, and man and governs their harmonization, then the next step is to admit its existence and powers *within* bodies and to then move to control, redirect, and manipulate its flow to enhance good effects (wellness) and minimize bad effects (illness).[1] If chi energy flows everywhere and permeates everything, then the skin is no barrier to its movement. This is the ontological or cosmological underpinning of TCM.

Traditional Chinese Medicine considers sickness or pain to be a result of chi blockage and/or unbalanced chi in the body. All TCM therapies – herbal concoctions, acupuncture, moxibustion, massage, diet, and qigong – are based on this fundamental philosophy and perspective. Get the chi balance right, and health (and for some, wealth and good fortune) follows. For example, the above quoted advocates maintain that:

> The *Kidney* consolidates and stores *Qi* that initiates and keeps life growing. ... Before we are born our parents endow us with *Essence Qi*, which following birth is fortified by *Air Qi* from the Lungs and *Food Qi* from the spleen. Both inherited and acquired *Qi* is collected within the reservoir of the *Kidney* to be dispensed as needed. (Beinfield and Korngold 1991, p. 121)

---

[1] On the traditional philosophical underpinnings of TCM, see Beinfield and Korngold (1991) and Porkert (1974, 1982).

5 Feng Shui and Traditional Chinese Medicine

And so on in similar fashion for each of the five body systems: liver, heart, spleen, lung, and kidney. Five being required by the Five-Phase Correlative Doctrine. They point to the daily rhythm of chi that needs be recognized in all diagnosis and treatment:

> During the day, *Qi* expands outward toward the surface of the body (*Yang*) and during the night *Qi* retreats into the body's core (*Yin*). *Yang* peaks at midday, *Yin* at midnight. Every two-hour period during the day and night clocks an alternating ebb and flow of *Qi*. (Beinfield and Korngold 1991, p. 92)

It is easy to appreciate the immense mental effort that is required to get on top of all of this, and to write authoritative books on the subject, and to practise chi-theory informed TCM. Despite all this mental effort by millions of commentators and practitioners, something so practical as a measuring instrument for chi has never in 3000 years been developed. Orthodox medicine has instruments for measuring blood pressure, heart beat, lung function, urine analysis, and most other things. Yet no instrument has ever been produced for chi measurement. Invisibility per se is not a barrier to instrumentation; we measure lots of things we cannot see (current, magnetic fields, etc.). The barrier is consistency and reliability of action by the invisible entity being measured. But if this is conceded as the explanation for no instrumentation, then what is left for invoking chi? Nothing.

Wang Fuzhi (1619–1692), the most prolific philosopher in Chinese history, embraced and developed the then millennia-old theory of chi, connecting it to personal health and well-being.

> What heaven endows in men is uninterrupted *qi*. If *qi* is uninterrupted, then *li* must also be uninterrupted. Therefore, as long as life continues, one's nature gets daily renewal. (Liu 2010, p. 365)

Traditional Chinese Medicine is concerned with the appropriate renewal of chi; 'recharging one's battery', so as to speak. This is to be done by diet (herbal medicine), exercise (the numerous qigong schools), acupuncture, astrological therapy, and other TCM regimes.

More fundamentally, in Chinese thinking since ancient times, the body was fashioned in the image of the cosmos; the first reflected the latter; just as harmonious chi movement was important for the cosmos, so to it was for the body and the body politic, the State. Chi was material, though material of an ethereal kind, it had a real existence; it was all-pervasive having no bounds yet moving in defined courses and meridians. For Chinese natural philosophers, the world was formed as a result of chi processes based on *yin* and *yang* cycles and interactions of the five elements, the *wuxing* of the five elements (soil, wood, metal, fire, and water). This is a deep-seated part of what commentators have labelled 'correlative thought' in Chinese culture. Ernest Eitel recognized this in his 1873 study:

> Observing the heavens, the constant change of day and night, the numbers and distribution of the heavenly bodies, moving on, hosts of them, each in swift course, and yet never interfering with each other …[and] observing our earth, with its constant revolutions of summer and winter, spring and autumn, growth and decay, life and death, the Chinese noticed, that here again the same mathematical order is repeated, that earth is but the reflex of heaven … (Eitel 1873/1987, p. 21)

Donald Harper has said: 'By the first century CE the paradigm of correlative thought was well established and remained the dominant paradigm in Chinese science throughout the pre-modern period' (Harper 2017, p. 45). Joseph Needham understood correlative thinking as:

> Things behaved in particular ways not necessarily because of prior actions or impulsions of other things, but because their position in the ever-moving cyclical universe was such that they were endowed with intrinsic natures which made that behaviour inevitable for them … They were thus parts in existential dependence upon the whole world-organism. (Needham and Ling 1956, p. 281)

There are correlations or parallels between the cosmos, nature, society, and individuals; everything is both interconnected and harmonized or has 'relatives' or parallels in other orders. Chinese correlative thought brought everything together; items in the Cosmic column correlated with items in the Nature column and in turn with things in the Personal column. This commitment is echoed in the standard Chinese saying: *Tian Ren He Yi* (Man and Heaven combined into One). This did potentially support a certain form of naturalistic thinking which is a presupposition of modern scientific thinking. In as much as the disposition and movement of internal chi was being attended to, then demonic possession, sin, and the like were not resorted to as a cause of illness. In this respect, the Asian tradition was in front of much in the West.

Consistent with, and flowing from, this correlative thinking, originally there were thought to be 365 points of intersection of the internal chi meridian lines. This gave one acupoint for each day of the year, making body structure correlate nicely with planetary structure and thus confirming the correlative view of cosmos-solar system-society-man. For different reasons, now 600+ acupoints are recognized. This, of course, does make falsifying the original acupuncture claims more difficult: when the same patient results are achieved by sticking needles into a non-acupoint, rather than admit the theory is falsified, just declare that particular insertion point a 'new', hitherto undiscovered, acupoint. And so the acupoint numbers keep going up.

The qigong (chi kung) component of TCM is practised daily by an estimated 20–30 million people in China and untold millions elsewhere in the world (Yan 2015). This correlative thought is nicely expressed by the International Yan Xin Qigong Association:

> The basic principle of Qigong is to coordinate the human body with the universe. It was assumed that all things in the world had spirit and intelligence. People were to keep in harmony with nature and absorb vital energy from outside the body to supplement their needs. The whole philosophy regarding the relationship of the human body with the universe gradually formulated the theory of Chinese Traditional Medicine. (Wozniak et al. 2001, p. 8)

Some commentators label chi as 'life force', and it is as a life force that it so easily enters medical science, discussion, and treatment. If the world is permeated by life forces, then the medical world should attend to them; not to do so would be irresponsible. A distinction commonly made is between internal chi (qi) and external chi. The former is the domain of qigong (chi kung), with many of its practitioners being sceptical about, or just giving lip service to, the latter. They cannot see that internal chi can be 'externalized', 'thrown', 'cast', or 'radiated' by qi masters or *qigongists* as they are more generally known. Nor of course can many 'outsiders' see these processes. It is also common to divide internal *qi* into nature and nurture varieties: the former we are born with; the latter accrues by qigong exercising which tops up and manages one's chi balance.

In this respect, chi has commonalities with 'imponderable fluids' widely proposed by 18th and nineteenth century European scientists (as will be outlined in Chap. 11) and *Élan vital* proposed in the early twentieth century by Henri Bergson (1859–1941) to account for the origin of life on earth and its evolutionary trajectory (Bergson 1911). Both constructs once had reasonable scientific support, and wide popular endorsement, but now they are found only in the scientific-relics room, though they linger in popular discussion and imagination. As many pointed out, the ideas occupied scientific space but paid no rent; they were idle notions. Julian Huxley (1887–1975) remarked: 'to think that *élan vital* explained the behavior of any organism, was like evoking *élan locomotif* to explain the movement of a railway engine' (Gillies 1996, p. 31). It did no such thing. The parallels with chi are telling and will be subsequently developed.

Chi-Based Medicine

Fritjof Capra gave TCM and chi enormous publicity in the West with his million-plus, best-selling *The Tao of Physics* (Capra 1984), that after its appearance in 1975, was published in 3 editions and 23 languages. He wrote:

> Traditional Chinese medicine, too, is based on the balance of *yin* and *yang* in the human body, and any illness is seen as a disruption of this balance. The body is divided into *yin* and *yang* parts. Globally speaking, the inside of the body is *yang*, the body surface is *yin*; the back is *yang*, the front is *yin*; inside the body there are *yin* and *yang* organs. The balance between all these parts is maintained by a continuous flow of *ch'I*, or vital energy, along a system of 'meridians' which contain the acupuncture points. …
>
> This system is elaborated in the *I Ching*, or *Book of Changes* …The *Book of Changes* is the first among the six Confucian Classics and must be considered as a work which lies at the very heart of Chinese thought and culture. The authority and esteem it has enjoyed in China throughout thousands of years is comparable only to those of sacred scriptures, like the *Vedas* or the *Bible*, in other cultures. (Capra 1984, p. 98)

In a following book, *The Turning Point* (Capra 1982), he elaborated on Chinese correlative thinking and its consequences for medicine:

> This correlative and dynamic way of thinking is basic to the conceptual system of Chinese medicine. The healthy individual and the healthy society are integral parts of a great patterned order, and illness is disharmony at the individual or social level. The cosmic patterns were mapped out by means of a complex system of correspondences and associations that was elaborated in great detail in the classical texts. (Capra 1982, p. 313)

And adds:

> The Chinese idea of the body has always been predominately functional and concerned with the interrelations of its parts rather than with anatomical accuracy. Accordingly, the Chinese concept of a physical organ refers to a whole functional system. (Capra 1982, p. 313)

These extracts are a fair condensation of the whole chi-based cosmology and cosmogony of China.

Ken Tobin, who before 'moving on from constructivism' (Tobin 2000), was a world leader of the constructivist movement in science education (Tobin 1993), the president of the National Association for Research in Science Teaching, and the recipient of numerous international education-research prizes, contributed an invited paper to the first volume of the new *Asia-Pacific Science Education* journal. In this paper he embraced chi-related medical practice, saying:

> The underlying theory relates to *Qi*, universal energy, and its flows through the body. In the case of humans there are 26 pairs of safety energy locks (SELs) through which *Qi* flows, providing the life source to the body … When a body is disharmonized, energy can be blocked at or close to the SELs, thereby disrupting one or more of the flows needed to distribute the life force to different parts of the body. (Tobin 2015)

And adds:

> The experience for an individual when blockages occur might be abnormal functioning of organs, illness, and different parts of the body being susceptible to damage (e.g., fracture of a bone). Using the principles of JSJ, an individual can use self-help techniques or a JSJ practitioner can use particular holds to restore harmony to the body and get various flows moving smoothly. In the past few years we have studied human conduct with JSJ as one of the components of a multilogical framework. (Tobin 2015)

That a person so prominent in the international science education community talks so easily, and without hesitation, about universal energy, chi flows, SELs, and blockages – underscores the degree of penetration of feng shui thinking beyond its Asian homeland. Such talk from established figures makes discussion of feng shui in science classes more natural and expected: if an acknowledged leader of international science education is endorsing and advocating the fundamental feng shui worldview, or the supposed scientific basis for it, then science teachers and students are justified in attending to it (Fig. 5.1).

These claims, that are commonly made by all feng shui advocates, are very specific, empirical, and should be able to be scientifically verified. But they have not been. The supposed fundamental cosmic and terrestrial 'energies' are unknown, undetected, and unmeasured by orthodox science. This clearly presents an educational problem, particularly for Asian science education. Are the above claims ignored, repeated, or are they appraised in classrooms? The argument of this book is that they should be critically appraised.

Chi is also the physical foundation for qigong, for the family of Prāṇa-based yoga regimes, for the many forms of Tai Chi Chuan (*tàijíquán*) exercises, and for the numerous other Asian martial, and not-so-martial, callisthenic routines that seek to marshal, enhance, and utilize this all-pervasive life force. One such practice is

Chi-Based Medicine

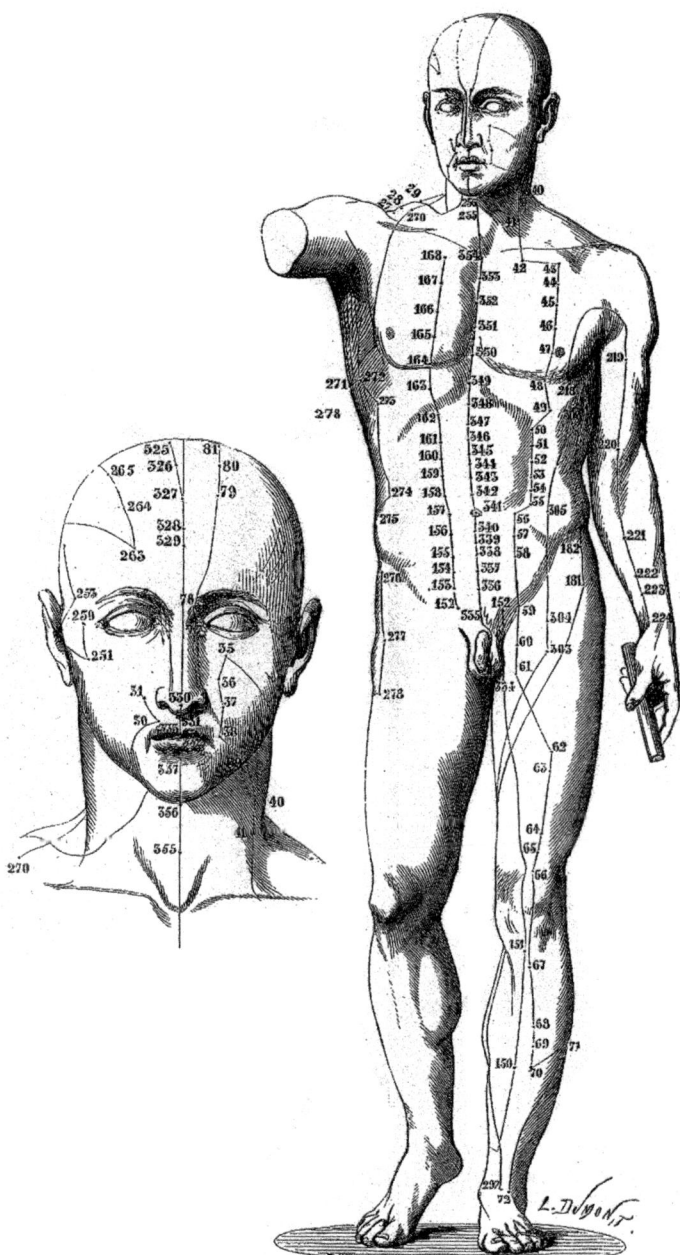

Fig. 5.1 Acupuncture meridians

Japanese Jin Shin Jyutsu (JSJ) therapy. This is the therapy and 'knowledge system' now promoted by Ken Tobin, who in the above extract refers to the 26 pairs of safety energy locks (SELs) that control the flow of chi within the body.

Reiki (pronounced *raykey*) is another Japanese form of chi-based medical therapy which is now practised universally. The name derives from Japanese *Rei* (soul) and *Ki* (energy). It had its origins in the early twentieth-century Japan where in 1914 Jatiji Kawakami developed a healing style he called Reiki Ryoho. There were other healers with comparable styles. Modern Reiki can be traced to 1922 when Usui Sensei, who had travelled in Europe and China, claims to have had a mystical experience on Mt. Kurama in which he was given Reiki energy and became the conduit for its transference to those who were, literally, touched, by him.

The Dopemystic Reiki Healing Institute is unrestrained in its homage to therapeutic universal energy, saying:

> There are fourteen great nerve centers in the physical body, in the astral body and in the body of the soul. And these centers are called chakras. Chakras are considered vortexes in which etheric energy can enter or leave the human energy body. They are energy centres that govern the physical, emotional, mental and spiritual aspects of the human form.[2]

The Reiki Training Centre estimates that four million people have taken the first stage of Reiki training and that Reiki therapy is offered in over 800 US hospitals, including at Yale University.[3] It substitutes pressure-manipulation of the chi meridians and nodes for the needle intervention of acupuncture or the heat application of moxibustion. The International Centre for Reiki Training says:

> Reiki is a Japanese technique for stress reduction and relaxation that also promotes healing. It is administered by 'laying on hands' and is based on the idea that an unseen 'life force energy' flows through us and is what causes us to be alive. If one's 'life force energy' is low, then we are more likely to get sick or feel stress, and if it is high, we are more capable of being happy and healthy.
>
> The word Reiki is made of two Japanese words – Rei which means 'God's Wisdom or the Higher Power' and Ki which is 'life force energy'. So Reiki is actually 'spiritually guided life force energy'.[4]

And the Centre goes on to claim that:

> the ability to use Reiki is not taught in the usual sense but is transferred to the student during a Reiki class. This ability is passed on during an 'attunement' given by a Reiki master and allows the student to tap into an unlimited supply of 'life force energy' to improve one's health and enhance the quality of life. (ibid)

---

[2] See https://www.dopemystic.com/dopemystic-blog/2018/2/21/the-chakras-continued-and-the-dopemystic-reiki-healing-process

[3] See Sacks (2014). And https://www.washingtonpost.com/national/religion/reiki-goes-mainstream-spiritual-touch-practice-now-commonplace-in-hospitals/2014/05/16/9e92223a-dd37-11e3-a837-8835df6c12c4_story.html?utm_term=.c4ab83e238cb

[4] See https://www.reiki.org/faq/whatisreiki.html

And reassuringly says:

> Because Reiki comes from God, many people find that using Reiki puts them more in touch with the experience of their religion rather than having only an intellectual concept of it. (ibid)

The US Wellness Institute says of its Reiki treatment that:

> The Reiki practitioner is the conduit between the patient and the source of the universal life force energy; the energy flows through the practitioner's energy field and through her hands to the patient. . . . [She] places her hands in specific energy locations . . . [the] length of time determined by the flow of energy through her hands. . . . The patient experiences the energy as sensations such as heat, tingling, or pulsing where the practitioner has placed her hands. Sometimes, the sensations are felt moving through the body.[5]

Another Reiki therapist describes her session with patient/client John, as:

> What we'll be looking for here, within John's auric field, is any areas of intense heat, unusual coldness, a repelling energy, a dense energy, a magnetizing energy, tingling sensations, or actually the body attracting the hands into that area where it needs the reiki energy, and balancing of John's qi.[6]

This description abounds with scientific words, but none of the extraordinary (at least outside of Reiki circles) claims has ever found support in a laboratory. The rallying cry is 'Science', with one book titled *How to Combine the Law of Attraction with Reiki Energy*,[7] but there is zero scientific support for the claims. Some therapists hold that the hands need not be on the patient's body, but merely over the body. Needless to say, there are thousands of sites selling Reiki Chakra pendants, stones, bracelets, rings, and anything else that will separate a gullible citizen from their hard-earned dollars. There is even Animal Reiki to assist with pet cats and dogs that might be a bit out of sorts. In the words of one commentator: 'Reiki is the hottest new Eastern healing practice making its way into the Western health industry' (Sacks 2014).

There are two questions: First, does the technique work? The answer given in a review of 205 relevant studies is: 'the evidence is insufficient to suggest that reiki is an effective treatment for any condition' (Lee et al. 2008). Second, does transmission of 'spiritually guided life-force energy' have anything to do with whatever effects might be found in Reiki therapy? The naturalistic default answer to the latter question is 'No' because there is no evidence for the existence of any 'life force energy'. A naturalist will always begin by looking for placebo effects or other consistently natural explanations of any therapeutic effects.

Despite the millions of people who have undergone Reiki training and the thousands of hospitals where it is practised, the *Oxford Handbook of Psychiatry* simply asserts that it is a pseudoscience (Semple and Smyth 2013, p. 20). Of course, it can be, as with so many other pseudosciences fanciful and nonsensical yet have some of

---

[5] See https://www.wellnessinstitute.net/ The Centre also, for a fee, offers instruction in Craniosacral Biodynamics, Polarity Therapy, NeuroEnergetic Therapy, and more.

[6] Brian Dunning, *Skeptoid* #411

[7] Available at www.silverlineddays.com

the effects claimed for it. All snake-oil merchants and faith healers depend upon this; they should be right some of the time. This concentrates attention on the identification of pseudoscience (something taken up in Chap. 13) and opens up the legal issue of medical fraud and malpractice. With mega-millions, if not billions of dollars involved in Reiki practice, these are both pressing and substantial issues. The more so when the cost of Reiki therapy is claimable on public health benefits.[8]

Some have tried to keep Reiki theory alive and naturalistic by proposing a 'biofield' that surrounds each individual and to varying degrees penetrates the individual. A group at the US National Institute of Health coined the term in 1994 to describe the supposed field of energy and information that surrounds and interpenetrates the human body. They say: 'It is composed of both measurable electromagnetic energy'; and then, without apology, add 'and hypothetical subtle energy, or chi'. It is claimed that Reiki, and other touch or energy therapies, alter the biofield. This is just a promissory note; one that has been constantly issued for 3000 years without payment. Tests have been done showing that there is no difference in outcomes between trained and untrained therapists, provided the latter do not divulge the pretence to patients. Hence whatever the Reiki effects, they are placebo ones.

In Australia, the 2018 cost is $50–100 per session. The question then becomes are there less costly, more beneficial and more efficient ways to get the placebo response?[9] In Australia there is no control over qualifications, whatever they could be; practitioners are unregistered; the whole field is set up for quackery and worse. In October 2018 in the State Supreme Court of New South Wales, Australia, one practitioner who operated a lucrative 'Universal Medicine' practice from his home, was found guilty of 'intentionally indecently touching clients'. His techniques included 'esoteric breast massage' and 'ovarian reading'. His clients included 10-year-old girls.[10] It was noteworthy that although one client successfully brought the charge, many others came to court as a cheer squad for the defendant. This moved the judge to comment that they were 'cult like'.

The engaging philosophical, and legal, question is: If Reiki theory – with its bioenergies and their touch guidance – is accepted, how is it possible to separate the deluded but harmless practitioner from the conniving, lecherous crook? There are two legal issues: one is fraud, selling something that is known not to work; the second is criminality, as in the Australian case. Scientific testing of the theory clears some of the legal air.

Qigong is a millennia old TCM practice and there are hundreds of thousands of commercial qigong health and 'healing' operations functioning throughout the world; it is a mega-million-dollar industry. Hundreds of clinics and businesses can be located with a few minutes of web search. One, picked-at-random operation, founded by 'International Qigong Master' Chunyi Lin, claims:

---

[8] On the legal-medical-economic issues, see: https://sciencebasedmedicine.org/

[9] See Centre for Scientific Medicine reports at: https://sciencebasedmedicine.org/reiki/

[10] *Sydney Morning Herald* 16 October 2018, p. 1

Chi-Based Medicine 101

> More specifically, disease or dis-ease is caused by energy blockages in the body. Usually this means too much or too little energy in one place. Either way, your energy is out of balance, creating an energy blockage. Energy blockages are caused by many things but most commonly by emotions or by stress. Energy blockages cause your body's natural healing system to breakdown and malfunction.
>
> Remove the blockage or blockages and restore energy balance, and the disease, sickness, or other problem goes away – not just the symptoms, but the root cause of the problem is removed. That is part of what is so beautiful and powerful about Spring Forest Qigong: It gets at the root cause of the problem, not just the symptoms.
>
> Spring Forest Qigong practice can help you remove the energy blockages in your body, restore your natural balance, and keep your energy flowing smoothly, enabling you to live the healthiest, happiest, most productive and rewarding life possible.[11]

Like King above, all major credit cards are accepted by Chunyi Lin.

It should be obvious that the flowing or non-flowing of energy in the body, much less chi energy, has nothing to do with the *causes* of stress, happiness, health, or leading a rewarding life. It is the height of individual narcissism to believe this; and it is the depth of commercial exploitation to charge credulous people for giving such advice and administering such treatment. Diseases can be immediately caused by infections, bacteria, fungi, parasites, viruses, worms, and a huge host of other organic and inorganic vectors. Malaria is not caused by any internal energy imbalance; it is caused by the plasmodium parasite most frequently delivered by a mosquito. Meningococcal disease, that kills multiple thousands of infants each year, is not caused by internal energy imbalance; it is caused by the bacterium meningococcus. Diseases are less immediately caused by defined and measurable environmental factors: water quality, air quality, lack of food, and so on.

Doubtless the diseases cause 'dis-ease' both to the sufferer and their families, but it is irresponsible to tell people that their illness can be cured by doing qigong exercise, and it is fraud to take money for doing so. A healthy lifestyle, exercise, and good diet are all commendable, but none of those prevents infection by meningococcus bacteria; what prevents the resultant disease is vaccination. Stress and anxiety can be caused by unemployment, relationship breakdown, and countless other things. Doing qigong will not generate work or prevent divorce. And time and money thus spent distracts from the main game of finding work, rescuing a relationship, or campaigning for state-supported vaccination programmes. The inwardness of the advice can misdirect people's actions away from dealing with the real cause to pondering and trying to marshal an imaginary one.

Psychosomatic diseases are real. The link between mental and physical states have been known since the first cave dweller's heart beat increased, their mouth dried, and they sweated when a lion approached. Fear, stress, and anxiety have physical effects. Grief or the shock of losing a loved one can immediately impair heart function. The 'upset' mind 'tells' the body to produce stress hormones such as adrenalin which then impact the heart. 'Broken heart' syndrome is a recognized

---

[11] See, https://www.springforestqigong.com/how-springforestqigong-works

medical condition – Takotsubo cardiomyopathy. There are standard manuals and handbooks of psychosomatic diseases; the subject is taught in medical schools. None of this has anything to do with chi energy, and everything to do with, speaking loosely, the mind-body interaction as manifest in recognizable biochemical pathways. Psychosomatic illness is explained by orthodox physiology, and rigorous experiments are conducted to isolate particular causal mechanisms. Austen Clark, a medical doctor, psychiatrist, and philosopher, has given an informed and comprehensive account of the subject, concluding:

> The danger in the holistic interpretation of psychosomatic disease is that it squanders these [scientific] hard-won insights. It promulgates a murky view of the role of the mind, embraces an occult view of the mind/body interaction, commits itself to contradicting current physical science, and abandons the enterprise of developing precise theories and rigorously formulated tests. (Clark 1985, p. 102)

Likewise, placebo effects are real. People thinking some agent or practice will work is often a condition of the medicine or practice working. Placebo or expectancy effects are a component of all medical treatment; people go to doctors or clinics because they expect to get better; this expectation needs be factored into causal accounts of treatment effects. A 2008 study of the effectiveness of anti-depressants found that none of the four most-prescribed drugs were any better than placebos (dressed-up sugar) in curing the condition. The occurrence of psychosomatic and placebo conditions should not be confused with feng shui or qigong effects; most of the latter are explained by the former.

The banner headline of the Spring Forest Qigong practice is: *A healer in every home and a world without pain and suffering*. This typifies the empty, but dangerous, sloganizing characteristic of feng shui. Why would anyone say something so stupid and vacuous? And why would so many people pay money to anyone speaking such drivel? In what world will breaking a leg, giving birth, losing a loved one, or being sacked from a valued job not cause pain?

The foregoing multitude of practices is unified in their 'theoretical' commitment to chi, but they diverge in their particular understandings of just what chi is and what physiological/mental processes are occurring when the practices are performed. For example, Wikipedia distinguishes qigong from tai chi on the grounds that:

> In qigong, the flow of *qi* is held at a gate point for a moment to aid the opening and cleansing of the channels. In *tàijíquán* [tai chi], the flow of *qi* is continuous, thus allowing the development of power for the use by the practitioner.

Of course, there is no instrumentation that would allow determination of whether the qi flow is continuous or discontinuous. This disagreement is comparable with theological ones about the flow, or otherwise, of grace.

A good proportion of the millions of daily practitioners of these exercises literally just 'go through the motions', and do so with little, if any, thought to the supposed underlying physical, much less metaphysical, foundations. Most people are satisfied if regular and disciplined performance of the prescribed dieting, breathing, stretching, exercising, disclosing, communicating, and related meditative practices, result in better physical and mental health, and lessened stress and pain. For many, the truth or otherwise of the theory behind the practice is secondary to their improved well-being; often it is a case of 'any port in a storm'.

This same pragmatic justification can be, and has been, evoked for a wide range of religious practices. If the individual and communal religious practices – doing communal good works, sharing a meal, disclosing problems or failings, meditation, prayer, singing in choirs, helping others – do no harm, and make people feel better, then that is sufficient justification for the religion and perhaps even its theology.

Chinese Government's Promotion of Traditional Chinese Medicine

Feng shui and qigong are being rolled together with Traditional Chinese Medicine (TCM) as components of the Chinese public health system that serves 1.4 billion people. The medical profession, scientists, philosophers, and educators cannot just look the other way; informed appraisal is required. The more so as China is actively exporting its medical operations to other countries as part of its 'soft power' strategy, and Western countries are variously adopting parts of TCM, usually under the guise of 'alternative' or 'complementary' medicine.

In 2016, following the National Conference on Hygiene and Health, the Chinese Government mandated support for TCM across all components of public health. This policy decision was conveyed in a report of the State Council Information Office (SCIO 2016). The SCIO report says that TCM 'represents a combination of natural sciences and humanities, embracing profound philosophical ideas of the Chinese nation' (SCIO 2016, p. 1). The report traces the history of TCM emphasizing the Chinese origin of herbal infusion in medicine, and of acupuncture, including moxibustion,[12] and rejoices in these practices having 'won popularity throughout the world'. It identifies Huangfu Mi who lived during Western Jin time (265–316) as the prime formulator of the theory of zangfu (internal organs) and jingluo (meridians) and their associated acupuncture and moxibustion practices. The acupuncture meridians are the movement lines for internal chi.

The SCIO report proceeds to make the progressive, but uncontroversial, claim that:

> First, setting great store by the holistic view, TCM deems that the relationship between humans and nature is an interactive and inseparable whole, as are the relationships between humans and the society, and between the internal organs of the human body, so it values the impacts of natural and social environment on health and illness. Moreover, it believes that the mind and body are closely connected, emphasizing the coordination of physical and mental factors and their interactions in the conditions of health and illness. (SCIO 2016, p. 5)

---

[12] Moxibustion is similar to medical acupuncture except that instead of needles, burning moxa wool is used to treat or prevent diseases. Moxa is often applied on the meridians and their acupuncture points. The SCIO claims that moxibustion is wonderful for simulating chi and blood flow and so ensuring that organs function well.

Most doctors, Western and Eastern, recognize this, and treat their patients accordingly. Treating the 'whole patient' is not a prerogative of alternative medicine. However, holism can, and has, covered a multitude of sins, and provided endless 'cover' against serious testing of claims in medicine, economics, sociology and much else; it can function as a 'get out of jail free' card for all kinds of dubious, ineffective and harmful practices. This needs to be recognized in reading the SCIO report. The more so as the report proceeds to say:

> Second, setting great store by the principle of harmony, TCM lays particular stress on the importance of harmony on health …of which the most vital is the dynamic balance between yin and yang. (SCIO 2016, p. 5)

This is, at best, high-grade metaphysics, and its connection to any specific practice is tenuous. If treatments succeed then yin and yang are in balance; if they fail, then yin is out of balance with yang. In either case there is no specific direction given for finding the cause of the success or failure. The yin-yang part is just a distraction from the task that needs be attended to; it is just a slogan or cloak waved over whatever might be going on. And, importantly, there have never been any quantitative measures of these 'factors'. While there are discipline-wide, agreed, calibrated measures of time, distance, velocity, blood pressure, heart rate, sugar content, resistance, voltage, and other science and health related factors, there are no such measures or instruments for chi variation caused by local yin-yang movement.

As soon as any factor or variable is thought to be influential, the immediate scientific response is to find an appropriate standard-unit and take measurements using some public instrument. This was the motivation and logic behind Galileo's creation of the pendulum-based pulsilogium to measure pulse rate, a known important factor in health care (Matthews 2015a, p. 215). There is not, and never has been, comparable, objective measures of yin, yang or chi. Fritjof Capra praises this:

> One consequence of this attitude is a distinctive lack of concern about quantification among East Asian doctors …doctors would not measure patients' temperatures but would note their subjective feelings about having a fever; herbal medicines are measured very roughly in little boxes without the use of scales, and are then mixed together. Nor is the duration of acupuncture therapy measured – it is simply determined by asking the patient how it feels. (Capra 1982, p. 319)

This is significant when considering the scientificity of TCM and its effectiveness. Claims for the latter depend upon measurement, something for which there are neither instruments nor agreed standard units upon which to base calibration. To equate the objective 'being well' with the subjective 'feeling well' is hardly a sound basis for evaluation of therapies or drug use.

Passing over such considerations, the Chinese government is clear that there will now be a dedicated and funded space in the nation's health system for TCM, saying that: 'Equal status shall be accorded to TCM and Western medicine in terms of ideological understanding, legal status, academic development, and practical application' (SCIO 2016, p. 7). This involves establishing a system to 'carry forward the theories and clinical experience of well-known veteran TCM experts, and efforts have been made to rediscover and categorize ancient TCM classics and folk medical experience and practices' (ibid).

In 2016 the World Education Service said the number of TCM outpatient clinics in China rose from 531 in 2010 to 5890 in the 5-year period to 2015; the number of dedicated TCM institutions rose tenfold from 4 in 1956 to 42 in 2016; and in 2015 there were 238 Chinese universities offering TCM programmes.

As well as being a multi-billion-dollar business in China, the SCIO report documents how TCM has spread to 183 countries and regions around the world. According to the World Health Organization, 103 member-states have given approval to the practice of acupuncture and moxibustion, 29 have enacted special statutes on traditional medicine, and 18 have included acupuncture and moxibustion treatment in their medical insurance provisions. There is a World Federation of Acupuncture-Moxibustion Societies, headquartered in China, that has 194 member-organizations from 53 countries. The government has funded the building and staffing of ten overseas TCM research and teaching institutions.

The 'Big Pharma' globalization of TCM is a conscious, government-funded, cultural parallel to China's huge 'Belt and Road' economic and transportation initiative. The SCIO report closes on a triumphant note:

China's economic development has entered a new historical period. TCM has come to play an increasingly significant role in the socio-economic development; it has become a unique resource in terms of healthcare, an economic resource with great potential, a scientific and technological resource with originality advantages, an outstanding cultural resource, and an ecological resource of great importance. The time has come for TCM to experience a renaissance. (SCIO 2016, p. 14)

With government backing, China's Pharma business is going to get bigger. The Yiling Pharmaceutical Company, that produces herbal remedies, has a market capitalisation of $3 billion. The Tianjin-based Quanjian herbal business was worth $2.8 billion in 2015; it had 10,000 staff in 110 countries; the owner, Shu Yuhui, took 6000 sales staff to France to celebrate a bumper selling season; it operated the largest cancer hospital in the world; it branched out into sales of 'negative-ion' sanitary pads and magnetic insoles for shoes. The treatment-related death of 155 users of the company's products did prompt a government investigation; yet when a doctor went online denouncing the treatments as ineffective and harmful, he was jailed for 3 months. Indicative of the current Chinese *Zeitgeist*, on his release he apologized publicly for 'not thinking clearly'.[13] In China, as in all authoritarian states, it has always been dangerous to not think clearly. In the Cultural Revolution millions lost their lives for not thinking clearly; which at the time simply meant thinking Chairman Mao thought. Not thinking clearly is very dangerous in the contemporary era of massive government oversight and monitoring of all print and online communications in China and among the Chinese diaspora.

Not surprisingly, Chinese schools are besting themselves to win government favour by flying the TCM flag. Zhejiang Province has been the first to announce that TCM will be, from 2018, added to the curriculum for the final year of primary

---

[13] See *The Economist* 16–22 March 2019, pp. 28–29.

school. Along with other aspects of TCM, 12-year-olds will have instruction in how to administer acupuncture. Already 700,000 textbooks have been printed. A particular challenge is that the course will be taught by science teachers. Fang Jianqiao, the president of Zhejiang University of Chinese Medicine, said that the course is being added to the school curriculum 'not just for the medical knowledge, but the culture and philosophy behind it and as a way to boost young people's confidence and pride in their country' (Hersey 2017). This makes Zhejiang science teachers the first in the world to formally have to weigh up at a systemic level the 'educational responsibilities and opportunities' of feng shui.

Appraising Acupuncture

The SCIO Report deals with all components of TCM each of which considers sickness or pain to be a result of chi blockage and/or unbalanced chi in the body. The focus here will be with an appraisal of the more limited explicitly feng shui informed part of the TCM spectrum, namely acupuncture.[14] The connection between acupuncture and chi is explicit. As two practitioners state the matter:

> The *Qi* courses through the body in perpetual motion similar to water in a riverbed. Like the matrix of waterways that cover the surface of the earth, these channels empty into one another, intersect, and have underground as well as surface streams, connecting the interior with the exterior of the body. (Beinfield and Korngold 1991, p. 236)

This reflects the underlying 'naturalistic' assumption of the Chinese philosophical tradition that forces governing the cycle of change in the external world also operate on human bodies and minds. Supposedly 14 major pathways traverse the body from the top of the head to the tips of the fingers and toes. Access to the streams are through acupuncture points (acupoint) or gates, many of the principal ones being below the elbow and knee where:

> the *Qi* changes its polarity from Yin to Yang and gathers force as it moves from the extremities to the core. Stimulation of these points exerts profound influence upon the equilibrium of the organism as a whole. (Beinfield and Korngold 1991, p. 236)

Stimulation of the points is not straightforward as the way the needle is manipulated depends on whether the aim is to consolidate or disperse chi. Nor is location of the acupoints for any given patient straightforward. The points lie near to anatomical landmarks, with the distance between these landmarks being measured in 'body units'. One body unit is equivalent to the width of the second joint of the thumb, with three body units being 'the width of the other four fingers when they are held together' (Beinfield and Korngold 1991, p. 255). This imprecision helpfully builds

---

[14] Numerous studies document the chi-basis of acupuncture and moxibustion practice. See Chang (1976) and Norris (2001). The latter affirms that 'qi is central to all aspects of traditional Chinese medicine … and the same can be said of the distinction between yin and yang' (Norris 2001, pp. 1, 6).

protection against falsification into the system: treatment failures are not failures of the theory, rather they result from the needle missing the acupoint, the 'body-unit' measurement was off.

The nationalism and triumphalism of the SCIO report does warrant examination. Acupuncture has not been the unchallenged part of the Chinese medical tradition (TCM) that the report makes it out to be. After a 1000 years of decline, in 1822, Emperor Dao Guang issued an imperial edict stating that acupuncture and moxibustion should be forever banned from the Imperial Medical Academy. It was re-banned a number of times, and finally by the Kuomintang government in 1929. Mao Zedong (1893–1976) brought it back from the dead in 1949 as part of his affirmation of Chinese tradition. It is noteworthy that Mao's personal physician revealed that Mao did not believe in TCM and refused its ministrations, insisting on Western medicine for his own ailments (Li 1996). Notwithstanding his own scepticism, Mao promoted TCM during the Cultural Revolution as a way of connecting the Communist Party to the anti-Western Chinese nationalist fervour consuming millions of youth and going close to destroying the country. While not using TCM, he spoke of it as a 'treasury of inherited knowledge'. Although speaking this way, Mao did not tangibly promote TCM. One estimate is that by 1978, of all Chinese doctors, just 10% practised TCM (Porkert 1982, p. 564).

The evidence for efficacy of acupuncture, much less for any inference to the yin-yang balance, or chi movement, as the cause of any such efficacy, is very problematic. One detailed study of decades of research into acupuncture, published in the journal *Anesthesia & Analgesia*, concludes:

> The outcome of this research, we propose, is that the benefits of acupuncture are likely non-existent, or at best are too small and too transient to be of any clinical significance. It seems that acupuncture is little or no more than a theatrical placebo (Colquhoun and Novella 2013, p. 1360)

One raft of studies that led Colquhoun and Novella to this conclusion was a review of genuine versus sham acupuncture trials for pain relief in Germany and the USA. In sham trials, needles did not penetrate beneath surface skin and that were placed randomly about the body, not near supposed SEL nodes. The trials revealed that:

> If, indeed, sham acupuncture is no different from real acupuncture, the apparent improvement that may be seen after acupuncture is merely a placebo effect. Furthermore, it shows that the idea of meridians is purely imaginary. All that remains to be discussed is whether or not the placebo effect is big enough to be useful, and whether it is ethical to prescribe placebos. (Colquhoun and Novella 2013, p. 1361)

This is exactly the result of the above reported authentic versus sham Reiki trials: there was no difference in outcomes in the two treatment groups. Some proponents of TCM, specifically acupuncture, simply say that modern science and TCM are incompatible as they make conflicting claims about the world, yet both are 'equally true' (Radner and Radner 1989, p. 152). This ecumenism extends to internal TCM disputes. Marc Estrin, the author of papers on holistic acupuncture, after surveying alternative and contrary practices, says that chi flow is affected differently by the thoughts of opposing therapists; hence contrary practices have the same outcome

(Radner and Radner 1989, p. 152). This goes a long way towards saving the vital fluid, or chi flow, theory of acupuncture from scientific appraisal, but it condemns it to being a pseudoscience.

The same effects were seen in a meta-analysis of 29 randomized controlled trials with 17,922 patients. In these trials, real acupuncture was better than sham but by a tiny amount that lacked any clinical significance. Colquhoun and Novella documented one notorious case that led to a flood of letters of complaint to the *British Journal of General Practice*. The authors in the case conducted an uncontrolled acupuncture/no acupuncture trial in which there was no difference in pain-relief outcome for the two groups. The research was published, but the authors highlighted acupuncture successes for a media release which was picked up by media and spread by the authors' university publicity department. The result is that the public read, again, of the 'success of acupuncture' (Colquhoun 2011).

In Colquhoun and Novella's survey of scores studies, involving thousands of patients, their conclusion is sobering, but not surprising:

> The best controlled studies show a clear pattern, with acupuncture the outcome does not depend on needle location or even needle insertion. Since these variables are those that define acupuncture, the only sensible conclusion is that acupuncture does not work. Everything else is the expected noise of clinical trials, and this noise seems particularly high with acupuncture research. The most parsimonious conclusion is that with acupuncture there is no signal, only noise. (Colquhoun and Novella 2013, p. 1362)

Their review of acupuncture studies is consistent with most major reviews of TCM.[15] America's National Institute of Health looked at 70 systematic reviews of TCM treatments. In 41 of them, the trials were too small or badly designed to be of use. In 29, the studies showed possible benefits, but problems with sample sizes and other flaws meant the results were inconclusive. Shu-chuen Li of Newcastle University in Australia, and formerly Head of the Pharmacy Department at National University of Singapore, found that only a quarter of the studies he looked at showed some benefits from TCM, but most of these were marginal. In contrast, nearly 100% of Chinese evaluative studies of TCM show its effectiveness.

The government has been embarrassed by this scholarly parallel to the routine '99.9% of voters endorsing the local dictator' in elections in all dictatorships. The government's own review of the published 1622 clinical trials of TCM revealed that in 80% of the trials the data were fake, a genuine case of 'fake news'. Publication bias is a serious issue in China and recognized as such by Chinese scholars. Yet on 23 October 2017, a posting on China's WeChat site that pointed to dangers in TCM remedies was removed by the government's web-oversight department.[16]

---

[15] See at least Cyranoski (2017), Huston (1995), and Skrabanek (1985).

[16] Because it cannot be controlled, Facebook is banned in China. State-controlled WeChat takes its place. It is no surprise that the International Journalist Association in 2017 ranked China 194 out of 197 countries for freedom of press.

Myron Wegman, the author of the Biomedical Research chapter in the *Science in Contemporary China* handbook (Orleans 1980), observes that:

> a classic weakness …of most Chinese research on therapy is the absence of true controls, essential if one is to draw secure conclusions about the efficacy of a particular treatment. (Wegman 1980, p. 273)

He notes that this is a particular problem for acupuncture research because the illnesses 'cured' are often ones where there is a high portion of spontaneous recovery or remission.

Consider the following description of clinical practice in Shanghai's Shu Guan Hospital Acupuncture Clinic:

> Rather than patients being isolated in private rooms, ten people at once were treated communally in a room crowded with tables and chairs. The atmosphere was neither frenetic nor chaotic, but friendly, vocal, and intimate. Some people had travelled long distances to be seen at this clinic renowned for resolving difficult cases. … Patients were helping each other by sharing both their individual stories and their common intention to make the most of this opportunity for treatment. Doctors were supporting each other with continual exchanges of advice and assistance. … People who had languished for years under other medical regimens changed dramatically in a matter of weeks. (Beinfield and Korngold 1991, p. 248)

Leaving aside just what proportion of patients changed dramatically in a matter of weeks and what the illnesses were that were so quickly cured, it should be obvious that for any individual patient, needle-effected change in chi flow is just one of a number of possible explanations of the cure. Placebo effects, group-dynamic effects, and expectancy effects all need to be controlled and investigated before attributing the cure to alteration of chi flow.

The lead investigator in a 3-year US acupuncture study, supposedly the largest ever done in the country, frankly admitted that 'the methods for collecting this information were not entirely objective and scientific' but nevertheless claims the study 'gives an indication of the effectiveness of acupuncture for treating the conditions listed' (Wensel 1980, p. 143). Such studies amount to collecting anecdotes, and a second anecdote counts as a replication study; such 'studies' give no indication of effectiveness and should never be funded.

Other studies identify the bulk of positive acupuncture outcomes as placebo responses (Sampson 1996, p. 193; Skrabanek 1985, pp. 187–188; Vickers et al. 2012). Andrew Vickers and colleagues looked at studies of acupuncture and pain relief for 18,000 patients. They concluded:

> Although the data indicate that acupuncture is more than a placebo, the differences between true and sham acupuncture are relatively modest, suggesting that factors in addition to the specific effects of needling are important contributors to therapeutic effects. (Vickers et al. 2012, p. 1414)

Others dispute even Vickers' modest effect. Edzard Ernst, Professor of Complementary Medicine at University of Exeter, wrote of the Vickers study:

This important analysis confirms impressively and clearly that the effects of acupuncture are mostly due to placebo; the differences between the results obtained with real and sham acupuncture are small and not clinically relevant. Crucially, they are probably due to residual bias in these studies. Several investigations have shown that the verbal or non-verbal communication between the patient and the therapist is more important than the actual needling. If such factors would be accounted for, the effect of acupuncture on chronic pain might disappear completely.

In addition, there are several problems with this technically well-done meta-analysis. For instance, its findings are strongly driven by the large German acupuncture trials; they received much press attention in Germany when they were conducted which, in turn, rendered their patient-blinding more than questionable. Moreover, we should be clear about the fact that, in all of these trials, the therapist knew whether he was administering real or sham acupuncture. Arguably, it is next to impossible to completely keep this information from the patient. In other words, a trial is either both patient and therapist-blind, or not blind at all.

Acupuncturists tend to tell us that therapist blinding is impossible, but this is clearly not true. I fear that, once we manage to eliminate this bias from acupuncture studies, we might find that the effects of acupuncture exclusively are a placebo response.[17]

Ernst and colleagues performed other large meta-analyses confirming the placebo-alone explanation of acupuncture effects (Ernst et al. 2011).

The complex methodological issues involved in appraising just the empirical base of acupuncture, that is, its effectiveness, are revealed in 'The Acupuncture Evidence Project: A Comparative Literature Review' conducted by a group of Australian professors from four universities and published in 2017. It found that 'acupuncture is of proven or possible effectiveness/efficacy for a very wide array of conditions'. But this newsworthy conclusion does not bear closer examination.

The Evidence Project was funded by the 'Australian Acupuncture and Chinese Medicine Association Ltd.' This, in principle, should cause alert, in the same way that tobacco company funding of lung cancer research does, or Coca-Cola funding of sugar-obesity relations, or coal-industry funding of anthropogenic climate change research. The project looked at 136 supposedly systematic reviews of acupuncture. The study was self-published; it did not appear in any orthodox medical or social science journal. The authors, between them, published ten of the reviews being examined; and they are all employed in Australian TCM contexts or teach alternative medicine courses. They acknowledge that some of the reviews considered were not randomized, controlled studies; and further that ten 'nonsystematic' reviews were included in the 136 sample. This raises the question of why noncontrolled and nonsystematic studies were included in an appraisal of supposed controlled and systematic studies? Were they included because they had positive results? Edzard Ernst, an Austrian medical researcher, described the project as 'an orgy in deceit and wishful thinking'.[18] Ernst was sceptical about the 'minor and residual'.

---

[17] See http://www.sciencemediacentre.org/expert-reaction-to-meta-analysis-of-studies-into-acupuncture-and-chronic-pain-2/

[18] See http://edzardernst.com/2017/12/a-new-comprehensive-review-of-acupuncture-turns-out-to-be-an-orgy-in-wishful-thinking/

The issues addressed above leave aside the warrant or otherwise of any inference about chi movement as the cause or explanation of acupuncture's effectiveness, even if there is any measurable effect. Even where judged successful, acupuncture studies, and more generally studies of chi-based medical interventions (Reiki, Jin Shin Jyutsu, etc.), simply do not entail any inferences to the reality of the evoked 'universal life force' of *chi*, *qi*, or *ki*. This is, again, the fallacy of 'affirming the consequence'; there can be perfectly acceptable mainstream explanations for any therapeutic success that is established. Lots of astronomical evidence supported Aristotle's crystalline spheres, but they were not there; an abundance of phenomena confirmed the phlogiston theory of combustion, but nevertheless phlogiston was not there; much electromagnetic phenomena pointed to the ether, but it was not there. All of these and, more, are just routine lessons in the history and philosophy of science. Together they are grounds for what Larry Laudan has called the 'pessimistic induction' from history of science against scientific realism.[19] Doubtless there is much to learn from TCM herbal infusion and various traditional practices, but continued and repeated mantra claims about life force only inhibit or confuse the learning; and they prepare people for more general mystification and obfuscation.

These effects were evident at the founding of the American Association of Acupuncturists and Oriental Medicine in 1981 at a Los Angeles convention. R.A. Dale in the keynote address announced that:

> Acupuncture is a part of a larger struggle going on today between the old and the new, between dying and rebirthing, between the very decay and death of our species and our fullest liberation. Acupuncture is part of a New Age which facilitates integral health and the flowering of our humanity. (Skrabanek 1985, p. 191)

This hardly prepares the field for sober, serious, scientific appraisal of its claims, and for careful extrapolation from the empirical claims to causes.

Conclusion

Despite billions of dollars being spent annually in China and around the world on TCM, and millions of people daily taking or practising one or other component of TCM – herbal medicine, qigong exercise, acupuncture, tai chi, and others – there is zero evidence to support the basic claim that TCM effects are due to good management of chi, much less that they establish the related yin-yang, five-element cosmological edifice. Where there are beneficial outcomes – pain relief, relaxation, stress reduction, and so on – many are shown to be placebo effects, psychosomatic effects, the result of standard biochemical processes, and so on. For each class of outcomes, the naturalist response is to look for scientifically established causal explanations.

---

[19] See Laudan (1981), and the numerous realist responses to this pessimistic induction, especially Brown (1994), Bunge (2006), Psillos (1999), and contributions to Agazzi (2017).

There is only a minimum of effects that cannot be so explained. But that does not mean that recourse needs be made to chi; all evidence so far is that it will explain nothing as there is nothing; it is feel-good invocation. Such invocation can be harmless, but it can also be distracting, time-wasting, scurrilous, and harmful. Better to keep looking for naturalist answers than.

Part III
Feng Shui: A Historical-Philosophical Narrative

Chapter 6
Matteo Ricci: A Sixteenth-Century Appraisal of Feng Shui

Three episodes in the history of contact between feng shui and Western science provide context for, and shed light on, the core argument of this book, namely, identifying feng shui as a pseudoscience. Each case provides material that can be utilized in science classes and in coordinated interdisciplinary teaching: first, the life and achievements of the Jesuit priest Matteo Ricci in the late sixteenth century; second, the observations and writings of the Lutheran pastor Ernst Johann Eitel in the mid-nineteenth century; and third, the landmark mid-twentieth-century studies of Joseph Needham on traditional, 'premodern' Chinese science. Each episode has historical, philosophical, political, cultural, theological, and scientific dimensions that can be examined. Each in their time distilled the give-and-take and mutual learning and accommodations that accompanied the historic encounter of Western worldviews and science with their Eastern, specifically Chinese counterparts.

The episodes can, to whatever degree is appropriate in the classroom and curriculum circumstance, be elaborated in science classes. They also provide ideal cases for cross-disciplinary cooperation among history, religion, philosophy, literature, cultural studies, and science teachers. Some aspects of these episodes will be outlined here. The historical texts, and debates occasioned by them, provide an opportunity for elaborating on current issues in history and philosophy of science, theology, and education. When used in class, the texts provide a common base for students and teachers in history, social studies, and science classes.

Famously, Marco Polo (1254–1324) travelled through China but his travel *Journal* deals with Chinese beliefs – Daoism, Buddhism, and Confucianism – only in the manner that any tourist might; his *Journal* exhibits no substantial engagement with any of the systems (Polo 1958). In the eighth century, there had been Nestorian Christians in China, but they left only fragmentary records of the Chinese philosophical or religious beliefs they encountered, and such records had no impact on European understanding of China (Palmer 2001). Later there were Franciscan missions and Macao-based Portuguese traders and missionaries who had engaged with

Chinese beliefs, but none left such a large, widely circulated, and influential trove of letters, reports, and manuscripts as did the Jesuits. Ricci's life in China is a justly celebrated 'first contact' between Western philosophy and theology and Eastern religion, metaphysics, and culture. Study of the contact can be something from which a better understanding of both traditions can emerge (Littlejohn 2014). Ricci's and other Jesuit writings informed Leibniz's and Enlightenment discussion of China and of non-European religion and worldviews (Cook and Rosemont 1994; Padgen 2007).

The Jesuit Mission

Matteo Ricci (1552–1610) was one of the first Europeans to give an informed and detailed appraisal of feng shui belief and practice in China and certainly one of the first whose account gained wide readership.[1] He was among the earliest, and foremost, of the Jesuit missionaries who were sent to Ming dynasty China in the late sixteenth century following the thwarted efforts of Francis Xavier.[2] As well as philosophical and religious training at Jesuit universities in Rome, Ricci had noteworthy mathematical, scientific, and technical competence. At the *Collegio Romano*, he studied mathematics and astronomy with the famed German Jesuit, Christopher Clavius, who was known as 'the Euclid of the sixteenth century'. Clavius was director of the Gregorian calendar reforms; and although he rejected the truth of Copernican heliocentrism, he taught the theory and embraced it for its utilitarian value in calculation, navigation, and map-making. Whether from conviction or pragmatism, he took the Bellarmine instrumentalist option. He was a confidant of Galileo, and they exchanged mathematical and astronomical papers. The Jesuits he trained were the conduits for bringing Galileo's arguments, telescope, and more generally the best European astronomy to China (Fig. 6.1).

As early as 1614, Jesuit astronomers were teaching and writing about Galileo's telescopic discoveries and his support for Copernicus.[3] And conversely, in 1687 the Jesuits provided the first translation into a European language, albeit Latin, of the *Analects of Confucius*, published as *Confucius Sinarum Philosophus*.

With 14 fellow Jesuit missionaries, Ricci left Lisbon in 1578 sailing to Goa where he spent 3 years completing theology studies and then sailed on to Portuguese Macao arriving in 1582. There he immersed himself in the study of Chinese lan-

---

[1] For studies of Ricci's life and influence, see Bernard (1935), Cronin (1955), Hsia (2012), Laven (2011), Littlejohn (2014), Rule (1972, 1986), Spence (1984), Tang (2015b), and Wright (2010).

[2] On the history of the Jesuit missions, and their precursors, see Brockey (2008), Dunne (1962), Laven (2011), Rowbotham (1942), and Rule (1986).

[3] The more prominent being Adam Schall von Bell, Manuel Dias, Johannes Schrick, and Ferdinand Verbiest. The latter reconstructed the Peking Observatory in 1673 replacing ancient Chinese instruments with European ones. On Jesuit scientific research, teaching, and practice in China, see d'Elia (1960), Spence (1980, Chap. 1), and Udías (1994, 2015, Chap. 4).

Fig. 6.1 Matteo Ricci

guage, literature, customs, and history. His quick mastery of so much completely novel material was abetted by his astonishing, and much commented-upon, memory. After one reading, he could repeat random lists of hundreds of characters. He had written an essay on memory during his studies in Rome.[4] With Xu Guangqi (1562–1633), his Chinese collaborator, friend, scholar, and Catholic convert, Ricci translated Clavius' geometry books *Geometrica Practica* and *Trigonometrica* into Chinese.

Importantly in 1607, they translated and published the first six books of Euclid's *Elements*. The translation was a work of huge historical importance for the development of mathematics, and much more, in China (Engelfriet 1998). Hitherto, Chinese mathematics had been exclusively practical, abjuring systemic, deductive, theoretical studies. Chinese astronomers had long had contact with Muslim astronomers in the Persian Marâgha observatory, some of the mathematical classics were brought back to China but they were not translated. Euclid's *The Elements* were brought to China in 1273, but not fully translated till 1856, almost 600 years after it appeared (Huff 1993, p. 240; Wang 2016, p. 67). To have *The Elements* in a country for 600 years without translation says a lot about the country's approach to mathematics. Court astronomers (there were no other kind) knew of Ptolemy, but they did not embrace a geometry-based astronomy.

While the Jesuits were establishing their mission, Johann Kepler published his Three Laws of Planetary Motion: all planets move in elliptical orbits with the sun at one focus; a line connecting a planet to the sun sweeps out equal areas in equal times; the square of the period of any planet is proportional to the cube of the semi-major axis of its orbit. These laws were the acme of geometric astronomy. Within 80 years, they would be used by Newton as evidence for his monumental universal

---

[4] On Ricci's astounding memory, see Spence (1984).

law of gravitational attraction. Kepler's laws could not be contemplated in the Chinese non-geometric astronomical tradition. A very small number of Chinese astronomers were moved to adopt the foreigners' astronomy and its methods, the majority preferring to persist with Chinese astronomy.

This 'failure to learn' circumstance could be traceable to China's well-documented elevated self-regard and xenophobia. China understood itself as being at the centre of the world and had for 2000 years regarded all those outside the empire as barbarians. It was this deep-seated cultural view that partly explained the failure of Chinese seafarers who, despite having compasses, did not explore the oceans in the way that Europeans did. Joseph Needham, in his study of premodern Chinese astronomy, remarked:

> In spite of so much accurate observation, Chinese study of planetary motion remained purely non-representational in character. Unlike that of the Greeks, in which the geometry of circles and curves was so prominent, it perpetuated the algebraic treatment of the Babylonian astronomers such as Naburiannu and Kidinnu, and never sought a geometrical theory of planetary motions. (Needham and Ling 1959, p. 399)

The main mathematical text for 1500 years had been the first-century AD Han dynasty's *Nine Chapters on the Mathematical Art* (Jiu Zhang Suan Shu) which consisted of 246 problems in cultivation, taxation, engineering, etc. (Needham and Ling 1959, pp. 25–27). The same utilitarian conception of mathematics is manifest in other famous and influential works – *Master Sun's Mathematical Manual* (Sun Zi Suan Jing), *Mathematical Manual of the Five Dynasties* (Wu Xao Suan Jing), and *Mathematical Treatise in Nine Sections* (Qin Jiushao).[5] As one historian concluded: 'traditional Chinese mathematics was distinctly practical' (Song 1996, p. 265). Tellingly, the remaining nine volumes of the *Elements* were not translated till the mid-nineteenth century by Li Shan-Lan (1811–1882), a Chinese mathematician, and Alexander Wylie (1815–1887), a Protestant missionary.[6]

Ricci also oversaw the Jesuit production of new and vastly more accurate and superior maps of China which considerably improved administration and tax collection, but this was at a cost to national esteem. Ricci's 'Comprehensive Map of the Geography of all Countries' caused outrage in court and literati circles as China was depicted as one country among many, not the largest, and not at the centre of the world; and the world was not flat. Subsequently, the introduction of the European pendulum clock was a great boost to astronomy, ritual, and culture. Along with all this, Ricci was tireless in translation, teaching, and preaching.

A dramatic astronomy-related episode in the early history of the Jesuit mission took place in 1644. Johann Adam Schall (1591–1666), a German Jesuit and successor to Ricci, sent on July 29 a formal petition to the emperor saying:

> Your subject presents to Your Highness predictions concerning an eclipse of the sun that will occur on September 1, 1644, calculated according to the new Western method, together with illustrations of the percentage of the solar eclipse, and the sun's reappearance as it may be seen in the Imperial capital and in various provinces. In some provinces the eclipse

---
[5] On this, see Guo (2014), Song (1996), and Wang (2016).
[6] On Li Shan-Lan, see Wang (1996).

comes earlier, in others later. The predictive data are listed and presented for examination. Your subject humbly begs from Your Highness a decree to the Board of Rites to test publicly the accuracy of the prediction of the solar eclipse at a proper time. (Spence 1969/1980, p. 3)

The test took place on the appointed date, and only Schall's predictions were confirmed; other submissions to the Imperial Calendrical Department based on extant Chinese astronomy fell short of the mark.[7] In recognition of this, he was appointed in 1645, by Emperor Shunzhi, director of the Bureau of Astronomy and the Tribunal of Mathematics. Fathers Giacomo Rho (1593–1638), Nicolo Longobardo (1559–1654), Johann Terrenz (1576–1630) were all appointed to Calendrical Bureau. Western astronomy, and soon mathematics, became, at least for 20 years, the court preference. Another solar eclipse test took place on 16 January 1665, and again Schall and his Jesuits got the timing exactly right; the Chinese astronomers were out by half an hour. Unfortunately, Shunzhi had died in 1661, and after court intrigues led by the 'demoted' Chinese astronomers, Schall was imprisoned and sentenced to death. Fortunately, there was an earthquake on the appointed day, the judges took this as an omen of heavenly displeasure, and Schall's head was saved (Udías 1994, p. 470). However, his health deteriorated, and he died in 1666 shortly after release. Nevertheless, his Jesuit colleague Ferdinand Verbiest succeeded to the Bureau Directorship in 1669. The centrality of science for the Jesuit Chinese mission has been recognized from the outset. This is captured in the much-reproduced 1747 drawing of Jesuit priests Ricci (left), Schall (centre), and Verbiest (right) all holding mathematical and astronomical instruments, along with a crucifix (Fig. 6.2).

Fig. 6.2 Three Jesuit scientists at work in China (Jean-Baptiste Du Halde 1747)

---

[7] Schall's experience in China is discussed in Spence (1969, Chap. 1). The all-important eclipse test is described in York et al. (2012, p. 284).

Ricci's China Travels and Journal

In 1583 Ricci began his 27-year sojourn in China. It was a country for whose culture and achievements he had the greatest admiration, incorporating many customs within the practice of the embryonic Chinese Catholicism that he and fellow Jesuits were establishing. He moved first in the south – Nanchang, Shao-chou, and Chao-ching – then finally, in 1600, gained admission to Peking where he spent the final decade of his life. He had the highest regard for Confucius and the social, political, and personal commitments of early Confucian followers, seeing them as compatible with Christian belief, against imported Buddhist and other Chinese belief systems. He purposely adopted the dress and demeanour of Confucian teachers.

Ricci died in 1610 in Peking where his grave is still maintained and respected. It was the site of the Communist Party's 1960 celebration of the 350th anniversary of Ricci's death – one of the few places in the world where the anniversary was celebrated. Ricci was a pioneer of European-Chinese cultural engagements and in so many ways embodied what today is regarded as the best kind of engagement between cultures – respectful and serious but critical where necessary. The *Encyclopædia Britannica* claims that 'Probably no European name of past centuries is so well known in China as that of Li-ma-teu (Ricci Matteo)'.

In 1615 Ricci's travel journal, *On the Christian Mission among the Chinese*, was published posthumously in Rome in Latin with the supervision of his fellow Jesuit Nicolas Trigault. Within a decade, four Latin editions appeared along with three French and one edition each of German, Italian, and Spanish translations. Interestingly, three centuries passed before the first English edition which was the 1953 translation by the Jesuit priest Luis Gallagher of Trigault's 1615 Latin text. Gallagher's 620-page translation is titled: *China in the Sixteenth Century: The Journals of Matthew Ricci, 1583–1610* (Ricci 1615/1953).[8] His many letters from China have also been collected, edited, and published (Ricci 2001).

Ricci's admiration of China, its culture and its Confucian traditions, is evident throughout his *Journals*. In the Preface to his translation, Gallagher wrote of Ricci that:

> The author of numerous works on science and religion, written in Chinese, Ricci was well known to the educated classes of China as a prominent professor of physics, mathematics and geography, as a learned philosopher of Chinese and of extraneous doctrine, as a prominent commentator on Confucius, and particularly as an eminent teacher of the Christian religion. Some of his Chinese compositions ... are included in the official index of the best Chinese writings of all time. ... His first Chinese mappamondi stood as a model for European cartographers for a century after its publication; and earned him the title, 'The Ptolemy of China'. His correction of the Chinese calendars was a preface to a century and a half of scientific advancement in China.

---

[8] There has been some debate about how much Trigault inserted himself into the 1615 Latin edition. But it seems that with some exceptions, it is a faithful translation of Ricci's original Italian text (Rule 1972, pp. 122–24). Gallagher rendered Ricci's given name as 'Matthew'; in this book the Italian 'Matteo' will be used.

> The nobility of character of the Chinese people, their love of liberty, of order, and of learning, their devout tendency toward religion, and their keen sense of justice and of ethical interpretation, were never more clearly revealed than when set forth in what Ricci calls his summary study of the customs, laws, institutes and government of the Chinese people. (Ricci 1615/1953, p. xxi)

Ricci's remarks in the opening of Chap. 10 of his *Journal* are repeated in many places:

> Of all the pagan sects known to Europe, I know of no people who fell into fewer errors in the early ages of their antiquity than did the Chinese. From the very beginning of their history it is recorded in their writings that they recognized and worshipped one supreme being whom they called the King of Heaven, or designated by some other name indicating his rule over heaven and earth. …They also taught that the light of reason came from heaven and that the dictates of reason should be hearkened to in every human action. … The same conclusion might also be drawn from the books of rare wisdom of their ancient philosophers. These books are still extant and are filled with the most salutary advice on training men to be virtuous. In this particular respect, they seem to be quite the equals of our own most distinguished philosophers. (Ricci 1615/1953, p. 93)

He was both astonished and impressed with the entrenched system of rigorous, impartial, countrywide Civil Service Examinations that were in place for selection into government service. Such arrangements were unheard of in Europe, where nepotism, family, marriage, cronyism, and purchase were the principal avenues to Royal or government service. The Jesuits, and other orders, had demanding education and exams, but they did not lead to, nor were they required for government service. In China the first-level exams were held biennially in every county; the second level were held triennially in each provincial capital; the third level, and most demanding exams, were held in Peking in the fourth year (Elman 2000; Laven 2011, pp. 132–139). A capable boy from a remote village could end up walking thousands of kilometres to Peking to succeed in the national exam then take a high place in court or in service.[9] The downside was that if the capable boy did not succeed, then he walked all the way back, and for 3 more years he memorized text and prepared model answers before once more walking to Peking to take the exams. And then perhaps doing it all again in another 3 years. This represented massive inefficiency in the state system: the boy could have been studying something else, observing nature, engaging with local technicians, doing rudimentary experiments, or just carrying on a business.

As a Jesuit missionary, Ricci's overwhelming preoccupation was with spreading the Christian Gospel to the Chinese and with making converts by establishing the wisdom of Catholicism in comparison to the esteemed Confucian belief, which was the belief of the *Literati* or the Chinese scholarly sect. His central evangelizing work, published in Chinese, and meant to place before the *literati* commonalities and differences between Christianity (Catholicism) and Confucian belief was *The True Meaning of the Lord of Heaven* (Ricci 2016). In a letter of 15 February 1609, towards the end of his life and his sojourn in China, Ricci writes:

---

[9] The examination system is discussed in Chap. 8.

> ... in the most ancient and authoritative works of the *literati*, they only worshipped Heaven and Nature and their common master. When making a careful study of all these works, we may find few things contrary to reason, but instead, most of them are corresponding to reason. And their natural philosophers are no worse than anyone else. (Tang 2015b, p. 180)

To further this mission Ricci wrote many studies, in Chinese, Latin, and Italian, on Confucian philosophy, history, and literature.

Ricci saw the European science he learnt at the *Collegio Romano* as preparing the ground for the religious conversion of China. In a 1605 letter from Peking to a high-ranking Portuguese Jesuit in Rome, he wrote:

> I really want to beg something of Your Reverence that for many years now I have been requesting without response. One of the most useful things that could be sent from Rome to this court would be a Father or even a Brother who is a good astrologer [astronomer]. And the reason why I say an astrologer, rather than an expert in geometry, clocks and astrolabes, is because regarding these other things I myself know a good deal, and I have sufficient books; but the Chinese take little account of this, compared with the course and true position of the planets and the calculation of eclipses; in short, we need someone who could draw up ephemerides [astronomical tables]. (Laven 2011, p. 130)

And in a 1609 letter, written almost at the end of his days, he emphasized the centrality of science to evangelism and conversion: 'If we can teach them our sciences ... it will be easy to persuade them to our holy law' (Laven 2011, p. 95).[10]

In one of his Chinese essays, *T'ien-chu shih-ihe*, he counselled the *Literati* on how best to support Confucius and oppose Buddhism and Taoism. This text displays something of Ricci's character and why he enjoyed contemporary and subsequent esteem:

> It is better to argue against them than to hate them, and better to convince them by reasoning than to argue against them. The followers of the two sects are all made by the Father of the Lord of Heaven, and so are our brothers. ... In my extensive reading of Confucian books I have frequently come across expressions of hatred for the two sects, ranking them with barbarians, and accusing them of heresy, but I have not seen the application of first principles in order to refute them. ...But if we both use reasoning in conducting our argument, then without harsh words the right and the wrong will appear, and the three schools will return to unity. (Rule 1972, p. 149)

Ricci was also occupied with the internal task of formulating a view about the appropriate 'style' or 'organization' of the Catholic Church in China – the 'Chinese Rites' issue as it has been called.[11] But his journal provided, along with much else of historical and cultural value, a much commented-upon account of the Chinese Daoist metaphysical and geomancy traditions. It is a generous and informed 'first contact' document of Western religion and science with Chinese religions and associated worldviews that warrants and rewards attention.

---

[10] On Ricci's scientific contribution to China, see Bernard (1935).

[11] Ricci argued with other missionaries and with the Vatican over how much of Confucian ritual concerning ancestors, burial, etc. and metaphysical beliefs could be maintained by converts to Catholicism. He wanted to maximize retention. Against Ricci's advice, the Vatican ultimately said that converts had to leave their cultural rituals and language about the 'Lord of Heaven' at the church door.

Astronomy

Astronomy in China was a serious enterprise but was conducted largely for furthering imperial (state) interests and for practical ends such as having a usable season-aligned calendar which itself was an indicator or manifestation of imperial competence and virtue. The court had to know and foretell the major astronomical events: lunar and solar eclipses, recurrent comet movements, 'blue moons', 'red moons', and so on. The court's prestige was significantly judged by its competence in these matters. To be unaware of the coming of these momentous events diminished the prestige of the emperor. One historian wrote:

> One of the most conspicuous characteristics of Chinese astronomy in ancient times is that it was an 'official' science closely connected with the government and government officials … This means that for many centuries astronomy was intended to serve as a science for supporting the Emperor's rule over the country. (Shinji 1996, p. 91)

All commentators point to the entrenched politicization of 'science' through Chinese history, including the present where everything is politicized (politics, law, journalism, sport, education, web access, and so on), and civil society beyond state control barely exists. Hua Shiping, writing on this topic, says:

> Politicization refers to the fact that science was influenced very much by politics. This is most clearly demonstrated in astrology, because the movement and formation of stars explain the mandate of the Emperor. Throughout China's feudal history, the development of astrology was not a process during which humans' understanding of this particular science progressed gradually. Rather, its development was a zigzag path; two steps forward, one step backward. New discoveries in astrology were subject to the ruler's political interests. (Hua 1995, p. 19)

When the Jesuits arrived, and long after, unless a person was in government service, they could not study astronomy; private study was banned or severely discouraged.

The Song dynasty (960–1279) devoted considerable resources to astronomical record keeping, planetary and stellar observations, building and refinement of astronomical instruments, creation and updating of libraries, periodic calendar reforms, and much else. Scores of astronomers/astrologers were employed in the Chinese Astronomical Bureau of the Ming dynasty (1368–1644) when Ricci and his Jesuit colleagues arrived, but they found Chinese astronomy under developed in comparison to European.

There was a long tradition of Chinese astronomical instrument construction, measurement, and record keeping. But theoretical understanding of the instruments and the readings was deficient. There was no awareness that earthly location, specifically latitude, impacted on astronomical observation and on the calibration of instruments. Chinese astronomy did not have a geometrical foundation. For instance, although the elevation of the polar star was different in Peking and in Nanking, and the length of summer and winter was different in both places, it was not until the end of the fifteenth century that these astronomical realities were recognized and built into record keeping and instrument design (Ho 1969). Huge and complicated instruments were moved from imperial capital to imperial capital – for instance, Nanjing to Peking – without realization that change in latitude (7°) affected the interpreta-

Fig. 6.3 Zhao Youqin's fourteenth century flat-earth cosmology (*A* Yangcheng, *B* Mt. Kunlun, *C* Sihai zhi Zhong)

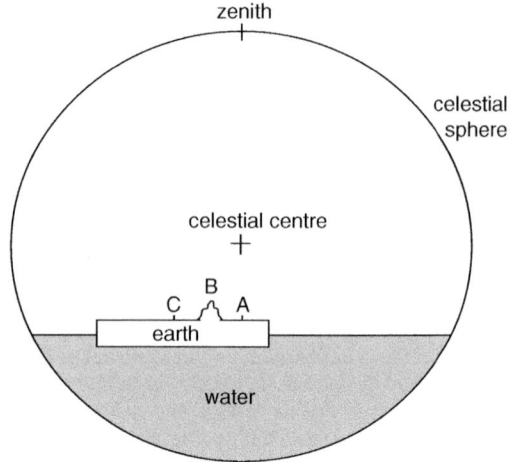

tion of the measurements taken. Ricci had noted in 1600 that the court instruments in Nanjing were set for latitude 36° which was inaccurate for both capitals.

Through trade and intellectual exchange, Muslim astronomy had some small impact on Chinese astronomical thought, but unfortunately through the Ming dynasty (14th – 17th centuries) this had been squeezed out in favour of traditional and inferior Confucian astronomy. A stark indicator of the difference between European/Arabic astronomy and Chinese astronomy was that, until the fifteenth century, Chinese astronomers believed in a flat earth. This is seen in the important early fourteenth-century astronomy book, *Ge xiang xin shu*,[12] written between 1324 and 1335 by the Daoist priest and astronomer Zhao Youqin.[13] In his version of neo-Confucian cosmology, the earth is likened to a flat board floating on water, with China in its centre, and surrounded by the heavens; the cosmos had a globular or egg structure. Zhao shared the dominant neo-Confucian view that the earth was stable and immobile, and the sun, five planets, moon, and stars moved daily from east to west; they were supported and propelled by the 'atmosphere' of *qi* (chi) that was omnipresent in the cosmos (Fig. 6.3).

China's highest mountain, Mt. Kunlun, lay directly under the zenith and so was at the centre of the earth. As the sun appears larger at the beginning and at the end

---

[12] There is disagreement about the title's translation. Needham renders it *New Elucidation of the Heavenly Bodies*; Alexeï Volkov says that this omits the title's all-important allusion to the *I Ching* with *Ge* denoting the hexagram 'Alteration'. So, the title should be *New Writing on the Image of the Alteration* (Volkov 1996a, p. 39). The latter is more reflective of the Daoist philosophical/theological commitments of Zhao and of the inextricable interlinking of 'science' with astrology, philosophy, and religion in the Chinese tradition. The science/non-science demarcation is even more problematic in the Chinese tradition than it is in the West.

[13] Arai Shinji has provided an extensive study of the book (Shinji 1996); and it is noted by Joseph Needham (Needham and Ling 1959, pp. 102, 208). Alexeï Volkov provides an account of Zhao's life (Volkov 1996a, pp. 34–39) and his calculation of π (Volkov 1996b, 1997).

of the day, it must be closer to earth at dawn and dusk, than at noon, when it appears smaller. Consequently, for Zhao 'there is much space above the earth, there is less below it' (Shinji 1996, p. 68). There was a dispute with astronomers who took the opposing view and said the sun was closer to earth at noon than at dawn and dusk because at noon it felt hotter and heat from an object varies inversely with distance. And there was dispute with Indian astronomers who moved the earth across to the east so that Buddha's birthplace in India would be under the zenith, and the Daoist sacred Mount Kunlun would be displaced off-centre; thus, the superiority of Buddhism over Daoism would be testified to by cosmology. This was, of course, the answer they had in advance of asking the question or conducting the investigation.

The arguments over Zhao Youqin's cosmology and astronomy demonstrate the role of politics, ideology, and religion in natural philosophy (science) and the limitation of instruments and measurements without a sound theory to underwrite what is being seen and recorded. Having instruments, and being able to measure, is not the same as having science. Phrenology, astropsychology, Nazi eugenics, Lysnkoists, and countless other such movements had instruments and measurements, but they did not have science.

All of this meant that Ricci's accurate predictions in 1601 of a solar eclipse in comparison to the court astronomers' hopelessly inaccurate eclipse prediction made such an impact on the Ming Emperor Wanli. The emperor allowed Ricci to construct the Cathedral of the Immaculate Conception in Peking and permitted him access to all the high court officials. Ricci and Jesuit colleagues taught the Ptolemaic system and published Chinese manuals for its calendrical use, but for his calculations, Ricci used Copernicus' heliocentric model of the solar system that he had been taught in the Jesuit college in Rome by Christopher Clavius. His Portuguese colleague Manuel Dias in 1614 published the first Chinese account of Galileo's telescope discoveries and his Copernican arguments; this at the time that Galileo was being condemned in Rome for his teaching of the theory.[14]

Ricci's mathematical, trigonometrical, astronomical, mathematical, cartographical, and chronological knowledge was taken up in a purely utilitarian way by the Chinese court and mandarins; it was seen to improve astronomy, astrology, map-making, and calendar calculations upon which so much of the functioning of the emperor's state apparatus depended. For the almost 200 years from Ricci's arrival to the expulsion of the Jesuit order in 1773, the Jesuits were the only source of Western astronomical knowledge in China (Udías 1994). However, and importantly for the overall contemporary educational and cultural discussion of science and worldviews, Ricci's 'natural philosophy', or his Western science, had minimal impact on Chinese science and no impact on Chinese culture.[15] Those who accepted the science held an early version of the later nineteenth-century traditionalist motto: 'Chinese learning for the fundamental principles, Western learning for practical

---

[14] Galileo was condemned not so much for actually teaching the Copernican theory, but teaching that it was true. Among countless sources, see Finocchiaro (1989).

[15] There was strong opposition to Jesuit science by Confucian traditionalists. This is elaborated in Wong (1963).

application'. The Chinese court anticipated the widely held contemporary educational 'border crossing' approach to the learning of modern science in multicultural environs, whereby science is learnt as just a technique without having an impact on cultural beliefs (Aikenhead 1996).

There was little distinction drawn between astronomy and astrology: they were simply two sides of the one coin. For this reason, the reach of astronomical study was severely restricted by the Imperial court because too much knowledge by too many people was too dangerous for Imperial rule. Through to the eighteenth century, the near-universal belief among astronomers and the educated elite was that astronomy simply was Chinese astronomy; there was nothing to learn from 'beyond the empire'.

Three centuries after Ricci, Alexis de Tocqueville, in his famous *Democracy in America* (Tocqueville 1835/1953), made comparable observations on the circumstance of traditional Chinese science:

> When Europeans first arrived in China, three hundred years ago, they found that almost all the arts had reached a certain degree of perfection there, and they were surprised that a people which had attained this point should not have gone beyond it. At a later period they discovered traces of some higher branches of science that had been lost. The nation was absorbed in productive industry; the greater part of its scientific processes had been preserved, but science itself no longer existed there. This served to explain the strange immobility in which they found the minds of this people. The Chinese, in following the track of their forefathers, had forgotten the reasons by which the latter had been guided. They still used the formula without asking for its meaning; they retained the instrument, but they no longer possessed the art of altering or renewing it. (Tocqueville 1835/1953, vol. II p. 47)

He concluded:

> The Chinese, then, had lost the power of change; for them improvement was impossible. They were compelled at all times and in all points to imitate their predecessors lest they should stray into utter darkness by deviating for an instant from the path already laid down for them. The source of human knowledge was all by dry; and though the stream still ran on, it could neither swell its waters nor alter its course. (Tocqueville 1835/1953, vol. II p. 47)

In the mid-twentieth century, as discussed in Chap. 10, this observation would be reformulated by Joseph Needham and become universally known as 'The Needham Question'.

Observations on Feng Shui

One of Ricci's most prominent converts, Li Yingshi, was a war hero and an esteemed feng shui practitioner who was an acknowledged expert on divination and geomancy. After he converted in 1602 and directed his family and retinue to do the same, he became Ricci's 'inside' source of feng shui knowledge providing information and a library of texts.[16] Chapter 9 of Book One of the *Journal* is titled

---

[16] Consistent with then current practice, all his 'heretical' library was eventually consigned to the courtyard flames. A blot on the reputation of the Catholic Church in China.

'Concerning Certain Rites, Superstitious and Otherwise'. Ricci's opening admonition is: 'I would request of the reader that he recognize in the two following chapters a reason for sympathizing with this people and for praying God for their salvation' (Ricci 1615/1953, p. 82). He then proceeds by noting:

> No superstition is so common to the entire kingdom as that which pertains to the observance of certain days and hours as being good or bad, lucky or unlucky, in which to act or to refrain from acting, because the result of everything they do is supposed to depend upon a measurement of time. This imposture has assumed such a semblance of truth among them that two calendars are edited every year, written by the astrologers of the crown and published by public authority. These almanacs are sold in such great quantities that every house has a supply of them ... and in them one finds directions as to what should be done and what should be left undone for each particular day, and what precise time each and everything should be done. In this manner the entire year is carefully mapped out in exact detail. (Ricci 1615/1953, pp. 82–83)

Ricci's observations on the embeddedness of geomancy in Chinese culture have been confirmed by all historians of China. Richard Smith in his work on *Divination in Traditional Chinese Society* writes: 'Although its particular manifestations and social significance may have varied from time to time, place to place, and group to group, divination touched every sector of Chinese society, from emperor to peasant' (Smith 1991, p. 9). And later that 'geomancy served remarkably well as shared lingua franca for Chinese at every level of society' (Smith 1991, p. 170).

Ricci says of Daoist geomancy that it is 'quite peculiar to the Chinese' and writes:

> In choosing a place to erect a public edifice or a private house, or in selecting a plot of ground in which to bury the dead, they study the location with reference to the head and tail and the feet of the particular dragons which are supposed to dwell beneath that spot. Upon these local dragons they believe that the good and bad fortune, not only of the family but also of the town and province and the entire kingdom, is wholly dependent. Many of their most distinguished men are interested in this recondite science and, when necessary, they are called in for consultation, even from a great distance. ... Just as their astrologers read the stars, so their geologists [diviners] reckon the fate, or the fortune of a place, from the relative position of mountains or rivers or fields, and their reckoning is just as deceitful as the reading of the stargazers. (Ricci 1615/1953, p. 84)

He is writing on the long feng shui tradition of 'siting'; of finding the appropriate location or niche where good chi energy is concentrated. He elaborates on divination, saying:

> These people worry a great deal about judging their whole lives and fortunes as dependent upon the exact moment of birth, and so everyone makes an inquiry as to that precise moment and takes an accurate note of it. Masters of this kind of fortune-telling are numerous everywhere ... (Ricci 1615/1953, p. 83)

And he observes, not unexpectedly, that:

> Fraud is so common and new methods of deceiving are of such daily occurrence that a simple and credulous people are easily led into error. These soothsayers frequently have confederates in a gathering who declare to a crowd that everything that was told to them by the performer came to pass just as he had predicted it. Sometimes, too, when strangers are brought in as confederates and relate marvels of the past, the followers of the local imposter

respond with loud applause. The result is that many deceived by this trickery, have their own fortunes told and accept what is predicted for them as the certain truth. (Ricci 1615/1953, pp. 83–84)

Fraud, hucksterism, deceit, and exploitation of people's credulity have remained a constant in the feng shui tradition down to the contemporary internet age as is manifest in the website extracts reproduced in Chap. 3.

Ricci believed that much of Daoism that he saw around him was a corruption of original Daoist belief and practice as was evident in the sixth-century BC founding writings of Laozi and the fourth-century BC writings of Zhuangzi a follower and expositor of Laozi. Ricci believed that this fundamental Chinese belief system or ideology had, over a 2000-year span, been intellectually and materially corrupted. He regarded ancestral Daoism highly as it was marked by pantheism, naturalism, and laudable beliefs about nature, holiness, responsibility, and social cohesion and improvement. But he thought that all of this was, over time, diluted and corrupted, doubtlessly beginning with the deification of Laozi but extending into the practice of geomancy and outright fraudulent behaviours where the 'priests got wealthy and the people got poor'. In this light, Chapter 9 of his journal, where the foregoing entries are found, might be more generously seen as combatting corruptions of Daoism rather than an example of European cultural arrogance. And doubtless at the time, many Chinese were thinking and saying the same as Ricci.

Ricci had grown up in a context in which he could observe at close hand the struggles of a corrupt tradition under attack. Just 30 years before Ricci's birth, Luther published his 95 theses detailing corruptions of Catholicism that animated the Reformation's march across Europe. Ricci was born (1552) in the Papal States acknowledged as one of the most corrupt, backward, and venal political-religious administrations in the history of Europe.[17]

Geomancy was thoroughly embedded in Ming dynasty (1368–1644) life, the practice being codified in the 1445 *Daoist Canon* that contained entire sections of geomantic charts for divination purpose. Few civil and military decisions were made without geomantic input; by law, no government building could commence construction without geomantic certification. Ricci regarded all of this as a *superstitio absurdissima* observing that:

> What could be more absurd than their imagining that the safety of a family, honors, and their entire existence must depend upon such trifles as a door being opened from one side or another, as rain falling into a courtyard from the right or from the left, a window opened here or there, or one roof being higher than another? (Ricci 1615/1953, p. 85)

Ricci had no hesitation in pronouncing feng shui beliefs as absurd, but the equally widespread and related belief in the power of devils or more generally of the interference of supernatural entities in earthly affairs presented a more complex problem for him. All three of the Abrahamic religions were committed to the reality of 'principalities and powers', or devils, or *Jin*, so Ricci could have no 'in principle'

---

[17] For the dismal later history of the Papal States, see Gross (2004) and McCabe (1935, Chap. 9).

objection to invoking devils as explanations of daily events.[18] This is well illustrated later in the *Journal* where Ricci recounts an instructive episode when a major Nankin city official offered him and his fellow missionaries a palace as a residence. Ricci writes of the official:

> 'A short time ago', he said, 'I built a palace, at public expense, for my colleagues; the staff of my tribunal. As soon as it was finished and given over to them, ghosts and devils took possession of it, and no one could live there without being harmed …if you are not afraid of ghosts, it is yours to purchase and there will be no wrangling about the price, because you may determine that yourself, as you see fit.' (Ricci 1615/1953, p. 345)

Apparently, all efforts of the Daoist masters to rid the new palace of ghosts had failed. Ricci recounts how the Jesuits succeeded:

> The first night they occupied the house, the Fathers recited appropriate prayers at an altar erected in the main aula, and went through the whole building, carrying a crucifix and sprinkling holy water, and from that time on, with God's grace favouring the spread of the faith in China, the evil spirits made no further appearance. He to whom all beings are subject had permitted the evil spirits to inhabit this house in order to prepare it for the coming of His servants, and when they came the spirits were driven out of it. (Ricci 1615/1953, p. 347)

It is not clear whether Ricci was doing this 'just for show' or whether he was engaging in genuine exorcism which was a commonplace Catholic rite of the time that was used to banish devils from people and places.[19] Whatever Ricci's inner beliefs, the city official was very impressed, saying:

> Now I can understand why the demons took possession of it. It was because Father Matthew's God commanded them to let no one live there, other than Father Matthew himself. (Ricci 1615/1953, p. 348)

Ricci knew that Chinese 'conversion' to European natural philosophy had consequences:

> And in truth it thus happened that many, having learnt our mathematical sciences [astronomy], laughed at the law and doctrine of the idols, saying that if they taught so much error in natural matters and those of this life, there is no reason to give them credit in supernatural matters and those of the other life. (Rule 1972, p. 166)

This comment was directed specifically at Chinese Buddhists, but it has become a perennial issue in all cases of contact between traditional belief systems and modern science: how to maintain traditional authority structures when authorities make claims about the world that younger people who have had some science classes can see are demonstrably false and without foundation? If respect for elders is predicated upon their being custodians of a culture's knowledge, then when that

---

[18] Paul's epistle to the Ephesians (6:12) 'For we wrestle not against flesh and blood, but against principalities, against powers, against the rulers of the darkness of this world …'. For elaboration and literature on this subject, see Matthews (2015a, pp. 361–63).

[19] Exorcism and belief in devil possession have continued to the present day in the Roman Catholic Church. In 2015 the Vatican hosted an Exorcism Convention. The Vatican's official exorcist, Father Gabriele Amorth, who has conducted 70,000 exorcisms, claimed that many paedophilia cases were the direct work of devils who had taken possession of the offending priests (Amorth 2010).

Contemporary Appraisal of Matteo Ricci

Ricci's journal comments might appear like a textbook example of insensitive missionary denigration of foreign cultures. But his story is more complex, and the lessons students can learn are more subtle. Ricci had a renowned appreciation of Chinese culture and custom, and he sought an 'accommodation' between Catholicism and Confucianism (Rule 1972, pp. 128–140). He thought his own historical studies of Confucian texts established that divination belief and practice was a corruption of original Confucian belief; it was unnecessary and unwanted baggage for Confucianism.[20] He fought the Vatican to allow the carry-over of many Chinese rites, specifically ancestor worship, by converts in the nascent Chinese Catholic Church. When finally, in 1773 the Roman Catholic Church banned these Chinese rites, the emperor, at the urging of traditionalist courtiers who wanted to reassert the authority of Chinese astronomy and natural philosophy, cast the Jesuits out of China. But expulsion of the order did not affect the high regard in which Ricci was held.

Upon Ricci's death in Peking on 10 May 1610, at age 57 years, his friend Li Chih-Tsao wrote to the court requesting an appropriate burial place, as in the Ming dynasty foreigners dying in China had to be buried in Macao. Li Chih thought that burial outside of China was a cruel slight on Ricci's commitment to the people and culture of China.

> Your servant, Li Ma-tou [Ricci] from the time he entered your court, began to absorb your brilliant culture, to read (Chinese) books and penetrate their meaning. Morning and evening, reverently and respectfully he burnt incense and prayed to Heaven, reciting your praises in poor return for your kindness. The loyalty of his heart is known to everybody, high or low, in the city, and we would not dare to embellish it. While he was yet alive he was reputed to be a lover of scholarship and a writer of no mean ability. In his earlier days across the seas he was known as a famous scholar, and when he came to this country he was praised by high officials who did not fear to liken him to the hermits who withdraw to lonely places. (Rule 1972, p. 119)

The emperor granted Li's request, designating a Buddhist temple for the purpose. Subsequently Adam Schall and other Jesuits were buried there, in what became the Zhalan Cemetery. In 2010 it was basically just the Chinese Communist Party and the Catholic Jesuit order who celebrated the 400th anniversary of Ricci's death. The

---

[20] The issue of what is 'corrupt' and what is not, what is authentic development, and what is misguided, deviationist, and heretical is something that all belief systems and ideologies face. The more so, of course, for party-based political movements and religions founded upon putative divine revelation. Communism, Judaism, Islam, and Christianity have all, famously and tragically, ruptured over this very issue of what is authentic and inauthentic development of a tradition.

Party approved a long Beijing newspaper account of his life, complete with photos of his tomb, portraits, and a listing of places with which he was associated.

Conclusion

The Jesuits, and other Catholic orders, were in China for evangelical purposes, but for nearly 200 years, the Jesuits, with full access to the court, had also been practicing, teaching, and promoting modern science in the centre of China. But modern science did not spread; it did not take root in China. Over the same period, 1600–1800, the New Science mushroomed throughout Europe and in colonies such as the American states. This 'failure to thrive' has been addressed as the 'Needham Question' and will be discussed in Chap. 10. Ricci's life and writings provide fertile material for cross-disciplinary study. Teachers and students of geography, history, religion, philosophy, culture, theology, and science can all collaborate in selecting threads of the rich Ricci tapestry and appropriately following through the many and diverse questions to which they give rise.

Chapter 7
Ernst Johann Eitel: A Nineteenth-Century Appraisal of Feng Shui

Two hundred and fifty years after Matteo Ricci's sojourn in China and his landmark accounts of Chinese natural philosophy and culture, another missionary, Ernst Johann Eitel (1838–1908), took the same path. Eitel was a sinologist and linguist and a German Lutheran missionary who graduated in 1860 as an MA in Theology from the University of Tübingen and whose thesis on Chinese Buddhism earned him a doctoral degree from the same university. In 1870 he went to Hong Kong and lived there for many years overseeing educational enterprises, before travelling widely within China.[1] He published a history of the colony *Europe in China: The History of Hong Kong* (Eitel 1895/1983), a Chinese-English dictionary, and studies of Buddhism (Eitel 1873). Eitel published a substantial appraisal of the history, metaphysics, and functioning of feng shui – *Feng Shui: The Rudiments of Natural Science in China* (Eitel 1873/1987). It could be the first Western book published with the title *Feng Shui*.[2]

Discussion of Eitel's book provides an opportunity for elaboration of wider issues concerning feng shui such as the interactions of science with society and culture, science and experiment, science in China, and science and worldviews. These are all issues that can easily and naturally be discussed in science classes and across school disciplines.

---

[1] From the 1870s, beginning with his administrative work in Hong Kong, Eitel preferred the Anglicized version of his name: Ernest John Eitel. For the life and work of Eitel, see Wong (2000).

[2] This book was originally published as *Fengshui: The Science of Sacred Landscape in Old China*, which recognizes the Chinese sense of landscapes being sacred: the form of the earth 'speaks to' an underlying 'heavenly' cosmology.

Feng Shui and Siting Practice

Eitel comments on the all-important role of feng shui in the siting of graves and dwellings. He mentions three things considered by the siting master: the topography of the location, the presence and flow of water, and the shape of local mountains. Concerning topography, he says:

> those elevations of the ground which indicate the presence of nature's breath, with its two currents of male and female, positive and negative energy, symbolically called dragon and tiger. The relative position and configuration of these two, the dragon and tiger, as indicated by hills or mountains, is the most important point, as regards the outlines and forms of the earth's surface. (Eitel 1873/1987, p. 45)

Concerning water flow, he says:

> Another important element in the doctrine of the outward forms of nature is the direction of the water-courses. We have had occasion to allude to this more than once, and the chief point is, that water running off in straight lines or forming in its course sharp angles is absolutely dangerous. A curved and tortuous course is the best augury of the existence of beneficial influences. (Eitel 1873/1987, p. 45)

And concerning the shape of mountains, he says:

> A third subject that calls for attention here is the form and shape of the hills, especially the outlines of their summit. I have remarked above that the summits of hills and mountains are the embodiment of certain heavenly bodies. It is therefore one of the first requirements of a geomancer that he should be able to tell at a moment's glance which star is represented by any given mountain. As to the planets and their counterparts on earth, the rules by which each mountain may be referred to the one or other of the five planets are very simple. (Eitel 1873/1987, p. 48)

He elaborates on how these simple linkages are made between celestial and terrestrial bodies:

> If a peak rises up bold and straight, running out into a sharp point, it is identified with Mars and declared to represent the element fire. If the point of a similarly-shaped mountain is broken off and flat but comparatively narrow, it is said to be the embodiment of Jupiter and to represent the element wood. If the top of a mountain forms an extensive plateau, it is the representative of Saturn, and the element earth dwells there. If a mountain runs up high but its peak is softly rounded, it is called Venus and represents the element metal. A mountain whose top has the shape of a cupola is looked upon as the representative of Mercury, and the element water rules there. (Eitel 1873/1987, p. 48)

Eitel notes how the European construction in the treaty ports that followed the mid-nineteenth-century Opium Wars was constantly embroiled in feng shui disputes about the work's expected positive and negative impacts on the flow and distribution of chi in the port area. He observes how much of the foreigners' building work in Hong Kong, including placement of European cemeteries, inadvertently turned out to be in 'good' feng shui sites (access to water, protection from high wind, away from malaria swamps, and so on) which led many Chinese to impute advanced feng shui knowledge, and their own concern for the well-being of the spirits of the dead, to foreigners.

This is the relatively harmless, near trivial face of feng shui – do what common sense and accumulated experience suggest and then call it 'feng shui guided'. For instance, all feng shui site selection manuals point out that invisible chi is blown away by wind and is accumulated and borne by water (Han 2001). Thus, attention to wind and water (the original meaning of feng shui) as indicators of chi is of paramount importance, and consequently ideal feng shui sites should not be wind-swept and should be located near water.

Maurice Freedman, in a 1974 study of construction in the Hong Kong New Territories, notes that desired construction and burial sites are ones that are:

> Protected from high winds by its hills. Places from which streams and rivers flow too directly and too fast are avoided. An ideal site is one which nestles in the embrace of hills standing to its rear and on its flanks; it is then like an armchair, comfortable and protecting. (in Bennett 1978, p. 11)

Such a site is called a 'lair'. In a seventeenth-century feng shui manual, favourable lairs were held to be places where 'a land configuration halts and *ch'i* collects' (ibid.), and so energy is focused for either the living who reside there or the spirits of the dead who are buried there. Domestic Chinese construction is often based on a lair design with protruding protective and collective wings.

No one needs cosmology and metaphysics to come to this conclusion. Water is beneficial in all ways: for drinking, plant growth, temperature control, transport, and so on. Good ecological and common-sense reasons suggest proximity to water and avoidance of wind tunnels make for good living (and conceivably dying) places. People claiming this conclusion as warrant for feng shui belief are committing the elementary logical mistake of affirming the consequent (*modus ponens*):

Feng Shui → site A will be healthy and beneficial.
Site A is healthy and beneficial.
∴ Feng Shui is proved/warranted/confirmed/established/vindicated.

But, of course, any number of other 'theories', including scientifically informed ones, can equally suggest site A. Further, if site A turns out not to be healthy and beneficial, then other theories, especially scientifically based ones, can suggest local reasons why this might be so and consequently can give guidance on rectification or remediation of the local problem. Feng shui theory on its own cannot do this. The theory is just an idle distraction when concrete, so to speak, decisions need to be made.

Feng shui cannot make considered adjustments in the way that scientifically informed predictors of site A can make adjustments in the event that A turns out to be unfavourable. Thus, ecological (acid rain), climatological (rising sea levels), geological (seismic activity), or economic (property values) theories can explain divergence from the expected favourable outcome. Feng shui consultants claim that, given appropriate soil, with water, sun, and chi, plants grow. A proper science education should lead students to inquire whether plants will grow with just two of these – water and sun. A classroom challenge will be to see how chi can be removed from the experimental set-up, as it is supposedly all pervasive. But if it cannot be removed, then equal gravitation or space could be identified as necessary for plant growth. It becomes apparent that chi is just an idle passenger; it is along for the experimental ride but does not pay for a ticket.

Quite predictably, two researchers report how feng shui-informed and feng shui-uninformed architects come to the same conclusion about the suitability of building sites and even floor plans:

> An empirical survey was conducted with architects in Sydney and Hong Kong, and the results show that the selection of surrounding environment for a building and interior layout as proposed by the [non-feng shui] architects generally concurs with the ideal Feng Shui model established more than two thousand years ago. (Mak and Ng 2005, p. 427)

Whether feng shui is just 'dressed-up' common sense and natural instinct has been an enduring question, with critics answering 'yes' and devotees answering 'no'. Also enduring has been the question of whether feng shui is a 'black art' made blacker by the deliberate or innocent connivance of those who materially benefit from its practice.

When feng shui advisors go on to advise on the purchase and placement of crystals, mirrors, wind chimes, three-legged gold toads with coins in their mouths, the five elements in each room, and all the other paraphernalia that supposedly brings luck and good fortune – then this suggests that the whole operation is just a piece of creative marketing and that the feng shui operation is indeed a 'dark art'.

Eitel recognizes that feng shui site selection and medical guidance go beyond harmless and benign commonsensical practices:

> Well, if Feng-shui were no more than what our common sense and natural instincts teach us, Chinese Feng-shui would be no such puzzle to us. But the fact is, the Chinese have made Feng-shui a black art, and those that are proficient in this art and derive their livelihood from it, find it to their advantage to make the same mystery of it, with which European alchemists and astrologers used to surround their vagaries. (Eitel 1873/1987, p. 1)

Eitel here articulates a standard criticism: what is good in feng shui practice is merely dressed-up commonsense: construct dwellings and plant crops in proximity to water and not in a wind tunnel or on a mountain top; have living areas orientated to the sun and sleeping areas away from the sun; avoid having front and rear doors opening in an uninterrupted line; do not have toilets opening into living areas; etc. An advocate for feng shui recently wrote in an American Institute of Architects web publication:

> Feng Shui is all about what nurtures a building's occupants and makes them feel comfortable in a space. As architects and designers, we instinctively do a lot of these things. Feng Shui provides the framework and the philosophy to support our instincts. (Knoop 2001)

Such practice has been labelled 'intuitive feng shui'. This is code for 'feng shui adds nothing to intuition and commonsense'. The argument of this book is that all feng shui's associated metaphysics, cosmology, and ultimately worldview is just hand-waving. As the saying goes: 'With hand-waving and a dollar you can ride the subway'. These issues are further elaborated in Chap. 13 when the distinction between chi as a hypothetical construct and as an intervening variable is discussed.

Chinese Protoscience and Experiment

Technology develops from experience, from day-to-day challenges, and from personal needs. Sophisticated technology can develop from science-informed challenges and needs. But there is something else needed for science to take root and grow, especially for a mature science. Every culture has had experience and needs, but not every culture developed modern science. Eitel is correct in identifying the crucial component of science as experiment: not 'kitchen' or pragmatic experiments dealing with how to improve production, transport, weapons, cuisine, or manufacturing, but mathematically formulated and controlled scientific experiments with explicit methodological principles.

For example, after 1000 AD, the common geomantic compass used for divination and then navigation had two rings staggered 7½° east and 7½° west to account for temporal change in declination (deviation of magnetic north from true north) which was something of which Europeans were not even aware (Needham 1963, p. 147).[3] But even the most sophisticated of this applied, practical experimentation fell short of the kind of controlled laboratory experiment in pursuit of theoretical illumination or goals that characterized the birth of modern science in Europe. The former is governed by immediate, experiential, or sensational considerations – the soup tastes better, the cake rises, the porcelain does not crack, and the iron is malleable. And this experiential input can be as narrow as 'it works for me' or 'it works here'. Scientific laboratory experiment is ultimately governed by the need for appraisal of theory and of *why* the empirical results might be as they are, and they need to be generalizable beyond one's own experience or beyond local experience.

In the opening pages of his book, Eitel wrote:

> Natural science has never been cultivated in China in that technical, dry and matter-of-fact fashion, which seems to us inseparable from true science. Chinese naturalists did not take much pains in studying nature and ferreting out her hidden secrets by minute and practical tests and experiments. They invented no instruments to aid them in the observation of the heavenly bodies, they never took to hunting beetles and stuffing birds, they shrank from the idea of dissecting animal bodies, nor did they chemically analyse inorganic substances, but with very little actual knowledge of nature they evolved a whole system of natural science from their inner consciousness and expounded it according to the dogmatic formulae of ancient tradition. Deplorable, however, as this absence of experimental investigation is, which opened the door to all sorts of conjectural theories, it preserved in Chinese natural science a spirit of sacred reverence for the divine powers of nature. (Eitel 1873/1987, pp. 3–4)

And towards the end:

> There is one great defect in Feng-shui, which our Western physicists have happily long ago discarded. This is the neglect of an experimental but at the same time critical survey of nature in all its details. (Eitel 1873/1987, p. 69)

---

[3] Amir Aczel cites a 1088 Chinese work of Shen Kua that recognizes declination. For historical studies of the compass, see Aczel (2001) and Fara (2005).

Eitel here echoes Kant's famous pronouncement in his *Critique* about the New Science only beginning when Galileo rolled balls down inclined planes and forced Nature to answer to questions that the natural philosopher asked: it was then that 'a light broke upon all students of nature' (Kant 1787/1933, p. 20). Eitel might well have learnt from Kant's commentary during his own doctoral studies at the University of Tübingen. His observation was not far off the mark. Nathan Sivin, the sinologist and historian of Chinese science, has written:

> Still we do not find anything corresponding to Galileo's conviction that physical realities must be isolated from the flux of sensation by measurement (as in experiments) so that physical truth can be made to manifest itself in straightforward quantitative relations. (Sivin 1984, p. 546)

Other contemporary Chinese philosophers have lamented that:

> … those experiments which had no practical purpose but were rather intended to verify scientific theory were relegated to an inessential position in ancient Chinese science and technology. This was even more true after the appearance of neo-Confucianism. (Jin et al. 1996, p. 147)

An economic historian makes the same claim:

> China fell behind the West in modern times because China did not make the shift from the experience-based process of invention to the experiment cum science based innovation, while Europe did so through the scientific revolution in the seventeenth century. (Lin 1995, p. 276)

Historians of Indian science concur for their own tradition:

> Finally, the tradition of experimentation was lacking in our ancient and mediaeval culture – even in Gautama's *Nyāya Śāstra* [6th century BC]. Experimentation to test a hypothesis based on observation or analysis of existing information, is the key to all modern scientific inquiry and progress, and an important step in the scientific method. (Bhargava and Chakrabarti 1995, p. 55)

Alexandre Koyré's account of the distinctiveness of the Scientific Revolution is relevant to these discussions about neglect or marginalizing of experiment in all cultures, including European, prior to the Scientific Revolution:

> … observation and experience – in the meaning of brute, common-sense observation and experience – had a very small part in the edification of modern science; one could even say that they constituted the chief obstacles that it encountered on its way. … the empiricism of modern science is not *experiential*; it is *experimental*. (Koyré 1968, p. 90)

Stephen Weinberg in his *The Discovery of Modern Science* (Weinberg 2015) makes the same point. He draws attention to the many 'air pressure' experiments of Boyle, Torricelli, and Pascal,[4] saying:

> No longer were natural philosophers relying on nature to reveal its principles to casual observers. Instead Mother Nature was being treated as a devious adversary, whose secrets

---

[4] Philosophical and educational aspects of these air pressure experiments are discussed in Matthews (2015a, pp. 117–128).

had to be wrested from her by the ingenious construction of artificial circumstances. (Weinberg 2015, p. 200)

Rolling balls down inclined planes, measuring the heights of pendulums after collisions, noting shadow dispersion lines around a stick, taking barometer readings as one went up a mountain, rubbing blocks of ice together in an isolated container to see if movement alone generates heat, etc. are all cases of early controlled, scientific experimentation; they are not kitchen experiments or Kant's 'random groping'.

Notwithstanding Kant's dramatic talk of Galileo causing a 'light to break upon all students of nature', there had been, as one might expect, anticipations of modern experimental practice in earlier European natural philosophy. Neither in politics nor in science do revolutions appear unannounced. Alistair Crombie (1915–1996), one of the foremost historians of medieval science, identifies the Oxford scholar Robert Grosseteste (1168–1253) as the first to begin moving natural philosophy away from Aristotle-induced experimental phobia that had constrained and thwarted European studies of nature for nearly 1500 years (Crombie 1953, 1955/1990).

Aristotle was a great natural philosopher, observer, and cataloguer of nature, but the rejection of experiment was intrinsic to his whole metaphysical system. Natural philosophy (science) was to reveal the operations and 'causes' of *naturally* acting nature; it was to describe and document *natural* motions in the heavens and on the earth. In principle, it did not bother itself with chaotic *violent* motions as these could reveal nothing of the nature, form, or essence that was 'driving', 'causing', and 'effecting' the visible motion or change of states. Aristotelian natural philosophers were directed to study freely falling bodies, but not bodies rolling on inclined planes or swinging at the end of string.[5]

Philosophers of science have long attended to what makes some convictions, preoccupations, and styles of inquiry important for some scientific traditions while unimportant or irrelevant for others. Alistair Crombie wrote of 'styles' of scientific thinking (Crombie 1994); Thomas Kuhn explained that it would only be a new paradigm that could make experiment, with its 'torturing' of nature, a legitimate part of natural philosophy; for Imre Lakatos, such legitimacy would only come with a new research programme.

Many scholars have pointed out that the 'New Science' of Galileo was not altogether new and that there had been precursors to Galileo.[6] There is, of course, agreement with Crombie that thirteenth-century Oxford natural philosophers had made a beginning on science. The question is how far did they go? Crombie is correct in focusing on experiment and maintaining that full-bodied modern experiment did not appear until Galileo:

---

[5] There is some affinity with the situation of contemporary anthropologists who recognize the inherent liability of participant observation methodology; their participation changes the dynamics they are there to study.
[6] See at least Moody (1951, 1966, 1975) and Wallace (1981).

The most important improvement made subsequently to this scholastic method was a change, general by the seventeenth century, from qualitative to quantitative procedures. Special apparatus and measuring instruments increased in range and precision, controls were used to isolate the essential factors in complicated phenomena, systematic measurements were made to determine the concomitant variations and render problems capable of mathematical statement. Yet all these were advances in existing practices. The outstanding original contribution of the seventeenth century was to combine experiment with the perfection of a new kind of mathematics and with a new freedom in solving physical problems by mathematical theories, of which the most striking are those of modern dynamics. (Crombie 1953, p. 10)[7]

Appraising Eitel's important claim about the centrality of experiment in modern science and its absence from traditional Chinese science is an occasion not just for historical study but for scientific/philosophical analysis of what constitutes an experiment, specifically a controlled experiment, and its role in science. This latter question has been widely and energetically addressed in recent philosophy of science, with the movement being labelled the 'new experimentalism' (Ackermann 1989). One issue is how theory-guided intervention in nature enables more justified appraisal of theory than does mere observation; another is the recognition that when experiments are controlled, their 'failure' allows the modification of assumed conditions and associated theories rather than immediate refutation of the theory being tested.[8]

Science and Metaphysics

Eitel's views were, for their time, a perceptive and informed account of Chinese natural science. But his comment that the absence of an experimental tradition 'preserved in Chinese natural science a spirit of sacred reverence for the divine powers of nature' (Eitel 1873/1987, pp. 3–4) did some injustice to the Western scientific tradition. Doubtless modern science is 'technical, dry and matter-of-fact', and the core of Western science was exactly that. But it needs to be remembered that up to, and including, the nineteenth century of Eitel's time, pretty well all European scientists were devout Christians and understood themselves as revealing bit-by-bit the handiwork of their God.[9] At least since St. Augustine (354–430) in the third century, to do natural science in the Christian West was the same as doing natural theology.

---

[7] Crombie's account of the working out of this new experimentalism in sixteenth- and seventeenth-century natural philosophy is given in Volume Two of his *Augustine to Galileo* (Crombie 1952) and in his essay on scientific methodology in the Medieval period (Crombie 1959/1990).

[8] Significant contributors to the philosophical understanding of experiment have been Franklin (1986), Galison (1987), Hacking (1988), and Mayo (1996). See also contributions to Gooding et al. (1989) and Radder (2003). Harré's *Great Scientific Experiments* (1981) has an informative introduction to experiment in the history of science giving accounts of 21 influential experiments.

[9] See, for example, Brooke (1991, 2001), Hooykaas (1999), Jaki (1978), Lindberg and Numbers (1986), and Mascall (1956).

No less than the Chinese did this Christian scientific tradition has 'a spirit of sacred reverence for the divine powers of nature'. And, of course, contributors to the Judaic and Islamic scientific traditions had the same sense of revealing the handiwork of God. As a philosopher and a physicist observe in a collaborative work:

> Newton's theology and physics were all of a piece. Believing in an omniscient God, it was natural to conceive of all points of space being alike immediately present to His consciousness; and only by the arbitrary *fiat* of an omnipotent God could he [Newton] explain the initial configuration of the atoms at the dawn of creation. (Lucas and Hodgson 1990, p. 241)

But having said this, there was no Catholicism within Copernican astronomy, Anglicanism within Newton's physics, Unitarianism within Priestley's chemistry, or Sandemanianism within Faraday's electromagnetic theory. God does not appear in any of the laws or equations of the founders of modern science. Though born in a religious home, science is not religious.[10]

Eitel marks down Confucius and his influential early disciples for not engaging with and correcting Chinese science, the proto feng shui systems of his age:

> It was in the power of Confucius and his disciples, Mencius and Sun-tze, who exercised a strong influence on the minds of their countrymen during this period [BC 500–200], to repress and rectify the superstitious notions already floating about among the people and tending towards a regular system of geomancy, by assuming a definite attitude, denouncing superstition and substituting an enlightened theory on the subject. But he and his disciples, though personally free from superstition, contented themselves with urging a reform of morality according to the pattern of the ancient sages, without venturing to grapple with the superstitions that were gathering around the ancient form of ancestral worship. In one word, they remained neutral, and the consequence was that superstition spread farther and farther. (Eitel 1873/1987, p. 54)

Many historians and commentators on Confucianism make the same point: Confucius was not engaged by or interested in how the world worked; he had little if any concern with 'natural philosophy' (Chan 1957). Joseph Needham commented that Confucians had no 'curiosity about Nature outside and surrounding man' (Needham and Ling 1956, p. 544). The very same point needs to be made about the other two major intellectual or ideological founts of Chinese thought: Buddhism and Taoism. All three were primarily concerned with right living (ethics), proper social arrangements (politics), and personal enlightenment. Investigating and explaining nature was not at the core of any of them.

Yijie Tang (1927–2014), a leading twentieth-century Chinese philosopher and cultural historian, concurs in Eitel's judgement of Confucius and his circle, writing:

> Confucianism is a humanistic mode of thinking. Confucianism does not proceed along the path of science nor does it subscribe to principles of science. (Tang 2015a, p. 76)

---

[10] A contentious claim when one considers the role of presuppositions, commitment, testimony, 'faith' in science, but that is another issue.

Other contemporary Chinese philosophers remark that:

> Confucianists used the model of induction based on personal experience (including social and psychological experience) to understand the world … Confucius' ethical centralism, in particular, gave scientific theory a tendency to conservatism and a lack of clarity. … In Confucianism, the understanding of nature served ethics. (Jin et al. 1996, pp. 145–146)

A senior historian of science makes a similar observation:

> Generally, there was plenty of technical knowledge in ancient Chinese civilization, but no independent scientific culture system existed. With the intelligence system hidden in the benevolence system, China developed its theoretical moral system in concurrence with the Confucian ideology; no independent knowledge system like that in ancient Greece emerged. (Wang 2014, p. 127)

That is, research or learning was for the purpose of leading a better life, being good. This is what justified the practice; learning and inquiry were not for the purpose of finding truth. Wang further claims:

> In ancient China, the learning culture had always been controlled or led by politics. More precisely, ancient Chinese politics had controlled and limited the development of the learning Culture. Besides, Confucianism with a cultural shielding had prevented the ancient Chinese from free explorations in science. (Wang 2014, p. 128)

Parallels with contemporary China, and many other states are obvious.

Romanticism in Chinese Science

Eitel sees a positive side to the Chinese neglect of experiment:

> Deplorable, however, as this absence of practical and experimental investigation is, which opened the door to all sorts of conjectural theories, it preserved in Chinese natural science a spirit of sacred reverence for the divine powers of nature. (Eitel 1873/1987, p. 4)

He adds:

> … yet I say, would God, that our own men of science had preserved in their observatories, laboratories and lecture-rooms that same child-like reverence for the living powers of nature, that sacred awe and trembling fear of the mysteries of the unseen, that firm belief in the reality of the invisible world and its constant intercommunication with the seen and the temporal, which characterise these Chinese groupings after natural science. (Eitel 1873/1987, p. 5)

And then:

> They see a golden chain of spiritual life running through every form of existence and binding together, as in one living body, everything that subsists in heaven above or on earth below. (Eitel 1873/1987, p. 4)

The Chinese approach to nature is manifest in a widely known nineteenth-century poem of Feng-shen yin-te who was a high-ranking state dignitary:

> With a microscope you see the surface of things.
> It magnifies them but does not show you reality.
> It makes things seem higher and wider,
> But do not suppose you are seeing the things in themselves. (Jaki 1974, p. 39)

Eitel's claims reflect the late-eighteenth- and nineteenth-century European Romantic criticism of science which he may have studied during his Tübingen years.[11] Such criticism led Goethe to reject Newton's theory of colours because it was mediated by, and dependent upon, optical instruments (Jaki 1969b). For romantics, the more things that got between nature and the observer, the worse was the apprehension of nature; the more nature was constrained in experiment, the more distorted was its understanding.

Eitel's Christian tradition had entrenched philosophical opposition to the perceived mechanical and materialist commitments of modern science, and this alone suffices for his endorsement of the Chinese 'sacred reverence for the divine powers of nature'. Atomism was unknown in China; neither Confucian nor Daoist metaphysics predisposed adherents to the 'mechanical worldview' of modern science. Qian Wen-yuan, a Chinese theoretical physicist and historian of science, sees Mohism of the second and third centuries BC as close as China came to an indigenous atomism but writes:

> ... it was a great loss to Chinese science that in the long history of traditional China, Mohism, as a staunch rival school of Confucianism, was almost completely left in oblivion. (Qian 1985, p. 51)

Eitel well knew that a 'sacred reverence' did not justify all epistemological, ontological, and scientific claims made in its name. Sacred, or otherwise, claims about the world needed to be justified by how the world was.

Eitel was correct in identifying in China the all-pervasive idea of a spiritual life or entity binding together nature, man, and cosmos: a 'golden chain' as he avers. This is a central plank in the traditional Chinese worldview and needs only to be stated in order to see how fraught is the project of accommodating such a worldview with the tenets and practice of modern science. Consequently, much thought, and anticipation of cultural consequences, needs to go into the teaching of modern science in China.

Eitel identifies Zhu Xi [Choo-he in Wade-Giles], the twelfth-century Song dynasty scholar 'whose commentaries are read in every school' as the principal expositor of the philosophical basis of feng shui (Eitel 1873/1987, p. 5). James Legge, another nineteenth-century missionary scholar, shared this estimation of Zhu Xi, saying: 'China has not produced a greater scholar. ... Nothing could exceed the grace and clearness of [his] style, and the influence which he has exerted on the literature of China has been almost despotic' (Legge 1887/2006, p. 20). Eitel relates how:

> According to Choo-he there was in the beginning one abstract principle or monad, called the "absolute nothing," which evolved out of itself the "great absolute." This abstract principle or monad, the great absolute, is the primordial cause of all existence. When it first moved, its breath or vital energy congealing, produced the great male principle. When it had

---

[11] On Romanticism and science, see at least contributions to Bossi and Poggi (1994) and Cunningham and Jardine (1990).

moved to the uttermost it rested, and in resting produced the female principle. (Eitel 1873/1987, p. 5)

And:

> When this supreme cause thus divided itself into male and female, that which was above constituted heaven, and that which was beneath formed the earth. Thus it was that heaven and earth were made. But the supreme cause having produced by evolution the male and female principles, and through them heaven and earth, ceased not its constant permutations, in the course of which men and animals, vegetables and minerals, rose into being. The same vital energy, moreover, continued to act ever since, and continued to act through those two originating causes, the male and female powers of nature, which ever since mutually and alternately push and agitate one another, without a moment's intermission. (Eitel 1873/1987, p. 6)

Eitel relates how the various movings and shakings in this primordial cosmology are not random but lawful:

> Now, the energy animating the two principles is called in Chinese Hi [Chi], or the breath of nature. When this breath first went forth and produced the male and female principles and finally the whole universe, it did not do so arbitrarily or at random, but followed fixed, inscrutable and immutable laws. (Eitel 1873/1987, p. 6)

Western science is predicated upon an a priori belief or commitment to the lawfulness or regularity of nature – what happens today, happened yesterday, and will happen tomorrow. Historians and philosophers distinguish two 'kinds' of lawfulness: the first coming from 'without' where a deity or primal creator is held to create or impose the laws on an otherwise chaotic, lawless mass of inchoate particles, objects, and organisms and the second coming from 'within' where all of the particles, objects, and organisms just move and interact according to their in-built natures. The former is associated with the mechanical worldview, and the latter is best exemplified in Aristotle's teleological worldview of natural places and natural motions. In both cases, laws state some uniformity about nature that can be ascertained by observation and/or experiment.[12] Early feng shui theory and practice had lawfulness with 'Chinese characteristics', as one might say. In Eitel's words:

> In short, the firmament of heaven is to a Chinese beholder the mysterious text-book, in which the laws of nature, the destinies of nations, the fate and fortunes of every individual are written in hieroglyphic mystic characters, intelligible to none but to the initiated. (Eitel 1873/1987, p. 10)

The central business of feng shui scholarship is to decode the mysterious textbook. Feng shui masters apply the deciphered textbook to local environmental, domestic, commercial, and personal circumstances.

---

[12] The basic text for this claim is Zilsel (1942). An orthodox account of causality is found in Agazzi (2014, Chap. 7), Bunge (1959, Chap. 4; 1998, Chap. 6), and Dilworth (1996/2006, Chap. 3). Nancy Cartwright (1983) and Ronald Giere (1999) dispute the orthodox view or at least claim that it needs to be considerably nuanced by recognition of, among other things, the omnipresence of idealization and abstraction in science. The latter are important considerations for methodology, but really do not impinge on the ontological question of causality.

Chinese Astrology

Given that the heavens were seen to clearly impact on *inanimate* earthly circumstances (the sun's movement on seasons, the moon's orbit on tides, etc.), it was a small step to see them impacting on *animate* human and personal circumstances. Indeed, the unity of heaven, earth, and man was at the centre of all Chinese cosmology whether Confucian, Daoist, or Buddhist; and the Christian claim that man was made in the image of God resonated with this worldview. Feng shui accepts and reinforces belief in a cosmic harmony that unites the heavens, earth, and man. Eitel remarked:

> Again, observing our earth, with its constant revolutions of summer and winter, spring and autumn, growth and decay, life and death, the Chinese notice, that here again the same mathematical order is repeated, that earth is but the reflex of heaven, the coarse material embodiment of the ideal mathematical problems, ethereally sketched on the firmament of heaven. (Eitel 1873/1987, p. 25)

The core of astrological belief is that the star constellation of the zodiac, against which the sun is located as seen from earth at the time of one's birth, affects an individual's personality, character, and much else. So, for instance, people born March 21 to April 9 are in the Aries 'star sign' and so have Aries characteristics and life fortune. Eitel spends many pages describing the detail of the complex Chinese solar astronomy: their division of the sun's ecliptic into 12 equal parts; the animal-naming of each – snake, monkey, rat, etc.; the identified asterisms (star groups) within constellations; the sun's defining of the 24 seasons, so that, for example, 'when the sun enters the 15th degree of Aquarius (February 5th) spring begins'; and so on.

A Chinese astrological treatise of 90 BC explains how the astronomer/philosophers discovered the unity of heaven and earth or at least the reflection of the former in the latter:

> Looking up they contemplated the signs in the sky. Looking downward, they found analogues to these on earth. In the sky there were the sun and moon, on earth yin and yang. In the sky there were the Five Planets, on earth the Five Phases. In the sky were the lunar lodges, on earth the territorial divisions. The Triple Luminaries [that is, sun, moon, and planets] are the seminal *ch'i* of yin and yang. The *ch'i* originally resides on earth, and the sages unify and organize it. (Yosida 1973, p. 73)

Not as important as the sun but still a significant heavenly entity with influence on terrestrial affairs, most obviously with the tides, is the moon whose monthly movement through 28 constellations, or abodes, was extensively studied. Aristotle attributed different kinds of madness to the effect of full moons. European astrologers right through to the eighteenth century and beyond identified lunar causes for mental disorders, hence the once common term 'lunatic' (Harrison 2000). By Eitel's time, such lunar beliefs had been banished from informed European thinking and medical practice; however they linger in *uninformed* opinion across all cultures. The term 'moon madness' is commonly used to label the supposed increased incidence of troublesome, odd, antisocial, or criminal behaviours induced by the full moon.[13]

Eitel relates of Chinese lunar theory that:

> These twenty-eight constellations are divided into four sections, one of which is called the azure dragon, located in the East, and comprising the first seven constellations. The next seven constellations are called the sable warrior, whose abode is in the North; the third seven bear the name of the white tiger, situated in the West, and the last seven are designated the vermilion bird, ruling the South. (Eitel 1873/1987, p. 12)

And somewhere back in the beginnings of the Chinese tradition, the different constellations that form the night-by-night abode of the moon were designated as auspicious or lucky. So 'Mercury when associated with constellations 7, 14, 20 and 28' was lucky and likewise Saturn with Nos. 3, 10, 17, and 24' (Eitel 1873/1987, p. 12). Knowledge of the nature and scope of science is gained, if along with standard teaching about causes of the moon's phases, something of this cultural history is also taught in science programmes.

Apart from the sun, moon, and planets, other groupings of stars were held to have significant impacts on terrestrial affairs: the seven stars of the Great Bear constellation and the nine stars of the Great Bushel. Of the latter, Eitel writes:

> But their position in the heavens is of little importance – some authorities even say they have no fixed place at all, but are moving about in the atmosphere – for they have each and all their counterparts or representatives on earth in the shape of mountains, and it is the business and art of the geomancer [feng shui practitioner] to determine which mountain-peak or hill corresponds to the one or other of these nine stars, each, of which has its own permanent relation to one of the five elements or to one of the above-mentioned eight diagrams. (Eitel 1873/1987, p. 14)

Newton's universal law of gravitation had brought the 'heavens down to earth' in the sense that the same law governing falling apples and swinging pendulums also governed the 'falling' moon and orbiting planets (Boulos 2006; Harper 2011, Chap. 6). Feng shui brought heaven down to earth in a very different, unscientific sense. With feng shui, the universal chi lines originated in the heavens, but their distribution and movement on earth governed human welfare. This was metaphysics without science; it was philosophical or literary metaphysics; metaphysics made easy.

Astronomical Problems for Feng Shui

Leaving aside any scientific or 'metaphysical' problem in translating from motions of the sun and moon to earthly and human affairs, Eitel identifies two major astronomical problems for feng shui theory: the precession of the equinoxes and missing planets.

---

[13] Arnold Lieber is one prominent proponent of 'moon madness' theory (Lieber 1978). But all rigorous tests of supposed moon-behaviour correlations show that there is none (Rotton and Kelly 1985).

For astrologers, stars influence our character and lives: more specifically, the star constellation that the setting sun is 'against' or 'sited' in at the time of our birth. As the earth orbits 360° around the sun, it appears that the sun moves through the background of stars along an ecliptic whose background is the 12 named stellar constellations. The number of days the sun spends in each constellation varies from 6 in Scorpio to 47 in Virgo; but for simplicity astrologers divide the ecliptic into 12 30° segments. So, for a good portion of people whose births are linked to specific zodiac signs, the sun was not 'in' the constellation at all but in limbo; at their birth, it was between constellations. But with some ingenuity, astrologers can adjust to this astronomical fact.

But there is another problem: the precession of the equinoxes from east to west, the effect of which has been to disconnect the astrological signs of the zodiac from their respective constellations. The Chinese astronomers/astrologers did not know of this. As the Earth 'wobbles' on its axis of spin, the axis making a complete rotation of its path every 26,000 years, the sun's sited background constellation at any given calendar date will slowly move. For those with an Aries birth sign, the sun was 'in Aries' between March 21 and April 19 2000 years ago when the astrological tables were created, but now for the same dates, the sun is in Pisces. So, the astrological star signs are now out of synch with astronomical reality. Some astrologers recognize this. Iona Miller, a clinical and Jungian hypnotherapist and lecturer, writes:

> Precession is the gradual shift in the orientation of the Earth's axis of rotation, which, like a wobbling top, traces out a conical shape in a cycle of The Precession of the Equinoxes was first noticed as a slow but steady slippage through the Zodiac of 1 degree every 72 years. A full cycle was a journey of Earth through all the signs of the Zodiac. The equinoxes move westward along the ecliptic relative to the fixed stars, opposite to the motion of the Sun along the ecliptic. To complete one cycle of the zodiac – a 'Great' or Platonic Year – requires 25,920 years. Dividing this sum by 360 yields the number 432, the root of the mythological count of 432,000 years.[14]

But this has little effect on the astrology business or on the countless millions of people who organize their affairs and interpret life's events, according to the star sign of their birth (Fig. 7.1).

Eitel comments:

> The Chinese, not knowing of the precession of the equinoxes, are rather perplexed by the discrepancy, but caring less for accuracy and more for ancient tradition, ignore the actual discrepancy, and still represent the twelve signs, not as they appear now, but as they appeared to their ancestors two thousand years ago. (Eitel 1873/1987, p. 11)

This is a matter that at least some contemporary feng shui consultants deal with by simply severing the connection of auspiciousness from the position of star groups in the galaxy. The above-mentioned Grand Master Skinner says of his Flying Star system that:

---

[14] Iona Miller, 2009: https://ancestorsandarchetypes.weebly.com/precession.html

Fig. 7.1 Precession of the equinoxes

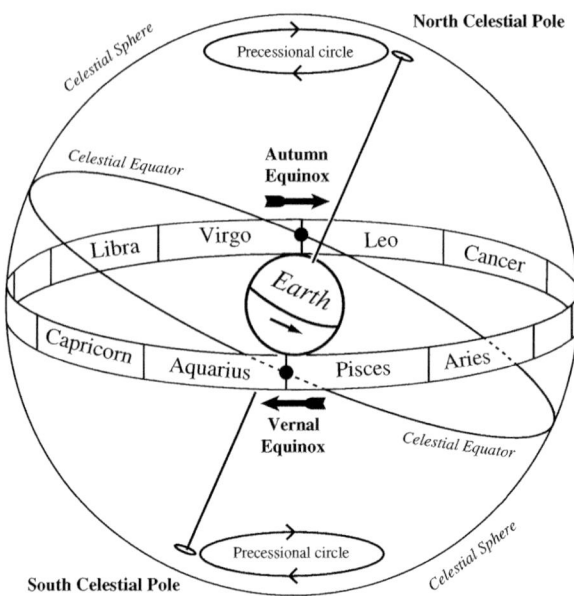

> This is the first book in English on Flying Star feng shui which explains how it compatibly relates to Eight Mansion feng shui, and provides 216 *lower kua* Flying Star charts with 24 Mountain direction indicator, as well as all of the variant Substitution Star (*ti kua*) charts. (Skinner 2015, rear cover)

But then he concedes that:

> Although the 'Flying Stars' were originally related to the stars of the astronomical Big Dipper asterism, in their feng shui usage they are simply terms for changing types of subtle *ch'i (qi)* energy present in our living environment, our homes, and our workplaces. (ibid.)

Needless to say, this is a major concession: either the galactic position of certain stars at one's birth has influence, but the tables of the original diviners are now out of synch with star movement; or there is no such influence, and feng shui practitioners can get out of astrology and just attend to the configuration of local chi. The latter option simplifies the task of consultants and puts a significant hole in the overall integrated cosmology that for 3000 years has underwritten feng shui. Many are not prepared to do this.

One dramatic way that some astrologers and feng shui consultants have chosen to overcome the seemingly fatal precession problem is to say that the 'signs remember the influence of the constellations that corresponded to them two thousand years ago' so that the ancient tables can still be used and so there is no need to go back to the astrological drawing board (Hines 2003, p. 209). This is a textbook example of the ad hoc alteration of a theory to save it from being falsified – the theory is adjusted precisely to account for the falsifying instance. Good science teaching can enable students to identify this all too-common manoeuver; its repetition is a sure indicator of pseudoscience.

Eitel identifies another major conceptual and evidential problem: Chinese astronomers of the time recognized only five planets – Jupiter, Venus, Saturn, Mars, and Mercury. Given the centrality of these in the whole Chinese cosmological picture and their agreement with the 'primacy of five' principle in Chinese metaphysical thought which is discussed above and below, it is easy to understand why Chinese astronomers did not seek to discover new planets.

The situation was comparable to the hold that Platonic-influenced spherical and circular motion had on European astronomers and natural philosophers, right up to and including Copernicus and Galileo. Until Kepler, European astronomers could not think 'outside the circle'; as so often happens, metaphysics constrained science (Zebrowski 1998).[15] Chinese astronomers may or may not have seen additional planets, but if so, they did not identify them as planets; they would have been regarded as comets, stars, or some aberration, just as Uranus was by those who happened upon it from antiquity till William Herschel's identification of it in 1781 (Miner 1998). This was the first of the three 'modern' planets to be identified: Uranus (1781), Neptune (1846), and Pluto (1930).

The spectacular 1846 discovery of Neptune by Johann Galle following its mathematically precise prediction by Urbain Le Verrier did not and could not have happened within the Chinese astronomical tradition. Astronomers knew that the orbit of the newly discovered Uranus did not accord with its Newtonian path. Twelve years before its discovery, the new planet was postulated as the means of reconciling the orbit of Uranus with Newton's classical mechanics. Le Verrier calculated its orbit and location, and it was found exactly where it should have been (Grosser 1962; Standage 2000).

For Chinese astronomy, there was, one might say, an epistemological obstacle impeding such investigations.[16] So much of their culture, governance and science were imbued with a five-planet astrological commitment that an alternative was inconceivable. These are merely examples of the commonplace reality that culture impacts on the conduct of science. The American Association for the Advancement of Science, in its 1964 report *The Integrity of Science*, well stated this nexus:

> Although science has its own history and tradition, it is not wholly independent or self-sustained. Scientists are human beings, and science is a part of culture. What is the influence of changing social conditions, particularly the growing importance of science to society and the rapid approach of scientists to positions of power, on the search for new knowledge and on the system of scientific discourse? Can the very success of science and its closer interaction with the rest of our culture lay it open to the influence of new and pos-

---

[15] Likewise, in Islamic countries, one does not expect serious research into authorship of the Koran or to find research into smoking and lung cancer supported by tobacco companies, or communist countries to maintain research into the economic functioning of open markets.

[16] This is the expression coined by Gaston Bachelard (1934/1984) to identify deep-seated conceptual barriers to scientific investigations. These categories blocked completely some lines of investigation and shaped the form of others. The notion was elaborated and utilized by Louis Althusser (1969).

sibly alien points of view which derive from other sectors of society: military, business, or political? (Holton 1967, p. 292)

Recognizing and being sensitive to such a nexus between science and culture are important outcomes of a good science education. The HPS-informed teaching of astronomy is one way that this sensitivity can be engendered.

A 'research problem' for the feng shui tradition was to determine whether the newly discovered Uranus and Neptune had or did not have auspicious influences and with what numbered constellations each should be linked. And whatever problems Western astronomy has had with the post-1930 planetary status of Pluto, these are magnified in the feng shui tradition which additionally has to ascertain the planet's impact, if any, on earthly affairs. Linda Goodman, a commercially popular and much-published astrologer, has an effective solution to this problem: planets have no astrological influence until they are discovered (Goodman 1971). The solution's effectiveness varies directly with the credulity of her readership. Fortunately for Goodman and fellow astrologers but unfortunately for society as a whole, credulity is little related to completion of school science programmes. Large numbers of people with science degrees are prepared to hold fantastic, unwarranted, and refuted beliefs – recall that 30–40% of US science teachers reject Darwinian evolution and believe in special creation (Otto 2016; Specter 2009).

The theory-driven growth of astronomy contrasts markedly with the static picture of astrology, feng shui, and classical Chinese astronomy. This dimension of change, growth, development, and empirical research is a feature of modern science which was launched in Europe in the seventeenth century. In contrast, static systems of ideas that do not change or transform are either ossified science or pseudoscience. Linda Goodman, the author of a shelf of best-selling astrology books, believes the stability of astrology is a good thing and indicative of its scientificity. She writes:

> Alone among the sciences, astrology has spanned the centuries and made the journey intact. We shouldn't be surprised that it remains with us, unchanged by time – because astrology is truth – and truth is eternal. (Goodman 1971, p. 475)

An outcome of good science teaching would be if students could recognise this stability 'ossification' and to regard it as marker of religious and political fundamentalism, not of good science; science does not stand still, it develops;[17] if it is not changing and growing, it is not science.

The Five Elements

Eitel relates how the principal agents through which heaven, and especially the five planets, act upon all living creatures are the five elements or 'agents' of nature. Though many, including Joseph Needham, caution that the Chinese term *wuxing*

---

[17] On account of its omnipublic presence, astrology is perhaps the most commented upon pseudoscience; it figures in every demarcation debate among philosophers since Popper. It is the most comprehensively tested of all the putative pseudosciences.

(*wu* meaning five, *xing* meaning agents, powers) should not be understood as 'elements' in the Western philosophical tradition of element, this is reading West into East (Porkert 1982, p. 565). Others have suggested 'agent', 'movement', 'processes', 'phases', or 'steps' would be more appropriate.[18] Whatever the translation, the Chinese 'theory' of five elements or phases is fundamental to feng shui.[19]

Eitel addressed the Chinese theory of elements and correctly cautioned that:

> By these however we must not understand five material substances, chemically indissoluble, but rather spiritual essences, each characteristically different from the other and forming the generative causes of all material substances. These five elements are wood, fire, earth, metal and water, the first of them being the agent of Jupiter, the second that of Mars, the third that of Saturn, the fourth that of Venus, the fifth that of Mercury. (Eitel 1873/1987, p. 14)

Belief in the defined powers of the five planets, acting in and through the five elements, was a confirmation of the primacy of five in Daoist cosmological/philosophical thinking. The feng shui or siting expert had to ascertain how well or poorly did the phases and planets 'align' or contest at any potential construction or burial site (Table 7.1).

In support of the five-element belief, expositors pointed to the fivefold division of the body (muscles, veins, flesh, bones, and hair), the five primary colours (white, black, red, blue, and yellow), the five fortunes (money, honour, longevity, children, and peaceful death), the five tastes (sour, bitter, sweet, acrid, salt), and so on. The fact of five fingers on a hand and five toes on a foot could also be called upon to bolster the primacy or foundational place of five in cosmology. So, marrying chi theory to 'five element' thinking was both natural and required for Tao thinkers.

Two contemporary expositors amplify the notion, saying:

> *The Five Phases* identify stages of transformation, patterns of expansion and contraction, proliferation and withering. Each *Phase* has an intrinsic primal energy, and ontological influence that shapes events. For example, human beings go through cycles in their lives similar to the seasons in nature – beginning in birth and ending in death, with stages of

Table 7.1 Daoist correlative thinking

5 elements (phases)	5 planets	5 seasons	5 directions	5 body systems	5 primary colours
Wood	Jupiter	Spring	East	Liver	Green
Fire	Mars	Summer	South	Heart	Red
Earth	Saturn	Late summer	Middle	Spleen	Yellow
Metal	Venus	Autumn	West	Lung	White
Water	Mercury	Winter	North	Kidney	Black

---

[18] On this, see Bennett (1978), Parkes (2003), and Porkert (1982, p. 565).
[19] The title of a recent expository book is *Feng Shui: The Chinese System of Elements* (Rolnick 2004).

growth, maturity, and decay in between. Within the life cycle, the power of each *Phase* can be observed. (Beinfield and Korngold 1991, p. 87)

Whether this particular amplification sheds light on the matter, or merely underwrites the essential mysteriousness, and hence non-scientific nature of 'element' or 'phase' category is an obvious question.

In the above table, the addition of 'late summer' to seasons and 'middle' to directions suggests that the 'facts' or 'nature' are being bent to fit the theory. Likewise, the ossification of the five planets of traditional Chinese astronomy in the table without adjustment to the reality of eight modern planets (with Uranus and Neptune) that have been in the solar system since the beginning and would have been exercising their influence when Taoist five-phase theory was first formulated suggests that the whole influential system and worldview is removed from science and indeed antiscientific in virtue of its failure to adjust.

Eitel's History of Feng Shui

Eitel provides an outline history of feng shui in which he maintains that while the leading principles of feng shui have their roots in antiquity and it did not rise as a distinct branch of study or professional practice till the Song dynasty (AD 960–1126), nevertheless, the 'history of the leading ideas and practices of Feng-shui is the history of Chinese philosophy' (Eitel 1873/1987, p. 51). Eitel notes that:

> With the rise of the Tang dynasty (AD 618–905), which is famous for its revival of literature generally and of poetical literature especially, which had hundreds of Buddhistic works translated from Sanskrit into Chinese, a new era opened, particularly favourable to the propagation of mystic and fanciful doctrines assuming, as geomancy had learned to do, the garb of national as well as Tauistic and Buddhistic philosophy. (Eitel 1873/1987, p. 61)

And:

> The notion of five planets (Venus, Jupiter, Mercury, Mars, and Saturn) influencing the earth and every living being, made its first appearance about this time, and was eagerly taken hold of by the professors of Feng-shui. (Eitel 1873/1987, p. 61)

Eitel observes that in his day the adherents of feng shui were divided into two schools, the Tsung-Miau (ancestral temple) school, which took its rise in Foh-kien (Fujian in Pinyin), and the Kwang-si (Guangxi in Pinyin) school. He says that the former school (Zumiao in Pinyin) of geomancers attributes the greatest importance to the doctrines of the order of nature (Li) and of the numerical proportions of nature (Su). They are therefore specially attached to the use of the compass. The second school placed the greatest stress on the doctrines of the breath (K'i) and outlines (Xing) of nature. He says that they use the compass also but only as a subordinate help in prospecting the country, for 'their principle is, first to look for the visible symptoms of dragon and tiger and of a good breath, and then to judge of the surrounding influences by consulting the compass' (Eitel 1873/1987, p. 63).[20]

Lillian Too, a modern feng shui master, affirms the long-standing division between the two main schools of feng shui: the landscape school and the compass school. She elaborates:

> The landscape school focuses on the terrain of the land, the configuration of mountains, the flow of rivers, the shapes of land masses, the strength of the winds, the quality of waterways … the techniques of landscape Feng Shui are very compelling.

> Meanwhile, different branches of the Compass school offer different ways of identifying good and bad compass directions. These are in turn calculated from formulas using one's four pillars of destiny (the year, the month, the day and the hour of birth), which are then interpreted according to a variety of formulas, some based on element analysis, and others based on the Pa Kua trigrams and the Lo Shu magic square. (Too 1994, pp. 15–16)

Amir Aczel well documents the priority of Chinese discovery of the lodestone, and the subsequent manufacture of 'south-facing' suspended pointers and thus the compass by at least 1040 AD (Aczel 2001, Chap. 7). He notes that the compass seemingly was not immediately used for navigation, but for divination:

> The feng shui practitioners used the magnetic compass as a divining tool. Their approach was animistic – the compass took the form of a fish floating on water, telling them how to make decisions, or it was a turtle whose head bobbed up and down as it stabilized, pointing in a favourable direction. The diviners saw the compass's response to a force acting on it at a distance as a magical sign about the nature of the land and its water and air, as well as about what was under the earth's surface. (Aczel 2001, p. 88)

The siting compasses have a central chamber, known as the 'celestial pool', which houses a magnetic needle. Concentric bands of characters (mostly cyclical) radiate from the celestial pool, and each band is related to some aspect of sky or earth. Once the disc is aligned with the needle, all that remains to be done is to find the coordinates of the site, and the most auspicious place for a house or tomb can then be established.

The Spirit World

Even all of the above does not exhaust the external influences, forces, or entities that bear down upon human life: the spirits of the dead reach beyond their graves and act powerfully in the world. Eitel comments that:

> This is a doctrine which seems strange to us, but which has nothing unreasonable in itself to a Chinaman accustomed to worship the spirits of his ancestors, whom he supposes to be

---

[20] As mentioned earlier, modern feng shui commentators and expositors refer to these schools as the 'compass school' and the 'form school' of feng shui. See, among many sources, Bennett (1978).

constantly hovering near, and to whom he therefore formally announces every event in his family, and offers sacrifices of meat and drink. (Eitel 1873/1987, p. 15)

Needless to say, many Chinese recognize the difference between respect for and even worship of ancestors and the need to reify their memory, that is, to give continuing existence to their ghost, spirit, or soul. The important early Chinese naturalist philosopher, Wang Chong (27–100 CE), made this distinction, saying in his *Discourses Weighed in the Balance* that if humans have ghosts, so too should animals, and given the countless number of once living humans, the ghostly realm is fully saturated so the notion of any individual ghost hearing, much less responding to, an earthly invocation, is fanciful. He advised just respecting the memory and not indulging in fanciful reification; the former did not require the latter.[21]

Eitel on the Educational Task for China

Eitel was writing almost 150 years ago, and his comments on the educational task confronting China need to be read and understood in their own context, yet his observations have current relevance and repay being read and appraised. He concludes his book by writing that while 2000 years ago the embrace of some of the fundamental ideas of feng shui was understandable and 'rational', such embrace no longer is:

> Feng-shui is at present a power in China. It is an essential part of ancestral worship, which national religion, neither Tauism nor Buddhism managed to deprive of its all-pervading presence. … Feng-shui is indeed the refined quintessence of Tauistic mysticism, Buddhistic fatalism and Choo-he's materialism, and as such it commands if not the distinct approval yet the secret sympathy of every Chinaman, high or low. (Eitel 1873/1987, pp. 65–66)

He notes that feng shui exponents:

> had studied nature, in a pious and reverential yet in a very superficial and grossly superstitious manner, but which, trusting in the force of a few logical formulae and mystic diagrams, endeavoured to solve all the problems of nature and to explain everything in heaven above and on the earth below with some mathematical categories. The result, of course, is a farrago of nonsense and childish absurdities. (Eitel 1873/1987, p. 69)

Eitel had no hesitation in pointing to the educational project, specifically the science-education project, that was needed in his day:

> The only powerful agent likely to overthrow the almost universal reign of Feng-shui in China I conceive to be the spread of sound views of natural science, the distribution of useful knowledge in China. … let correct views be spread regarding those continually interchanging forces of nature, heat, electricity, magnetism, chemical affinity and motion; let these views be set forth in as forcible and attractive, but popular a form as Choo-he employed, and the issue of the whole cannot be doubtful. (Eitel 1873/1987, p. 69)

---

[21] Wang, in contemporary terms as will be elaborated below, was a naturalist, not a materialist. He regarded chi as a genuine, existing all-pervasive 'fluid' that acted lawfully and had real effects.

Although a Lutheran clergyman, he spoke with a voice informed by science and the common Enlightenment hope and expectation:

> The fires of science will purge away the geomantic dross, but only that the truth may shine forth in its golden glory. (Eitel 1873/1987, p. 69)

But he was informed and realistic enough to recognize that:

> Feng-shui is, moreover, so engrafted upon Chinese social life, it has become so firmly intertwined with every possible event of domestic life (birth, marriage, housebuilding, funerals, etc.) that it cannot be uprooted without a complete overthrow and consequent re-organisation of all social forms and habits. (Eitel 1873/1987, p. 65)

Not surprisingly, this contention has drawn down on Eitel the ire of some who have read his work. It is the kind of statement that is put up as 'exhibit A' to demonstrate cultural insensitivity, Western chauvinism, and objectionable scientism and positivism. Notwithstanding all of this, Eitel's recognition of the material embeddedness of feng shui beliefs, and the consequent social and cultural adjustments needed if they were to be abandoned or significantly changed, is both realistic and commendable. Further, Eitel's view has been independently adopted by many contemporary Chinese scientists, philosophers, educators, and cultural critics concerned with the modernization of Chinese culture (Tang 2015e).

It is not a simple matter to 'just' change belief from Islam to Christianity in Saudi Arabia or Malaysia or to reject the 'ontology' of the caste system in India or the unquestioned authority of elders in Australian aboriginal cultures. There are some cultural beliefs and commitments that cannot be changed as easily as shirts; feng shui is one of them, and Eitel recognized this.

Eitel's contemporary, the famed Dutch sinologist and ethnographer Jan de Groot (1854–1921), shared Eitel's evaluation of feng shui and wrote in the fourth volume of his monumental *The Religious System of China*:

> Fung-shui is a mere chaos of childish absurdities and refined mysticism, cemented together, by sophistic reasoning, into a system, which is in reality a ridiculous caricature of science. But it is highly instructive from an ethnographic point of view. The aberrations into which the human mind may sink when, untutored by practical observation, it gropes after a reasoned knowledge of Nature, are more clearly expounded by it than by any other phenomenon in the life and history of nations. (de Groot 1901, p. 938; in Bruun 2008, p. 85)

This is an illuminating lesson that can emerge when feng shui is included in a science programme, or when it is studied across the curriculum (see Fig. 11.3).

Conclusion

Eitel was not a naïve idealist who thought that important core beliefs of a person or culture can change without any collateral social or ideological dislocation; worldviews are not changed like singlets. This idealist outlook is, unfortunately, common in multicultural and indigenous science education discussion. It does need to be recognized that the changing of some deep and substantial beliefs entails large-scale changes in personal and cultural epistemology; the status and power of traditional authority figures and sources; relations between men and women, between parents and children, and between teachers and students; in many cases the legal system of countries; and so on. All of this is a common experience for communities and cultures adjusting to the dominant intellectual and commercial realities of the modern world. Witness the dynamic of cultures embracing modern science, technology, human rights, and democracy and the effort expended in resisting such embrace. The examination of Eitel's writings in science classes and in coordinated cross-subject teaching allows students to engage with fundamental issues in cultural history and philosophy of science. Such engagement makes a positive contribution to their education.

Chapter 8
Science, Westernization, and Feng Shui in Early Twentieth-Century China

Since the 1911 overthrow of the Manchu dynasty and the creation of modern political China, each generation has grappled with issues about the maintenance of Chinese culture, specifically its mixed Confucian/Daoist traditions, alongside the modernization of China's economy, politics, and culture. In 1985 President Deng Xiaoping announced the four modernizations required for contemporary China – industry, agriculture, national defence, and science and technology (Suttmeier 1980). Yijie Tang (1927–2014), a Peking University philosopher, spoke for many when he said: 'we felt that without the modernization of thought, the modernization process will come to naught' (Tang 2015c, p. 282). Science, especially science education, plays a central role in this 'dialectic' or conversation. What are the essential requirements of science to which all cultures in which it takes root need to adjust? What are the essential features of science that need to be conveyed and learnt in science education programmes? And concerning the argument of this book, is modernization of Chinese culture compatible with widespread practice of feng shui and belief in its chi-based cosmology? Underneath these questions is the long-standing realist/instrumentalist philosophical question: Does science tell us how the world is, or does it merely provide a means of successfully manipulating the world?

As with all serious questions, it is useful to put this 'feng shui and modernization' question into historical context. Along with all religions, ideologies, and worldviews, Confucianism and Daoism have been subject to external influence and have variously adapted and changed over time under these influences and have of course changed due to internal 'reformist' influences within the traditions (Berthrong 1998). All substantial and vital ideologies grow, adapt, and change. Saying that they should not do so, or that all such change is inauthentic, is just fundamentalist primitivism – an unfortunately too common occurrence in the Judaic, Christian, and

Islam traditions (Schimmel 2008). Education is essential for cultural maturation; it both responds to it and drives it.

There was, and is, a significant internal Chinese tradition that views feng shui in the same light that outsiders such as Matteo Ricci and Ernst Eitel did and see the comparable social and cultural changes required for remediation of its detrimental impact on the country that they did. This Chinese cultural 'reformist' tradition has both philosophical and political dimensions.

The Chinese Naturalist Tradition

The 'dissident' naturalist tradition in China goes back at least to the third century BC from which time Needham quotes Hsün Chhing:

> If (officials) pray for rain and get rain, why is that? I answer, there is no reason at all. If they do not pray for rain, they will nevertheless get it. When (officials) 'save the sun and moon from being eaten [during eclipses], or when they pray for rain in a drought, or when they decide an important affair only after divination – this is not because they think that they will in this way get what they want, but only because it is the conventional thing to do. The prince thinks it is the conventional thing to do, but the people think it supernatural. He who thinks it is a matter of convention will be fortunate; he who thinks it is supernatural will be unfortunate. (Needham and Ling 1956, pp. 365–366)

Hsün goes on to discuss 'ominous' signs, saying:

> But when human ominous signs come, then we should really be afraid. Using poor ploughs and thereby injuring the sowing, spoiling a crop by inadequate hoeing and weeding, losing the allegiance of the people by government bent on evil – when the fields are uncultivated and the harvest is bad, when the price of grain is high and the people are starving, when there are dead bodies on the roads – these are what I mean by human ominous signs (ibid.)

There is nothing supernatural about these signs; they are all natural, human-made, and repairable by man. No outside agency needs be invoked; doing so is just a distraction.

Four hundred years later, the first century AD philosopher Wang Chong (Chhong) lamented the grip of irrationalism and superstition on his countrymen; Wang was a rationalist in that he appealed to facts of the matter to settle disputes and completely rejected the idea that either nature or the gods would bend themselves to human wants and desires – humans existed in a world that was indifferent and unresponsive to their needs (Wang, Chong ca.80/1963). Needham calls this the 'Sceptical Tradition' in ancient Chinese thought (Needham and Ling 1956, pp. 346–394). Such beliefs were, at the time and in the present day, at odds with the man-centred core of the Chinese Confucian worldview and philosophy and, indeed, with the central beliefs of nearly all past and present cultures.

The non-naturalist, ego-centric subjective tradition is well exemplified by Yang Hsiung [Xiong] (53 BCE–18 AD), a foremost Confucian poet-philosopher and nationalist of the Han dynasty. As well as the *Analects of Confucius*, he absorbed and internalized the 'text' of the *I Ching* which he updated and sought to improve.[1] He came away from this philosophical formation with a very ego-centric worldview, and understanding of the task of natural philosophy, writing:

> Someone who understands the nature of heaven, earth, and man is a universal man of learning; someone who understands the nature of heaven and earth, but not of man, is an artisan. (Wilhelm 1977, p. 131)

Yang's step up from 'artisan' to 'man of learning' is very high and demanding. Transported to another context, it reduces Galileo, Newton, Darwin, Einstein, and perhaps all of the great European scientists to artisans; no longer are they 'men of learning' as they did not study, much less claim to understand, the nature of man. They looked out, not in. But the same consequences and judgments held for natural philosophers in the Chinese tradition: on Yang's account, they would not be invited into the circle of the learned unless they demonstrated knowledge of the nature of man. Yang further strengthens this ego-centric worldview when he contends that it is the human spirit that establishes the harmony of the world (Wilhelm 1977, p. 132). He is not saying that the human spirit studies or apprehends the harmony of the world – which was the explicit motivation for Christian and Muslim natural philosophers, as such harmony was God's doing – but rather it establishes the harmony, a truly idealistic, species-centric, even ego-centric position. Some 2000 years after Yang, a prominent science educator stated this ontologically idealist position as:

> For constructivists, observations, objects, events, data, laws, and theory do not exist independently of observers. The lawful and certain nature of natural phenomena are properties of us, those who describe, not of nature, that is described. (Staver 1998, p. 503)

Subjective idealism has a long and deep reach.

By the late-nineteenth century, there had been a number of Chinese translations of European scientific and mathematical works which were read by ever-increasing numbers of scholars. These translations began with Ricci and his Jesuit colleagues, they continued with later Protestant missionaries, and then in the nineteenth century again with Jesuits who had been re-admitted in 1844 and with Chinese scholars who had studied abroad. In 1865 courses in science were introduced into the Imperial College of Peking, the first such university-level courses in China. In 1898, Yen Fu (1853–1921) translated Thomas Huxley's *Evolution and Ethics* which was widely read and even adopted in many schools. Many of Huxley's phrases – 'survival of the fittest', 'struggle for existence', and 'natural selection' – entered into popular discussion and were used for describing China's political and military struggles. Additionally, Yen translated works of Spencer, Mill, Hume, Adam Smith, and Montesquieu. They were widely sold and read (Brière 1956, pp. 19–20). Others translated more core works of European science and philosophy.

In 1865, the balance of Euclid's *Geometry* was translated and published by Li Shanlan who wrote in the Preface: 'the inherent defect in Chinese mathematics lay in

---

[1] He replaced the classic idiographic hexagrams by tetragrams in which the single line can not only be undivided and divided but also twice divided; he rigorously coordinated the position of single lines with administrative and social institutions (Wilhelm 1977, p. 135).

its complete absorption in specific practical functions resulting in a failure to discover the underlying principles [laid out by Euclid]' (Peake 1934, p. 181).[2] The revered Chinese mathematical treatise from the first century AD that had been studied and examined for 1500 years, *Nine Chapters on the Mathematical Art* (Jiu Zhang Suan Shu), consisted of 246 problems in cultivation, taxation, engineering, and so on, with the problems being largely treated in isolation; there was no Euclidean attempt at ascertaining 'first principles' or axioms.[3] Arabic versions of Euclid had been brought to China but ignored. The first six books were not translated and published until Ricci and Xu Guangqi's publication in 1607, with the remaining books being translated by Li Shanlan in the mid-nineteenth century.[4] The mathematicians concern with technique and technology instead of theory was endemic in Chinese culture.

Nineteenth-Century Reckoning

In the three centuries following Ricci's death, Western science had made precious little impact in China. Knowledge of Western science was confined to a miniscule portion of the mandarin class, and there it was understood in utilitarian terms and pursued mainly for technological advantage. Dai Nianzu, the Chinese historian of science, put it plainly: 'From the end of the Ming Dynasty to the end of the last century [19th], the field of modern physics in China was a wasteland' (Dai 1996, p. 207). Others have documented the long and deep-seated conflict between the worldviews of most traditional Chinese cultural movements, especially the Confucian tradition, and the worldview and practices of modern science.

Qian Wen-yuan, the Chinese-educated theoretical physicist and US-educated historian of science, wrote one such account: *The Great Inertia: Scientific Stagnation in Traditional China* (Qian 1985). He saw all the multifaceted dimensions of Confucian-supported, conformist, authoritarian, centralized, imperial state power as restricting the seeds of modern science and suffocating shoots that might, against all odds, take root.[5] He, as with others, regarded the Mohists of the Warring States period (403–222 BC) as protoscientists from which something good could have developed. They were both artisans and philosophers; they formulated naturalistic accounts of engineering, optics, mechanics, logic, and economics; they were comparable to the Hellenic traditions of the time. However:

> … it was a great loss to Chinese science that in the long history of traditional China, Mohism, as a staunch rival school of Confucianism, was almost completely left in oblivion. (Qian 1985, p. 51)

---

[2] On the above historical matters, see Kwok (1965, chap. 1), Needham and Ling (1959), Peake (1934), Qian (1985, chap. 3), and Wang (1996).

[3] See Needham and Ling (1959, pp. 25–27).

[4] See Wang (2016).

[5] Derk Bodde commented: 'The Confucians succeeded in China where the Legalists failed, because of their realization that the only lasting unity is one based upon homogeneity of ideas and culture, and not on a forced obedience to a common body of law' (Bodde 1938, p. 8).

Karl Popper and many others have made the same observation about the eclipse of pre-Socratic atomism in the West by Platonic and Aristotelian traditions. Wallis Suchting concurred: 'Atomism carried on a basically marginal existence, ...till it was recuperated by Galileo' (Suchting 1994, p. 45). That is, a naturalistic metaphysics that could have supported or directed a proto-modern science was suffocated in Classical Greece and then given oxygen by the seventeenth-century European natural philosophers. Stephen Greenblatt has provided a book-length treatment of Lucretius' famed naturalist poem *De rerum natura* (*On the Nature of Things*) written in the first century BC. Lucretius, a Roman poet and philosopher, put into a six-book poem the naturalist and materialist worldview of the atomists who were writing three to four centuries earlier but who did not have a popular press (Greenblatt 2011). Confucius [Kong Fuzi] wrote and taught in the fifth century BC, the same time as the Greek atomists, and his thought and 'system' became the Imperial norm, or official state ideology, in the Han dynasty of the second century BC. It dominated through the Yuan (1280–1367), Ming (1368–1643), and Ch'ing (1644–1911) dynasties. There is a certain parallelism between the fate of materialism in the West and in the East.

Neo-Confucianism emerged in the Song dynasty (960–1279) and lasted energetically through to the nineteenth century and less energetically into the twentieth century. In the nineteenth century, the main preoccupation of neo-Confucian scholarship was 'evidence-based philological approaches to ancient texts' (Angle and Tiwald 2017, p. 209). The 'evidence-based' hermeneutics was progressive and enlightened scholarship, just as it was in nineteenth-century European biblical studies; nevertheless it was clearly out of step and irrelevant to the material and intellectual challenges to China in the period; a better understanding of ancient texts was no protection against European and Japanese guns. John Berthrong, a historian of Confucianism, well summarizes the dominant scholarly opinion, namely:

> The main focus of Ruist [Confucian] concern was neither in proto-science nor descriptions of the natural world. Above all else, Confucians sought to cultivate ethical life and reform society. (Berthrong 2003, p. 375)

This began to change dramatically with defeats in the Opium and Trade Wars (1839–1842, and 1858–1860),[6] the Sino-Japanese War (1894–1895), and the Boxer 'uprisings' (1900). In each of these conflicts, China again came into direct conflict with the 'front line' of Western science in the form of modern navies, armies, munitions, transport, railways, medicine, surgery, communications, and telegraph. The humiliating national defeats in the field were followed by more humiliations:

- The forced opening of the trade ports (Canton, Amoy, Foochow, Ningbo, and Shanghai).
- The creation of foreign zones which legally were part of their respective European countries.
- The loss of Hong Kong to Great Britain (1842), Annam to France (1885), Qingdao to Germany (1898), Tianjin to Italy (1901), Hankou to Russia (1896), and Ryukyu Islands (1879) and Taiwan (1895) to Japan.

---

[6] On the Opium Wars and the end of China's last Golden Age, see Platt (2018).

- Granting the right of foreign ships to traverse Chinese rivers unmolested and uninspected. They were floating pieces of 'extra-territorial sovereignty'.
- Paying massive financial indemnities to the invading powers – Great Britain, Japan, Russia, Germany, and France – for the cost of their invasions. This amounted to 21 million dollars to Great Britain when the Treaty of Nanking was signed in 1842 ending the First Opium War and a billion dollars to the allies when the Boxer Protocol was signed in 1901.
- Having to send envoys to the government of each invading country to make formal apologies for Chinese resistance to the invasions.

The 1915 Japanese demands for a 'Protectorate' over huge portions of northern China was the last straw in national ignominy. Anti-imperialists divided between ultranationalists who wanted to purify and reaffirm Confucian tradition against the immorality and bestiality of the foreigners and those that wanted to adopt the worldview and science that made the invaders so powerful.

In the early twentieth century, an entirely different 'New Confucianism' movement began. Angle and Tiwald remark that what distinguished this movement was its 'engagement in dialogue with Western philosophy', and they cite Mou Zongsan's (1909–1995) translations of Kant as an exemplar (Angle and Tiwald 2017, p. 211). The argument of this book is that what is needed for genuine modernization is engagement with *scientifically informed* philosophy, not just with Western philosophy. There are many 'schools' of scientifically ill-informed Western philosophy where engagement brings little intellectual benefit and even less social benefit, indeed on the contrary.

After the first defeats, a 'Westernization Movement' sometimes called 'Self-Sufficiency Movement' began in China and continued through the late nineteenth century. With each national humiliation, more people inside and outside of the government embraced the 'modernization' cause, but the embrace was partial. Western technology and industry were embraced, with steel mills, paper mills, textile plants, shipyards, railways, military arsenals, telegraph, and much else being built. The arsenals, in particular, beneficially had education, translation, research, and publication facilities.

For example, the Kiangnan Arsenal was established in 1865 and through to 1905 was the largest Chinese Translation Centre, with 178 Western scientific and philosophical works translated and published. Peking University was established in the 1890s. But many commentators have pointed out that Westernization did not extend to large-scale embrace of theoretical science, unified science, the scientific worldview, or the scientific habit of mind. Nathan Sivin observed that historically while the Chinese had sciences, they did not have science; until modern times, they did not have any overreaching notion of a science of the world, a science where components were answerable to each other and were to be consistent with each other (Sivin 1984, p. 533).

The historian Dai said of the Westernization Movement that: 'they merely adopted some of the most advanced Western technologies and had no intention of seriously studying natural science' (Dai 1996, p. 207). None of this 'big picture of science' found its way into the school system and little, if any, into higher education. The depressing national situation was well captured at the time by the American physicist Henry Rowland (1848–1901) in an address given to the American Association for the Advancement of Science:

> To have the applications of a science, the science itself must exist. Should we stop its progress, and attend only to its applications, we should soon degenerate into a people like the Chinese, who have made no progress for generations, because they have been satisfied with the applications of science and have never sought for reasons in what they have done. The reasons constitute pure science. They have known the application of gunpowder for centuries; and yet the reasons for its peculiar action, if sought in the proper manner, would have developed the science of chemistry, and even of physics, with all their numerous applications. By contenting themselves with the fact that gunpowder will explode, and seeking no farther, they have fallen behind in the progress of the world; and we now regard this oldest and most numerous of nations as only barbarians. (Rowland 1883, p. 242, cited in Guo 2014)

China's crushing defeat in the Sino-Japanese War of 1894–1895 signalled the end of the superficial technology-only Westernization Movement. Many saw that the modernization of China would mean more than buying and operating technology; modern science, and all that it entails, had to be embraced. Beiyang University (Tianjin University) was established in 1895 and the Imperial University of Peking in 1898 each with a remit to teach modern science and mathematics. This was, and still is, a challenge in China. Millions of technicians, engineers, and applied scientists have been trained, but far fewer theoretical or 'big picture' scientists, and there is little encouragement at school or university for acquiring any philosophically informed 'big picture' of science. What space there is is taken up with compulsory 'learn-and-repeat' courses on Marxist *Dialectics of Nature* (Gong 1996a; Wan et al. 2013).

The Imperial Examinations

Of fundamental importance was the late-nineteenth-century introduction of European science and mathematics as optional subjects in the 2000-year-old system of Civil Service Exams.[7] Since the beginning, these exams were based on the Confucian classics; they demanded decades of study of the literary classics, with 'study' mostly meaning hagiography, memorization, and identifying cross-text relations. The content of the exams did not change for 2000 years; the exams defined the Chinese cultural canon; and until their scrapping in 1905, they were gatekeeper for advancement in the Chinese state bureaucracy. A question in the Ming Examination of 1487 was typical of all the questions that went before for hundreds of years and that would follow for 500 years till the system was abandoned in 1905. Examinees had to write a strictly structured comment on a fragment of Mencius: 'He who delights in Heaven,

---

[7] Miyazaki (1976) offers a classic discussion of the practice and impact of these imperial exams. See also Franke (1963), Ho (1967), and Huff (1993, pp. 275–285).

will affect with his love and protection the whole empire'. The answer had to have antithetic pairs such as pro and con and shallow and profound, and these had to balance each other in diction, imagery, and rhythm (Huff 1993, p. 280).

Exams were taken first at the district level, with top students going on to prefectural exams, and then the top students from there going to provincial exams, and the topmost on to court for examination by the emperor and the most senior civil servants. Materially they were a massively inefficient operation, with hundreds of thousands of scholars walking hundreds of kilometres to sit exams, answers being recopied for sending to the Directorate of Education for marking, and so on. Those that failed might 'study' for a further 3 years to sit in the next examination cycle. Culturally they were hyperefficient: they guaranteed the production of a docile, narrow-minded, loyal, inward-looking bureaucracy for the administration of the imperial state. The admission of Western science into this canon, even as an option, was of the greatest symbolic and practical moment.

Today, through its stress on political correctness and Marxist ideology in the National College Entrance Examination (*Gaokao*), the Chinese Communist Party is repeating the social conformity and selection procedure of the imperial examinations. Any deviation from approved CCP positions rules out university entrance or at least entry to elite universities: no honest assessment of the Cultural Revolution, Mao's 'Great Leap Forward', the 1989 Tiananmen Square 'incident', or numerous other matters can be given. This naturally feeds back down into all high school instruction because so much of this is geared to the gaining of entry into the coterie of high-status universities.

Early Twentieth-Century Adjustments

In the middle and late nineteenth century, the conservative and traditionalist option was to ignore the West, along with its science and philosophy. But there was a reformist tradition which was clearly expressed by the Chinese governor and scholar Zhang Zhidong [Chang Chih-tung, 1837–1909].[8] He wrote in his 1898 *An Exhortation to Learning* that: 'Western learning should be adopted for practical purposes, but Chinese learning should remain the guide' (Peake 1934, p. 13). This work received imperial endorsement, with a million copies being printed and distributed throughout the realm. A variation of Zhang's pronouncement became the rallying cry of the Self-Strengthening Movement: 'Chinese learning as the substance and Western learning for application'. The idea was that Confucian learning should continue as the basis of social life and understanding, while Western learning be utilized in technology and manufacturing. The limitations of this position and more particularly Zhang's limitations were exposed when, as governor, he ordered a complete metal smelting plant from England without knowing what kind of metal

---

[8] He was a precocious scholar who at age 13 passed the first degree of the imperial exams, completing all degrees by age 26. His grave was ransacked by Maoists in the Cultural Revolution.

ore China had available and not placing the imported plant near any ore mine. This was akin to 'cargo mentality' in some Pacific islands after World War II. Nevertheless, Zhang's political career moved on. He introduced large-scale school reforms, sent hundreds of university students to study in Japan and Europe, was scrupulously honest, and died as a model Confucian without money or possessions (Bays 1978).

Instrumentalist Science became the most common option among the educated classes. It is an option still taken. In multicultural science education, it is called 'border crossing' (Aikenhead 1996). It has many consequences, including as noted in 2014 by the Chinese philosopher, Yualin Guo: 'the Chinese public does not distinguish science from technology. The science that newspapers, TV, radio, networks, etc. propagate is mostly technology' (Guo 2014, pp. 1835–36).

China's staggering defeats in the 1894–1895 war with Japan, a neighbour that in all living memory was regarded as inferior, accelerated the utilitarian turn to the West, with more arsenals, factories, railways, telegraphy, and shipyards established. Chinese officials and intellectuals could see that the military defeats were one outcome of the 1868 Meiji Restoration in Japan in which there was a wholesale embrace of Western technology and manufacture. There was a move to do the same.

At the end of April 1895, about 1200 of the candidates who had come to Peking for the highest-grade Civil Service Examination signed a petition imploring the emperor not to sign the Treaty of Shimonoseki with Japan and to embark with all speed on modernization consistent with China's Confucian tradition and values. The petition was drafted by Kang Youwei (1858–1927), a poet and political writer, who looked to the constitutional monarchy of Japan as a model for China.[9] This movement was given the unlikely name of 'Confucian Radicalism' (Zarrow 2005, chap. 1). It saw the tools of democracy – elections, councils, and parliaments – not as a way of governing the country or determining national policy but as a way of better and more informedly bonding the people to the emperor. This was thoroughly Confucian. Despite the number and eminence of the signatories, the petition was not handed to the emperor; the court officials were aghast that it was written, much less signed by anyone. Kang merely wanted a constitutional monarchy for China; it was hardly a 'storming the barricades' revolutionary call.

All of these military, social, economic, and cultural forces coalesced in the famed 'Hundred Days Reform' (11 June to 22 September 1898) initiated by youthful Emperor Guangxu (1871–1908) of the late Qing dynasty. The emperor was led to this liberal, modernizing initiative by an inner group of advisors, 'the Six Gentlemen', who were affected by Kang Youwei and had varying degrees of commitment to Enlightenment ideals. For some, these ideals were sourced to European writings and for others, to liberal traditions and writers in the Chinese Confucian and Daoist heritage. Crucially, in taking these policy initiatives, the emperor by-passed and ignored the court's conservative Grand Council.

---

[9] In this advocacy, Kang was 'on the same page' as the conservative wing of the European Enlightenment. On this period see Bays (1978), Furth (1983), Karl and Zarrow (2002), Spence (1982), and Zarrow (2005).

The reform initiative brought in significant industrial and manufacturing changes. An important component was an Educational Reform Act that gave an opening for modern science in the Chinese school system. However, the reforms were squashed after just 100 days by the Empress Dowager Cixi supported by a cohort of Grand Council traditionalists. Cixi in effect staged a coup, putting the emperor under effective house arrest. The 'six gentlemen' were arrested on 22 September 1898 and beheaded shortly after without trial. The enabling letters from Peking for the education reforms would barely have reached the outer provinces before they were countermanded. The Manchu (Ch'ing or Qing) dynasty collapsed in 1912, the Republic of China was established, and disarray, warlordism, and invasions prevailed. But European Enlightenment ideas had limited, but nevertheless relatively wide, exposure among educated Chinese, and this would have a lasting influence. Along with these 'foreign' ideas, there was a limited affirmation of liberal, naturalist, secular ideas in the divergent Chinese traditions. Although suppressed in 1898, these ideas and ideals surfaced again just 20 years later in the 'May Fourth Movement' of 1919.

Through it all there was debate and agitation about the modernization of Chinese thought and the role of Western science in China's Confucian identity. Scholarly groups seriously engaged with the task of seeing how Confucian thinking might be augmented to allow China to both compete technically with the foreigners and to master the foreign knowledge that had eluded Chinese mandarins formed in the Confucian tradition.

As in the West, orthodoxy in China had its defenders. Some of the better minds worked in the evidential studies tradition where the classic texts were examined in a 'scientific' manner to see if, where, and when textual corruption and foreign incursion might have occurred. One question asked was: Is traditional Confucianism really Confucian? Beginning with the work of Baruch Spinoza (1632–1677), the West had become accustomed to critical scriptural studies; finding the 'Historical Jesus' beneath the accretions of textual tradition became a routine task for scholars and for educated clergy (Schweitzer 1910/1954). Conservative Chinese embarked on a comparable study: 'Who was the historical Confucius?' 'Did he exist?' The historical irony is that rebellion against Song and Ming Confucianism began in a conservative and fundamentalist spirit – to recover the original meaning of the classics.

Those who had travelled overseas to be trained in science, some taking PhDs at the very best Western and Japanese institutions, expected that Chinese scholars would contribute to the advance of science and further, that a scientific outlook would take root and grow in China.

The 'New Thought' and 'May Fourth' Movements

Following the collapse of the Manchu imperial reign, and among the international and local turmoil of the Great War, the learned strata of China was, not surprisingly, swept by a 'New Thought' movement, of which there were two significant markers.

First was the May Fourth Movement of 1919, which was commonly known as the 'Mr Science and Mr Democracy' movement; second was the 'Philosophy of Life Debate' of 1923, sometimes called the 'Metaphysical Debate'.

Hundreds of thousands of students, teachers, professors, and journalists participated in these agitations, with countless thousands of books, leaflets, newspapers, and posters being printed and avidly read and debated. All of this intellectual debate occurred during the chaotic and deadly Warlord period (1916–1928). It is useful to elaborate a little on the issues then disputed as they recurred in China during Deng Xiaoping's post-1978 reforms that attempted to roll back the economic and cultural disasters of the Mao years and have surfaced again during Xi Jinping's tightening of the Communist Party (CCP) rule. The social, philosophical, political, and educational issues debated in China almost 100 years ago have a certain perennial quality; they are still debated both in the West and the East.

A prime instigator of the New Thought and May Fourth movements was Ch'en Duxiu [*Ch'en* Tu-hsiu] (1879–1942), a noted champion of science, critic of traditional Confucian philosophy and culture, and founder of the Chinese Communist Party (CCP).[10] He was an ardent 'public intellectual' and political reformist. He was one of thousands who looked to Western thought, especially science, in order to modernize Chinese thinking and strengthen the society so that the endless humiliations of the late nineteenth and early twentieth centuries would not recur. He took the view that Confucianism had dominated Chinese thinking, culture, and politics since the Han emperor Wu Di's decree of 134 BC. It had not cultivated or encouraged natural philosophy; it was synonymous with conservative, backward-looking, if not reactionary thought; it was incompatible with any Chinese embrace of modern science. He saw that China's modernization, and adoption of science, was a systemic problem; there was no isolated switch that could just turn on science. He gave this damning account of the Confucian-dominated imperial examination system:

> In spite of the many good things that may be said to the credit of Hung Wu [1328–1398], he will ever be remembered in connection with a form of evil which has eaten into the very heart of the nation. This was the system of triennial examinations, or rather the form of Chinese composition, called the 'Essay', or the 'Eight Legs', … The writer could not express any opinion of his own, or any views at variance with those expressed by Chu Hsi [Xhu Xi] and his school. All he was required to do was to put the few words of Confucius, or whomsoever it might be, into an essay in conformity with the prescribed rules. …. But absurd as the whole system was, it was handed down to recent times from the third year of the reign of Hung Wu, and was not abolished until a few years ago [1905]. No system was more perfect or effective in retarding the intellectual and literary development of a nation. (Russell 1922, p. 45)

His estimation of the examination system was underscored by the floggings he received from his grandfather when between the ages of 6 and 9, he failed to correctly recite passages from the Confucian classics and by the fact that his own father spent a lifetime studying for the second degree in the exams without ever obtaining it.

---

[10] For his life, teachings, and activities, see Kwok (1965, chap. 3).

After republican agitation, Ch'en fled China for Japan but did not support Sun Yat-sen's campaign against the Manchus because of its unacceptable and mistaken national chauvinism which appalled him. He travelled on to France in 1907 and there found the Enlightenment commitments that guided him for the remainder of his career: democracy, science (especially evolutionary theory), and socialism. Each of the 'rights of man', the scientific method (experimental, empirical, quantitative, and naturalist), and the outlook of science (anti-authoritarian, sceptical, tentative, universal, internationalist) had to be embraced; and none of this was compatible with Confucianism and probably not even with neo-Confucianism. He became a Francophile. Anticipating later critiques of CCP rule by noted scientists (Miller 1996), Ch'en maintained that science required a democratic republic in which to flourish, not an authoritarian monarchy sustained by a subservient culture.

In 1910 Ch'en returned to China and took senior positions in different high schools, until his republican writings had him expelled from the country by President Yüan Shih-kai. Once more he went to Japan, returning in 1915, whereupon he became a founding and influential member of the 'New Thought' (sometimes referred to as 'New Culture') movement. Ch'en wanted to modernize Chinese thinking, not just Chinese industry and technology. This 'progressive' movement was locked in conflict with a 'traditional' or 'conservative' one, both vying for influence in the early life of republican China following the 1911 collapse of the 300-year Manchu (Ch'ing) dynasty (Furth 1983). 'Liberal' conservatives espoused the foregoing motto of Zhang Zhidong: 'Chinese learning for fundamental principles and Western learning [science] for practical application'. Amid all the debate and national self-examination, in 1914 President Yüan issued an edict saying:

> The doctrines of Confucius and the classical literature are without equal among mankind. The offering of incense and sacrifice is historic, and it is appropriate for the Republic to follow the old customs. (Kwok 1965, p. 72)

Revelation of such entrenched antimodernist conservatism at the top of the Chinese political structure dismayed participants in the New Thought movement. In 1915, in response they launched their own journal *New Youth* (*La Jeunesse Nouvelle*). This was edited by Ch'en who saw its purpose being 'introducing western ideas, opposing Chinese ideas and eliminating Confucianism' (Schwartz 1951, p. 62). The journal was immediately embraced by all the modern intelligentsia of China; with a circulation of 15,000 per issue, it had great educational, cultural, and ideological impact (Wang 1966, p. 309). Between 1915 and 1918, articles were published on Adam Smith, John Stuart Mill, Darwin, and many other key figures of European liberalism, but no articles on or discussion of Karl Marx (Wang 1966, p. 315). Ch'en published his own essay 'The French People and Modern Civilization' arguing that both politics and science contributed to the supremacy of Europe and that both European liberal, democratic politics and Western science had to be embraced by China. There could be no return to the old ways.[11] Two articles on science appeared in 1919 and 1920.

---

[11] On the history of 'modernization' in Chinese education of the period, see Wang (1966, chap. 11) and Zarrow (2018).

The first issue of *New Youth* contained an editorial essay 'My Solemn Plea to Youth' in which Ch'en wrote:

> Our men of learning do not understand science; thus they make use of *yin-yang* signs and beliefs in the five elements to confuse the world and delude the people and engage in speculations on geomancy ... The height of their wondrous illusions is the theory of *Ch'I* [primal force] ...We will never comprehend this *Ch'I* even if we were to search everywhere in the universe. All of these fanciful notions and irrational beliefs can be corrected at their roots by science, because to explain truth by science we must prove everything with fact. Although this is slower than imagination and arbitrary judgment, every progressive step is taken on firm ground. It is different from those flights of fancy which in the end cannot advance one bit. (*New Youth* 1915, vol. 1, p. 1. In Kwok 1965, p. 65)

Every sentence of the editorial raises issues in philosophy, philosophy of science, history of science, cultural criticism, politics, and more. One century later these issues still warrant class discussion, especially in any coordinated, cross-disciplinary approach to curriculum planning.

Ch'en, as with many in the movement, was a secular liberal and shared much of the philosophical, cultural, and political New Thought programme of the European positivists – he called for 'Mr. Confucius' to be replaced by 'Mr. Science' and 'Mr. Democracy'. In January 1919, before the May Fourth demonstrations, he editorialized:

> In order to foster Democracy, we must oppose Confucianism, chastity, old ethics, and old politics. In order to foster Science, we must oppose the old arts and religion. In order to foster both Democracy and Science, we must oppose our national heritage and our old literature. (Wang 1966, p. 311)

Mao Zedong called him 'the Commander-in-Chief' of the May Fourth Movement (Gong 1996b, p. 13). But Ch'en's call for the Chinese adoption of modern science *and* Enlightenment politics was, and remains, a challenge to the Communist Party. The party wants the first, but not the second. This position has only hardened since potentially President-for-Life, Xi Jinping took control of the Party.

The New Thought movement promoted the 1919–1921 China lectures of John Dewey (Wang 2007) and the 1919 lectures of Bertrand Russell (Russell 1922). Ch'en and colleagues fully embraced Dewey's characterization of the 'scientific outlook' and the necessity of its cultivation among all classes of Chinese citizens. Dewey's position, as noted in Chap. 2, was expressed in all his education writings and later as:

> In short, the scientific attitude as here conceived is a quality that is manifested in any walk of life. What, then, is it? On its negative side, it is freedom from control by routine, prejudice, dogma, unexamined tradition, sheer self-interest. Positively, it is the will to inquire, to examine, to discriminate, to draw conclusions only on the basis of evidence after taking pains to gather all available evidence. It is the intention to reach beliefs, and to test those that are entertained, on the basis of observed fact, recognizing also that facts are without meaning save as they point to ideas. (Dewey 1938, p. 31)

Bertrand Russell, in a book following his China sojourn, wrote of the dangers besetting any modernization programme in China:

> The first danger is that they may become completely Westernized, retaining nothing of what has hitherto distinguished them, adding merely one more to the restless, intelligent, industrial, and militaristic nations which now afflict this unfortunate planet. The second danger is that they may be driven, in the course of resistance to foreign aggression, into an intense anti-foreign conservatism as regards everything except armaments. This has happened in Japan, and it may easily happen in China. The future of Chinese culture is intimately bound up with political and economic questions; and it is through their influence that dangers arise. (Russell 1922, p. 3)

Understandably, the New Youth group embraced the widespread agitations of the May Fourth (1919) Movement[12] which grew from Peking University student demonstrations on 4 May 1919 against the government's servile acceptance of the Versailles Treaty requirement to, among other things, hand over the territories of Shandong to Japan. It became nationwide and fanned modern Chinese nationalism. This movement was in all ways progressive. Along with everything else, they established and maintained schools, which by one estimate had 50,000 pupils in just Peking. Gong Yuzhi (1929–2007) wrote:

> Progressive people who supported science during the May 4th Movement primarily had in mind its function as an ideological weapon in the struggle against the darkness and ignorance of feudal, philistine ideology. On the other hand, at that time people also emphasized the function of scientific development in vitalizing industry, hoping to cure the long-standing weakness and poverty of old China by means of science and industry. (Gong 1996b, p. 14).

Ch'en saw Confucianism as the chief impediment to China's advancement and modernization and science as the effective and only dissolver of its associated superstitions and idealisms. Not only was he condemned by the traditionalists and imprisoned by the Chinese Republican government for his science; he suffered the same fate at the hand of Chinese communists. By 1920 Ch'en lost faith in the possibility of China becoming a democracy; and inspired by the example of Lenin's Communist Party overthrowing the Czarist regime and old culture and religion of Russia, he became a communist. But his Enlightenment and liberal politics were seen as Trotskyist Deviations; he was labelled a 'counter-revolutionary', a 'right deviationist', a 'bourgeois democrat', and worse. Ch'en died lonely and friendless at age 62.

Ch'en's crusade was supported by Fung Yu-Lan (1895–1990) (also rendered 'Feng') the noted philosopher and historian of Chinese philosophy.[13] In his much-cited 1922 publication written during the dark depths of Chinese warlordism and national humiliation, Fung maintained that:

---

[12] See Chow (1960), Schwarcz (1986), Wang (1966, chap. 10), Zarrow (2005, chap. 8), and contributions to Schwartz (1971).

[13] Fung studied philosophy at the University of Peking, then in America at Columbia University where he was influenced by Dewey's pragmatism and neo-realism. His two-volume *History of Chinese Philosophy* was written in 1934 (Fung 1952–1953). He was a founder, in 1935, of the Chinese Philosophical Society and later became its president. His philosophical writings are in Fung (1947, 2008).

> What keeps China back is that she has no science. The effect of this fact is not only plain in the material side, but also in the spiritual side, of the present condition of Chinese life. China produced her philosophy [Confucius (551–479 BCE), Mencius (372–289 BCE)] at the same time with, or a little before, the height of Athenian culture. Why did she not produce science at the same time with, or even before, the beginning of modern Europe? (Fung 1922, pp. 237–38)

Twenty years later, as discussed in Chap. 10, this question was repeated by Joseph Needham and became 'The Needham Question' which launched hundreds of books and thousands of articles.

Fung's answer was the domination of subjectivist and idealist thinking in the Chinese tradition: 'China has not discovered the scientific method, because Chinese thought started from mind, and from one's own mind' (Fung 1922, p. 260). He breaks with the dominant, millennia-long Chinese tradition of a living or 'vital' cosmos:

> We cannot admit the existence of a cosmic soul. For us, the universe is only a general term, a *universal*; it does not call for a real structure necessitating a soul. Only the superior animals have a soul. The existence of a cosmic spirit is impossible in our system. From this point of view we sympathize with materialism. (Brière 1956, p. 52)

This observation recognizes the huge cultural and worldview reconsiderations that modern science requires of societies where it is embraced and normalized. Variants of Fung's question and answer need to be engaged with. Without such engagement, there can be utilitarian and technological science, but hardly the energetic enthusiasm for science that is required in a modern society. Education is typically the battlefield between contending 'schools' in this science and modernization and culture and tradition contest.

After the 1912 collapse of the Manchu (Qing) dynasty, and in the early years of the Republic, thousands of Chinese left the country to study abroad, with hundreds completing science degrees and many being supported by missionary societies.[14] The example of Li Yaobang is illustrative. He studied electronics at the University of Chicago, being the first Chinese person to receive a PhD in modern physics. His 1914 doctoral dissertation was on 'The Determination of the 'Value of 'e', by Millikan's Method Using Solid Spheres'. He was a participant in the then frontier of physics. But when Li returned to China, there was nothing for him to do, nowhere for him to teach, and much less to continue research in atomic physics. He gave lectures on physics to a church group and began a business. There was little, if any, scientific infrastructure in China.[15] An immediate institutional challenge for the commencement of modern science in China was the establishing of such infrastructure.

Fung, along with everyone else, recognized that long before the sixteenth century, China had all the material apparatus for science (writing, printing, books, technology, a scholarly class, and well-functioning state institutions), but he maintains

---

[14] Wang (1966) provides detailed information on numbers, destinations, and fields of study.
[15] For this and other such examples, see Dai (1996).

that it was cultural factors and philosophy that thwarted the rise of a modern science. And he linked this to the syncretism that Ricci noted:

> After the Chin Dynasty [220–200 BC] ... Soon came Buddhism, which again is a 'nature' philosophy of the extreme type. The Chinese mind oscillated among Taoism, Confucianism, and Buddhism for a long time. It was not until the tenth century AD [Song Dynasty] that a new group of men of genius succeeded in combing these three, Taoism, Confucianism, and Buddhism, into one [neo-Confucianism], and instilling the new teaching into the Chinese national mind, which has persisted to the present day. (Fung 1922, p. 256)

His answer to what would become the 'Needham Question' was:

> China has not discovered the scientific method, because Chinese thought started from mind, and from one's own mind. ...... Chinese philosophers loved the certainty of perception, not that of conception, and therefore, they would not, and did not translate their concrete vision into the form of science. (Fung 1922, p. 260)

Fung Yu-Lan says of his paper that it is an attempt to answer this question 'in terms of China herself' (Fung 1922, p. 238) and presages that: 'At the end of this paper I shall venture to draw the conclusion that China has no science, because according to her own standard of value she does not need any' (Fung 1922, p. 246). Fung is correct in realizing that any attempt to embrace, incorporate, or absorb modern science was going to require a change in Chinese values which are intimately connected to Chinese worldviews. For any society and culture, there was no quick and easy route to modern science. Science is an ensemble with ontological, epistemological, ethical, sociological, educational, and, it needs be said, political parts. All of these need to be coordinated and in harmony for a society to embrace and nurture science.

An alternative is to 'have' science without embracing it. This can be simply 'buying in' science and technology in the manner pursued by the USA where overwhelmingly its science and technology research and teaching faculty come from elsewhere; they are first bought and then brought in. The same 'rootless' result occurs when science is taught in a purely technical manner and where there is a degree of scientific competence developed, but the 'ethos' or worldview of science is not cultivated. This is the situation in many Muslim and Asian countries and pertains in traditional societies where the popular educational principle of 'border crossing' reigns.[16] It is what leads to 40% of US biology teachers having biology degrees but believing in special creation or even in the young earth doctrine.

Without using the term, Fung was laying the blame for China's failure to move forward to modern science on the dominating impact of empiricism on the thinking of its natural philosophers. Many western philosophers have identified commitment to empiricism as the chief intellectual obstacle to the birth of modern science in seventeenth-century Europe. Alexandre Koyré's observation on the retarding role of empiricism on the Scientific Revolution has been noted earlier: 'the empiricism of modern science is not *experiential*; it is *experimental*' (Koyré 1968, p. 90).

---

[16] Students are told to leave their culture at the door of the science class and put it on again after the lesson (Aikenhead 1996). This is the anthesis of the Enlightenment's expectation of education.

Other contemporary Chinese philosophers identify the same impediment to the development of modern science in China:

> .. in those areas of science which could be explained by daily experience and direct induction, Chinese scientific theories produced wonderful descriptions, such as the explanation of meteors, aerolites, fossils, and rainbows, but once outside the area which could be mastered by direct perception or induction, ancient Chinese scientific theories became ambiguous. (Jin et al. 1996, p. 146)

Ch'en and Fung affirmed Enlightenment principles. The above-quoted *New Youth* editorial touches on most of the core scientific, philosophical, cultural, and educational issues surrounding feng shui and how it is best treated in a school system. These were momentous issues for China at the beginning of the twentieth century. The same issues recur in all societies and cultures as they 'come to terms with' science and with Enlightenment political beliefs especially secularism (separation of state and church), liberalism (freedom of speech, association, and press), and democracy (ultimate power belongs to citizens, not to kings, emperors, parties) in politics. Just as these 'big issues' are prompted by and involve science, so too they should find some appropriate place and exposition in science education. It is unfortunate to allow students to complete science education yet remain ignorant of the movement of these science-related tectonic cultural plates.

The Philosophy of Life Debate

In 1923 the neo-Confucian scholar Chang Carsun (Zhang Junmei 1887–1969) of Tsing Hua University exhorted Chinese youth to turn away from the science-based philosophy of life promoted by the *New Youth* movement (Chang 1957). This was the first shot in the 'Philosophy of Life' or 'Metaphysical' debate that spread far and went deep into Chinese teaching, learning, and public discussion.[17] One student of the period has said 'all the influential thinkers of China took part in the debate' (Kwok 1965, p. 149). A published 'Proceedings' of the debate amounted to 250,000 words.[18] Hu Shih, the prominent Chinese pragmatist and empirical philosopher,[19] wrote in *The Effort Weekly* of 1923 that:

> Everyone joined in this discussion of Science and the Philosophy of Life. The length of the battlefront, the large number of its warriors, and the duration of the conflict all contributed to make this the first great battle in the thirty years since China's contact with Western culture. (Kwok 1965)

---

[17] On this debate see Brière (1956), Chang (1957), and Kwok (1965).

[18] *The Battle of the Philosophies of Life,* 2 volumes, Tái-tung Book Co., Shanghai

[19] Hu as a teenager was influenced by Darwin and Spencer; he went to Columbia University in 1915 and completed a PhD under John Dewey with a thesis on 'The Development of the Logical Method in Ancient China'. He said that Dewey 'had taught me how to think … to understand the nature and function of the scientific method' (Kwok 1965, p. 89). During World War II he was Chinese Ambassador to the USA. After 1949 he was hounded and denounced by the CCP.

With the horrors of World War I and its mass killings and city-wide destructions so clearly underwritten by Western science and technology, constantly before the educated classes, Carsun's call was heard by receptive ears. The much-touted modern science seemed not to provide an enviable philosophy of life for European people, much less for Chinese people who had their own Confucian tradition to give meaning to life and stability to society.

Liang Chí-cháo (1873–1928), an enthusiast for science in the 1898–1912 period and who was a member of the Chinese Observers' Team after the war, was so disturbed by what he saw and what was related, that he turned on science and supported Zhang's campaign against Western science and culture. His 1922 appraisal was the same as countless commentators, social critics, philosophers, and religious figures in the West:

> Because of the development of science which in turn created the industrial revolution, the external life of modern man has experienced rapid and numerous changes, while his inner life has faltered and become unstable …Those materialist philosophers, sheltering themselves in the respectable aura of science, have established a purely materialistic and mechanistic view of life. … Religion and traditional philosophy have been defeated and are in utter confusion, but this Mr. Science barges in and wants to build the great new law of the universe through experimentation. … This World War is but a consummate outcome of it all. … The Europeans have had a grand dream of the omnipotence of science, but now they are talking about its bankruptcy. (Kwok 1965, pp. 137–38)

The 'science and modernization' side was championed by many, including three prominent Western-educated Chinese scientists – geologist Ting Wen-chiang (V.K. Ting 1887–1936), mathematician Jen Hung-chün, and psychologist T'áng Yüeh. Ting labelled Chang and the traditionalists 'Metaphysical ghosts'. He and fellow scientists defended the reach of science into the sphere of values and worldview. They advocated what subsequently would be called a scientific worldview and what pejoratively has been called 'scientism'. In 1923, in a special issue of the *China Institute Bulletin*, Ting stated the realist view that: 'The object of science is to eliminate personal subjective prejudices and to search for the truth which is general and universal' (Brière 1956, p. 30). In defending science against the claims of Liang, he wrote:

> As for the main responsibility for the war, it is politicians and educators who should be blamed. The majority of these two kinds of people are still not scientific … Science is not only not external, it is the best tool for education and cultivation of the personality; the reason is that through the constant search for truth and the incessant desire to do away with prejudice, the men of science not only gain the ability to look for truth but also acquire the sincerity about truth. No matter what they are confronted with, they can study and analyze with dispassion, seeking the simple in the complicated and order in disorder. … How can this state of mind ever be dreamed of by those who sit meditatively talking about Zen, and by those who speculate metaphysically? (Kwok 1965, p. 147)

He was relaxed in saying 'Science is all-sufficient, not so much in its subject matter as in the manner of procedure' (Brière 1956, p. 30). As has already been seen, and as will be further developed in Chap. 13, this 'all-sufficient' claim is an overreach if taken too literally and simply. There are legitimate ontological, methodological, epistemological, and ethical issues within science that require *philosophical* elaboration and defence; science by itself does not provide this guidance; informed philosophical thinking is required. Ideally, scientists themselves have this philosophical competence, then the 'all-sufficient' claim is strengthened, but cross-competence is rare, and conceptually the science/philosophy distinction needs be drawn and both dimensions need be attended to in the education of scientists and science teachers.[20]

T'ang Yüeh, a psychologist and translator of many of William James's works, argued in an article on 'Psychological Phenomena and the Law of Causation' that all phenomena, including psychological or mental phenomena, are subject to the law of causation and hence are in the domain of scientific investigation. Everyday experience such as love, jealousy, aesthetic response, and so on can also be investigated by science. His view was that:

> In matters relating to sentiment, we must rely on the extent of our reason and solve them with the scientific method. Our idea that sentiment is suprascientific comes from the immediacy of experience with sense data; but this experience is no different from all experience, and it is the starting point for science. (in Kwok 1965, p. 151)

The mathematician Jen Hung-chün held that a scientific philosophy of life is possible and was being progressed by the best scientifically informed philosophers and thinkers. What was important about, and for Jen distinctive of, science was its spirit and attitude. In an essay 'On the Scientific Spirit' he writes:

> What the scientist knows has factual data as its basis, experiment as its corrective, application as example, and experimental proof as its claim to finality. It does not accept subjectivism, established teachings, and words of ancients. (in Kwok 1965, pp. 123–24)

He made the standard points that 'the more the knowledge of the material world improves and becomes scientific, the more the philosophy of life will become scientific and be proportional to it' (in Kwok 1965, p. 153). And:

> Science studies relations among matter. Only when the relations are understood can laws be discovered. Such studying of relations and discovering of laws all give men an attitude of causation [sequential thinking] … Thus this philosophy of life aims at whatever conforms with reason and the demand for proof. (in Kwok 1965, p. 125)

The 1923 'Philosophy of Life' debate mirrored the Romantic critique of science that had earlier occurred in Europe (Cunningham and Jardine 1990; Holmes 2009); it anticipated the post-World War II debate prompted by the same revulsion at science-enabled mass horrors; it was a foretaste of the anti-Enlightenment campaigns of the later twentieth century.

---

[20] Two well-informed elaborations of philosophical questions arising within the practice of scientists are Gauch (2003) and Johansson (2016).

Conclusion

The early twentieth-century China-wide debates about science, tradition, and culture can be recast as a debate about worldviews: What are the essential components of a worldview? Does the practice of science require a worldview? Is there a single scientific worldview or a range of them? Does science have a worldview? Can science engage with and criticize components of worldviews? And so on.[21] The argument of this book is that a worldview that comfortably accommodates feng shui, and its related chi-based practices and cosmology, cannot be reconciled with the practice and growth of serious science. That is with the epistemology, ontology, methodology, and ethics of modern science. The feng shui worldview does not fit into the 'ecology' of science as will be outlined in Chap. 13. It is an outlier, at best a distraction, but routinely a hindrance to the development of an intellectually healthy culture.

---

[21] On the subject of science, worldviews, and education, see the 14 contributions to Matthews (2009a).

Chapter 9
Feng Shui, Science, and Politics in Contemporary China

The modernization of thought called for by contemporary Chinese scholars echoes the Enlightenment project launched in Europe in the eighteenth century (Israel 2001, 2011). Assuredly today 'modernization' means recognizing and engaging with the core political (universal rights, anti-absolutism, freedom of speech, and assembly), philosophical (primacy of reason, autonomy of philosophical investigation, and its freedom from religious, party, or state oversight), and scientific (primacy of observation, experimental testing, and commitment to methodological naturalism) arguments advanced by the Enlightenment thinkers. For most 'modernizers' in China, the task was to formulate 'Enlightenment with Chinese Characteristics'. This was not an easy task; the millennia-engrained Confucian tradition does not easily, if at all, accommodate the epistemology or politics of the European Enlightenment as outlined in the preceding sentences.

Education, the Enlightenment Tradition, and the Modernization of China

A minority of Chinese citizens rejected the modifier of 'Chinese Characteristics': they held that science and related Enlightenment commitments were universal. Hence Fang Lizhi said:

> Words like 'Eastern' and 'Western' were useless. … The keys to modernization are neither Chinese nor foreign, neither Eastern nor Western. They are the ability to absorb advances – advances called science and democracy – that are available to anyone, anywhere. (Fang 2016, p. 193)

Modernization does not mean the embrace of the historical Enlightenment's actual understanding of law, government, commerce, religion, politics, or science; it is not a matter of going *back* to the eighteenth century but being *challenged* by the best of that Age. The Enlightenment thinkers were products of an outstanding eighteenth-century European milieu; they were not a monolithic cabal; and the difference between radical and reformist wings in the Enlightenment has long been recognized. There were differences among them on all of these core subjects, and assuredly there were aspects of each that they got wrong in their day. They were right to embrace the New Science, but their understanding of its methodology and epistemology needs be refined. Famously, philosophical enthusiasts for the New Science – for instance, Francis Bacon, John Locke, and Robert Boyle – had, by modern standards, an inadequate grasp of the methodology of the New Science being carried on around them by their heroes. Bacon famously rejected Copernician astronomy.[1] Recognition of such failures, inadequacies, and short-falls is more consistent with a proper understanding of the Enlightenment not as a set of ahistoric principles or findings, but rather an openness of inquiry in all fields. There is no Enlightenment catechism; for the contemporary world, there is more an enlightened direction than an Enlightenment destination. It confounds the spirit of the Enlightenment to say that Enlightenment thinking cannot be revised; as with all intellectual movements, a 'hard core' and a 'protective belt' can be recognised, and legitimate argument can be had over where to draw the boundary.[2]

It is to be expected that the Enlightenment took different forms in different national contexts (Porter and Teich 1981), and so it did, and will do, in China. Whether such Enlightenment is authentic, or is a government-permitted sham, is something for Chinese citizens and scholars to ascertain.

The work of Yijie Tang (1927–2014) is illustrative of contemporary Chinese concerns with Enlightenment thinking and for expanding modernization beyond the economic sphere to the modernization of politics and of thought. Tang was one of the most prominent Chinese philosophers of recent decades, a Peking University professor, and director of the huge National Confucian Project involving 400 scholars working on 5000 Confucian works.[3] He was sympathetic to the European Enlightenment project and expressed his sympathies in an essay titled 'The Enlightenment and Its Difficult Journey in China' (Tang 2015c). In 1958 he had been expelled from the Communist Party, and in the 1966 Cultural Revolution, he lost his teaching position at Peking University and was not allowed to resume it till 1980. He and those around him were engaged on the task of identifying what it was to be a 'modern' Chinese citizen. They saw this, in part, as requiring a second Enlightenment in China.

---

[1] For detailed accounts of these failures by philosophers to correctly identify the methodology of the New Science, see at least Agassi (2013).

[2] An account of the core Enlightenment positions is given in Matthews (2015a, pp. 23–27).

[3] Twenty-five of his essays are collected in Tang (2015e). They include essays on 'New Confucianism', 'Matteo Ricci', 'Daoism', and 'The Enlightenment in China'.

Tang well stated the importance of this 'cultural appraisal' task:

> If Chinese people want to make contributions to the 'coexistence of civilizations' in contemporary human society, they must first know their own culture well, which means they must have a cultural self-consciousness. The so-called cultural-consciousness refers to the fact that people in a certain cultural tradition can give serious consideration or make earnest reexamination of their own culture's origin, history, characteristics (including both merits and weakness), and its tendency of progress ... we must analyze the weak points of our own culture as well to better absorb other cultures' essences, and to give a modern reinterpretation of Chinese culture, so that it can adapt to the general tendency in the development of modern society. (Tang 2015d, p. 299)

This cultural appraisal can be advanced if science education is informed by the history and philosophy of science. It will be further advanced if feng shui is brought into the science programme as a topic for discussion. Tang spoke for many in saying:

> We felt [in 1985] that Comrade Deng Xiaoping's promoting of economic development is central and that intensely realizing the four modernizations was correct, and we were totally in full support. At the same time, we asked if the problem of modernization is only to be understood as the modernization of industry, agriculture, science and technology, and national defense [the four modernizations], because we felt that without the modernization of thought, the modernization process will come to naught. (Tang 2015c, p. 282)

The task of 'cultural self-consciousness', as outlined by Tang, Wei, Fang, and so many other progressives, is an educational project that requires science-informed historical and philosophical input. It is a task with which *all* societies need to engage. Responsible science education will contribute to this task in all countries. The more explicitly the HPS dimensions of science are presented in classrooms, then the more fruitful will be such contribution to cultural self-consciousness. This holds for China as well as other countries. A narrow and purely technical science education makes no contribution to this task and indeed thwarts it. A class of 'experts' is created who know only their own special slice of the scientific spectrum and so are ill-informed or ignorant about wider natural and social realities and are unable to contribute to mature and progressive policies concerning those realities.

Science, Liberalism, and the Modernization of Chinese Politics

In 1978 the dissident, Wei Jingsheng, was sentenced to 18 years in jail[4] for calling for a 'fifth modernization', namely, a modern political system.[5]

---

[4] The CCP released him from jail one week before the International Olympic Committee met to decide the venue of the 2008 summer games. Beijing was awarded the games.

[5] This particular 'modernization' still cannot be voiced in China. Many of those who voiced it in Tiananmen Square in 1989 were either shot, jailed, lost their jobs, left the country, or lived cowed lives (Kristof and Wudunn 1994). Many of the individuals repeating it in Hong Kong in 2015 simply disappeared. In September 2018, the independence-minded Hong Kong National Party was declared illegal. The massive Hong Kong demonstrations of June 2019 were similarly motivated.

> We want to be the masters of our own destiny. We need no gods or emperors and we don't believe in saviors of any kind…we do not want to serve as mere tools of dictators with personal ambitions for carrying out modernization. We want to modernize the lives of the people. Democracy, freedom, and happiness for all are our sole objectives. (Wei 1997, p. 208)

Since the late nineteenth century, there has been a constant scientific liberal, dissenting thread in Chinese culture. This was proclaimed on banners of the May Fourth Movement and its promotion of 'Mr Science' and 'Mr Democracy'. Fifty years later, it was restated in the writings, achievements, and tribulations of Fang Lizhi (1936–2012), the prominent cosmologist, astrophysicist, and Peace Prize recipient who shared the same jail fate as Wei Jingsheng.[6] In 1956, during Mao's brief '100 flowers' interlude, Fang spoke out about the purpose of education and how education was inconsistent with sloganizing, indoctrination, and repetition of party dogma that Maoism was enforcing. He regarded freedom of association and opinion, democracy, and human rights as both the natural consequence of and also prerequisite for the conduct of science (Fang 1990/1992). Shortly after, in 1957, he was imprisoned and tortured as a 'Rightist' in the beginning of the Cultural Revolution.

In 1987, 20 years after rehabilitation, Fang was again expelled from the Communist Party and dismissed from his position as Vice President of the University of Science and Technology of China. His crime, and that of other liberal scientists such as Xu Liangying (1920–2013), the physicist, historian, and translator of Einstein, and Li Xingmin, was to reject Deng Xiaoping's policy that 'intellectuals row the boat, the Party holds the tiller'. The scientists thought that intellectuals should be involved in policy determination, not merely researching what they were told to research; they should have a hand on the tiller. This held especially so for scientists and science policy. In 1990, in the aftermath of Tiananmen Square, Fang was expelled from the country; in the early 1990s, after a brief period in England, he was appointed a physics professor at The University of Arizona where he published and taught till his death in 2012 (Fang 2016; Kelly 1990).

In 1990, in an interview conducted in the US embassy, Fang said: 'I came to my own views about democracy through my conception of science' (Fang 1990/1992, p. 295). And in his autobiography:

> The keys to modernization are neither Chinese nor foreign, neither Eastern nor Western. They are the ability to absorb advances – advances called science and democracy – that are available to anyone, anywhere. (Fang 2016, p. 193)

Xu Liangying (1920–2013) was candid in saying:

> … the most essential requirement [for science] is freedom of speech which, in turn, is only possible under a democratic system in which people can enjoy democratic rights. (Xu 1981/1996, p. 7)

---

[6] See his autobiography (Fang 2016) and his essays on 'Science, Culture, and Democracy in China' (Fang 1992).

Fang and the other liberal scientists were making the same political demand as made by the eighteenth-century European and American Enlightenment thinkers. This was unacceptable to the Communist Party and led immediately to his second expulsion.

Michael Polanyi (1891–1976), the Hungarian-British chemist-philosopher, refugee from Nazism, and defender of Liberalism, wrote of the scientific passion for truth, pointing out:

> Articulate systems which foster and satisfy an intellectual passion can survive only with the support of a society which respects the values affirmed by these passions, and a society has a cultural life only to the extent to which it acknowledges and fulfils the obligation to lend its support to the cultivation of these passions. ... Love of truth and of intellectual values in general will now reappear as the love of the kind of society which fosters these values, and submission to intellectual standards will be seen to imply participation in a society which accepts the cultural obligation to serve these standards. (Polanyi 1958, p. 203)

Lyman Miller's *Science and Dissent in Post-Mao China* (Miller 1996) provides an informative, though disheartening, history of liberal-democratic dissent by scientists in China from the May Fourth Movement of 1919 to the end of the twentieth century. He writes:

> For scientists such as Fang [Lizhi] and Xu [Liangying], the anti-authoritarian norms of science translated easily into a classically liberal politics. The message these scientists carried into the larger political arena defended above all the sanctity and worth of individual autonomy and conscience above the claims of state and society. They advanced a pluralist politics rooted in appeals to reason. They called for all of the freedoms attendant to liberal politics – freedom of speech, assembly, the press, and so forth. Above all, they placed sovereignty squarely among China's citizenry, not in the state itself. (Miller 1996, p. 4)

Philosopher and Nobel Peace Laureate, Liu Xiaobo (1955–2017) was four times jailed and finally in 2008 jailed for 11 years where he died in 2017. His crime was instigating the Charter 08 petition which was initially signed by 313 Chinese intellectuals. Its opening paragraph read:

> This year is the 100th year of China's Constitution, the 60th anniversary of the Universal Declaration of Human Rights, the 30th anniversary of the birth of the Democracy Wall, and the 10th year since China signed the International Covenant on Civil and Political Rights. After experiencing a prolonged period of human rights disasters and a tortuous struggle and resistance, the awakening Chinese citizens are increasingly and more clearly recognizing that freedom, equality, and human rights are universal common values shared by all humankind, and that democracy, a republic, and constitutionalism constitute the basic structural framework of modern governance. A 'modernization' bereft of these universal values and this basic political framework is a disastrous process that deprives humans of their rights, corrodes human nature, and destroys human dignity. Where will China head in the 21st century? Continue a 'modernization' under this kind of authoritarian rule? Or recognize universal values, assimilate into the mainstream civilization, and build a democratic political system? This is a major decision that cannot be avoided.[7]

Since the Enlightenment there has been a science-based, epistemological argument for liberalism: given the goal of science is to find truths about the natural and

---

[7] See https://www.nybooks.com/articles/2009/01/15/chinas-charter-08/

social worlds; given truth-finding requires open and public debate where all views are welcomed independently of race, gender, class, religious, or political affiliation; then science requires an open society with freedoms of association, publication, and speech. This argument was first stated by Locke in the seventeenth century; it was advanced by Priestley in the eighteenth century, by Mill in the nineteenth century, and by Dewey, Popper, Polanyi, and countless others in the twentieth century. All science students, indeed all students, should encounter and come to terms with this argument as it is equally applicable to intellectual advance in history, economics, mathematics, psychology, and every other truth-seeking field, including theology. This argument is, of course, an embarrassment to constructivists and postmodernists who reject the very idea of 'truth finding' as a goal of science.

In the first decades of the twenty-first century during the presidency of Xi Jinping, liberalism has been more tightly squeezed, and what small spaces there were between the party, the government, and civil society have further constricted. More correctly, civil society has all but been extinguished. There are social and economic costs in doing this:

> China's reform-era trajectory is being reversed. ... Beijing's failure to deepen political reform when it had the chance to do so – during the last two decades of the twentieth century and in the first decade of the twenty-first – is now leading the entire system to cannibalize itself and its own prior efforts as political institutionalization. ... Under these pressures, China's tradition of coherent bureaucratic rule ... is beginning to wobble. (Minzer 2018)

The 'wobbles' are depressingly documented (Shirk 2007): two-thirds of Chinese millionaires had, or were planning, to immigrate and have all purchased property abroad where they have also sent their children to school; the gap between haves and have-nots, winners and losers in China's state-capitalist system, is widening, and the gap is more and more obvious as Rolls Royce and Bugatti dealerships keep opening alongside Rolex shops; 100+ of China's 1000 multimillionaires hold high-ranking CCP positions and were in the Party congress appointing and reappointing President Xi; the budget of the Internal Security Department of the CCP rivals that of the army; when an *Economist* or *Time* or any other magazine reports any of the wobbles, sale of the issue is banned and its website closed.

In China, each 200 or so Internet-using households are monitored by a security manager via a 'grid management' system. Each Internet provider employs up to 1000 censors to ensure that nothing on the site violates government dictates; the government itself employs between 20,000 and 50,000 Internet police (*wang jing*); perhaps 300,000 auxiliary personnel (*wumao dang*) are employed at all levels of government to monitor communications and net activity. In one extensively documented study, apart from the to-be-expected censorship of the Tiananmen Square 'incident', the Democracy Wall, the Great Leap Forward, Hong Kong Independence, Fang Lizhi, and the Cultural Revolution, other topics deemed dangerous to state security, meaning CCP control of China, were protests in Inner Mongolia, collective action over lead poisoning in Jiangsu, and students throwing shoes at Fang BinXing (King et al. 2013). These researchers conclude:

> The size and sophistication of the Chinese government's program to selectively censor the expressed views of the Chinese people is unprecedented in recorded world history. (King et al. 2013, p. 1)

In 2013 the government instructed universities, schools, and media that seven topics were off limits and in-class or on-campus discussion of them prohibited. The forbidden topics were constitutional democracy, universal values, judicial independence, economic neoliberalism, free media, disputing the CCP account of party history, and questioning the primacy of socialism with Chinese characteristics. Such prohibitions demean Chinese citizens. They also violate the very idea of a university, much less a modern university. In the nineteenth century, John Henry Newman (1801–1890), a prelate of the Roman Catholic Church, wrote in his 1872 *Historical Sketches* of a university:

> It is the place to which a thousand schools make contributions; in which the intellect may safely range and speculate, sure to find its equal in some antagonist activity, and its judge in the tribunal of truth. It is a place where inquiry is pushed forward, and discoveries verified and perfected, and rashness rendered innocuous, and error exposed, by the collision of mind with mind, and knowledge with knowledge. (Tristram 1952, p. 64)

It verges on the unreal to think that while Tsinghua University in 2018 is the world's leader in the table of most cited maths and computing publications, the government does not allow its students, or its stellar faculty, to discuss, debate, or appraise questions commonly found in first year programmes in universities and even high schools around the world. The same constraint applies at the other great Chinese universities in the C9 group and at all the lesser ones.

In the opening days of 2019, thousands of the highest-quality Chinese physicists, astronomers, mathematicians, and technicians combined to successfully land the Chang'e-4 spacecraft on the far side of the moon, something hitherto never achieved, yet the same scholars are prohibited from openly appraising the contribution of Mao to China or any of whatever number of other topics the CCP deems unpatriotic or disturbing of public order. That the very best and brightest of Chinese school leavers cannot openly discuss the Cultural Revolution, the freedom of religion, independence of the judiciary, treatment of the Uighurs, John Stuart Mill's *On Liberty*, or any other topic they choose is bizarre and an insult to the intelligence of students and staff. A pretext for the suffocation of dissent and debate is invocation of Article 1 of the Chinese Constitution: 'Disruption of the socialist system by any organization or individual is prohibited'. In China, debate and dissent means disruption, punishable by fines, loss of job, and prison. In liberal-democratic countries, constitutionalism means that constitutions can be criticized, revised, and amended; in China to advocate constitutionalism is regarded as advocating Western multiparty democracy. Human rights lawyers by the hundreds, if not thousands, have lost employment for such an interpretation of constitutionalism. In January 2019 the lawyer Wang Quanzhang was sentenced to four and a half years in prison for such 'subversion'. While free discussion and open debate, the defining features of science and a liberal society, are prohibited, many universities are opening Xi Jinping Thought Centres, named for the very person who amended the constitution to allow himself to remain president for life.

Marxism as Official Philosophy

The philosophical/political backdrop to the history of science and dissent in China is the adoption of Marxism as the official philosophy of the People's Republic of China which was established in 1949 after the victory of Mao's Communist Party in the Civil War. Marxism had been part of the New Thought movement from the 1920s but had never dominated politics or Chinese thought (Knight 2005). With the establishment of Communist China, Engels' *Dialectics of Nature* (Engels 1875–82/1934) and Lenin's *Materialism and Empirio-Criticism* (Lenin 1908/1970), as expanded upon by Mao Zedong in various essays, became the philosophical standard against which science, putative sciences, and all philosophical endeavours were judged (Kelly 1985). In this, China followed by 30 years the Soviet Union's experience of Marxism as a state philosophy (L. R. Graham 1973).

The crucial point is that in both countries, the state imposed an official philosophy which enabled orthodox/heterodox, acceptable/unacceptable categorizations of beliefs and the researchers affirming them. With unemployment, prison or execution is being the fate of those in the latter category. There are other examples of the marriage of philosophy to power. None of them is inspiring.

In the Roman Catholic tradition, philosophy was the handmaiden of the Church. In 1879, Pope Leo XIII (1810–1903) promulgated his encyclical *Aeterni Patris* (*On the Restoration of Christian Philosophy*) that gave the name *philosophia perennis* (perennial philosophy) to Thomism and directed Catholic universities, seminaries, and schools to base their philosophical and theological instruction upon it. In 1914 Pius X (1835–1914) issued his *Doctoris Angelici* decree, stating that:

> We desired that all teachers of philosophy and sacred theology should be warned that if they deviate so much as an iota from Aquinas, especially in metaphysics, they exposed themselves to grave risk. (Weisheipl 1968, p. 180)

Thomas Aquinas was elevated to near apostle rank in the Catholic pantheon. It was only in the final years of the twentieth century, with Pope John Paul II's 1998 encyclical, *Fides et ratio*, that the Catholic Church relaxed its attachment to Thomism as official Church philosophy. This was a contra-philosophical policy, notwithstanding that Aquinas was one of history's great philosophers and intellects and that Thomism was a serious philosophical programme advanced by countless great philosophers.

The same dynamics have played out in the Islamic world where the medieval view of philosophy as the servant of the Koran still holds. It is impossible for a Muslim holding any official position to commit to any philosophical system that cannot be reconciled with the assumed ontology, epistemology, politics, and ethics of the Koran.

Outside of China, wherever Marxist parties came to power (Soviet Union, Albania, Bulgaria, Czechoslovakia, East Germany, Cuba), philosophy became the handmaiden of the State; philosophy did not have, and was not allowed, an independent existence. Even if the official philosophy is the best there is, there are good educational, political, and philosophical arguments why it should not be enforced by the State (or Church). Promoted yes, enforced no. This is the fundamental liberal argument advanced by Joseph Priestley in the eighteenth century, John Stuart Mill in the nineteenth century, and Michael Polanyi in the twentieth century. But *Dialectics of Nature*, China's official philosophy, falls a long way short of being the best available philosophy.

The first ever Chinese national science congress was held 30 years after the revolution, in March 1978, and there the Party repudiated and denounced the Gang of Four and its anti-intellectualism. Vice Premier Deng Xiaoping's four modernizations – of agriculture, industry, national defence, and science and technology – were announced as Party policy, and the 6000 attendees were ordered to contribute to the modernization of science.[8] A broader science training was envisaged, but there were limits. Fang Yi, a politburo member, future Vice Premier, and President of the Chinese Academy of Science, in his speech to the congress said:

> We actively advocate the study of Marxist philosophy by scientific and technical workers, and we should encourage and help them to do so. It is necessary to hold different kinds of forums regularly, begin publishing journals on the dialectics of nature, carry out research on the history of natural science, and encourage scientific and technical personnel to guide their scientific research with Marxist philosophical concepts. (Orleans 1980, p. 554)

Fang's programme of the forced marriage of science with Marxist philosophy has continued to the present day, with *Dialectics of Nature* being a compulsory course in all science graduate programmes and 'dialectical materialism' being a mandatory subject in most, if not all, university programmes (Gong 1996a; Guo 2014).

To be expected, there has been significant debate about, and reappraisal of, Marxist philosophy and specifically *Dialectics of Nature*. A good deal of this debate has occurred in the pages of the *Bulletin of Natural Dialectics*, a journal established in 1956 by the physicist and historian Xu Liangying, whose subtitle is 'A Comprehensive Theoretical Journal of the Philosophy, History, and Sociology of Natural Science'. It was, of course, shut down during the Cultural Revolution but resumed publication in 1980 under the editorship of Fan Dainian also a physicist, philosopher, and historian. In 1983, Qian Xuesen, a star CCP scientist, branded the journal a source of 'spiritual pollution' in China and moved unsuccessfully to again have it closed down along with other such sources of pollution – religious publications, liberal periodicals, and so on.[9] Despite, or perhaps because of, these attacks, it has continued to publish and grow in influence.[10] It is little wonder that many Chinese intellectuals lament that a fifth modernization, namely, the modernization

---

[8] Deng Xiaoping and Fang Yi's congress addresses are reproduced in Orleans (1980, pp. 535–556).

[9] In December 2018, 45 Chinese journalists and contributors to the online religious magazine *Bitter Winter* were arrested, imprisoned, and tortured. Such repressive action is routine. Reporters who have shot film of Re-education Camps have simply disappeared, never to be seen again. By signing the September 2018 Accord with China, the Vatican is now complicit in this repression. Cardinal Zen of Hong Kong called it an 'incredible betrayal'. See https://www.theaustralian.com.au/news/world/catholic-church-ignores-chinese-atrocities/news-story/c46a4d9e5bc3282b8fed3a6cb0b6b767

[10] Thirty-four articles published in the *Bulletin* from its refounding in 1980 are republished in Dainian and Cohen (1996).

of culture, was not added to Deng's four modernizations. But this would have deleterious consequences for Communist Party rule: as demonstrated in all the former Soviet Republics and East European nations, one-party rule is inconsistent with the modernization of culture. Modernization is inherently liberal.

Mario Bunge visited China in 2011 to give lectures to the CCP's School of Marxism. He reaffirmed his published position that Marx and Engels 'were serious, important social scientists; they pushed liberalism towards the left … they were materialists on the whole; and they wrote clearly except about dialectics' (Bunge 1994, p. 30). And then proceeded to say:

> However, I have not come to flatter you. Instead, I came to offer some constructive criticisms of Chinese philosophy. Allow me to start by telling you brutally that your philosophy is primitive and unscientific, and that you should update it in the light of science, logic, and mathematics. The reason for carrying out this task is it that it is unreasonable and even politically hazardous to remain stuck in philosophy's past while advancing so quickly in modernizing the rest of society. After all, social policies are designed on the basis of a handful of philosophical principles about the nature of the world, human beings, society, and knowledge, as well as ideas about what is worth and just. If any of these guiding principles are wrong, the social policies will fail or worse: they will bring misery. (Bunge 2011)

The following year, a version of these Beijing lectures appeared as a book chapter 'Marxist Philosophy: Promise and Reality' (Bunge 2012a, chap. 9). There he concluded:

> Given the attachment of Marxist scholars to writings they regarded as infallible and forever topical, it should not be surprising that nearly all the great advances in the natural and social sciences during the last century occurred outside the Marxist box, and that some of them, the most revolutionary, were criticized in the name of Marxism. Thus, when ossified, Marxism became a serious obstacle to the advancement of knowledge. (Bunge 2012a, p. 93)

Feng Shui and the Chinese Communist Party

Given the dominance of the Communist Party in Chinese society, culture, and education, it is warranted to examine separately the Party's position on feng shui theory and practice.[11] In 1940 Mao Tse-tung declared for science and against superstition, including feng shui, saying:

> The culture of this New Democracy is scientific. It opposes all feudal and superstitious thought; it advocates practical realism, objective truth, and the union of theory and practice. From this point of view, the scientific thought of the Chinese proletariat, along with the comparatively progressive material monists and natural scientists of the capitalist class,

---

[11] The Chinese Communist Party does not so much dominate Chinese politics; rather it constitutes Chinese politics and dominates civil society. University student societies cannot form and exist without a CCP-approved patron. In November 2017, Zhang Yunfun was arrested for organizing an illegal Marxist Study Group in a Guangzhou university. He was sentenced to 6 months in jail for 'disturbing public order'. Likewise the CCP insists on bringing Christian churches and Islamic mosques under Party control.

must unite to oppose imperialism, feudalism, and superstition; [they] must not ally themselves with any reactionary idealism. (*Selected Works*, vol. 2, p. 700; in Kwok 1965, p. 19)

When the Chinese Communist Party came to power in 1949, it outlawed feng shui as a form of backward superstition and as being incompatible with Marxist theory and its associated materialist ontology. The Party said feng shui was one of the 'Four Olds' (Shapiro 2001). The clamp down on feng shui was tightened during the disastrous Mao-led and CCP-organized 'Great Leap Forward' (1958–1962) and the following intellectually more disastrous and sadly longer Cultural Revolution (1966–1976).[12] In both destructive episodes, the materialist part of the ruling 'dialectical materialism' ideology was emphasized. Mao's efforts to outlaw and ban the practice, and to systematically refute the associated cosmological beliefs as being incompatible with Marxist dialectics and materialism, were Canute-like in their effect: the feng shui tide kept rising in China.

The Party's eradication of feng shui was always going to be difficult as its sibling qigong was practiced daily by millions of citizens as part of a healthy lifestyle. Belief in the reality and powers of chi (qi) links feng shui and qigong. With the jailing and discrediting of the Gang of Four, both practices flourished. At this time, the distinction between 'internal' and 'external' chi was made. Scientists at the Shanghai Institute of Nuclear Physics published papers in the Chinese *Journal of Nature* (vols. 6, 10, 1979–80) identifying the external chi generated and radiated by qigong masters with microwave energy, electromagnetic waves, and other carriers (Lin et al. 2000, pp. 51–53). This blunted the superstition charge; a materialist basis for the practice was seemingly identified.

Part of the reason for the continuation of feng shui and qigong after the communists came to power was that Mao's embrace of science was the embrace of a stunted science; a purely technological science that ignored science's wider social and philosophical implications, much less embracing its connection with the Enlightenment's critical and 'open society' traditions. In the early years of the Cultural Revolution, Mao's Education Directive of 21 July 1968 laid out his vision for a stunted, verging on brainless, Chinese education, saying:

> It is still necessary to have universities; here I refer mainly to colleges of science and engineering. However, it is essential to shorten the length of schooling, revolutionize education, put proletarian politics in command and take the road of the Shanghai Machine Tools Plant in training technicians from among the workers. Students should be selected from among workers and peasants with practical experience, and they should return to production after a few years' study. (Bloembergen 1980, p. 104)

Because the Party's Directive so closely tied the university curriculum to practical outcomes, there was no course in General Physics; rather physics was divided into subspecialties – lasers, acoustics, magnetism, radio-physics, etc. Once admitted to a specialty in first year, students could only with great difficulty change. During this period, China produced hundreds of thousands, if not millions, of science grad-

---

[12] On the first episode, see at least Bachman (1991). On the second there is a tsunami of biographical, historical, and political publications. See at least MacFarquhar and Schoenhals (2006). Neither episode can be critically discussed in China.

uates who had no inkling of the tradition and culture of modern science, or its philosophy, much less had cultivated a scientific habit of mind, or a social-historical critical facility.[13] Students graduated with what amounted to bachelor's degrees in radios, internal combustion engines, water pumps, and so on. Countless hundreds of thousands of radio graduates knew how to assemble and fix vacuum-valve radios, but after William Shockley's (1910–1989) invention of the transistor and then its commercialization, all of this technical competence was rendered useless, as were the degrees. The radio engineers became shop assistants or taxi drivers. Science graduates were more technicians than scientists. Little in their education would prepare them for intelligent engagement with the fundamentals of their supposed discipline, much less with feng shui theory or party ideology; they knew not much about very little. They were the sad products of a wretched, narrow, and stunting science education.

This 'stunted worldview-free' science is favoured by proponents of 'border-crossing' in multicultural science education. Such proponents maintain that indigenous communities should be taught enough science to cope with modern technology – toasters, refrigerators, and motor cars – but not use the scientific tradition and method to appraise their own culture's worldviews or metaphysical commitments.

The Communist Party, after rejecting Confucianism as one of the 'Four Olds', belatedly recognized, as Ricci had in the sixteenth century, that there were multiple accretions on original Confucian teaching and it endorsed 'restricted' or 'original' Confucianism as official policy because this stressed tradition, filial duty, obligations to the State, social harmony, and good behaviour. Such Party endorsement was a way of showing the Chinese identity of the party.[14] But the Party decried, as Ricci had earlier, 'extended', or 'augmented' Confucianism. It regarded the latter as corrupted.

---

[13] As with other countries, there are blind spots or 'no-go' areas in Chinese history education. Serious appraisal of Mao, and the policies of the CCP during Great Leap Forward and Cultural Revolution, is simply not included in the curriculum. The CCP position is that the 'the Great Helmsman was 70% right and 30% wrong' and no further elaboration or discussion is required. The 30% wrong recognizes the 45 million citizens needlessly killed in the Great Leap Forward and the millions killed in the cultural and educational havoc of the Cultural Revolution.

[14] The Party now exercises its 'soft power' by, among other initiatives, funding hundreds of international Confucian Institutes in schools and universities around the world (16 in Australia in 2018). Some have been closed when the Party's control of staffing and curricula has become too overt. So as not to offend the CCP and thus endanger the flow of government-sponsored students, a number of universities have prevented the screening on campus of the Canadian film critical of the Institutes – *In the Name of Confucius*. In December 2018, the University of Victoria in Australia joined the ranks of many others around the world in preventing screening of the film.

Feng Shui Rehabilitated

Not surprisingly, as communism has ebbed as the nation's dominant ideology, feng shui has risen, a case of 'Good-bye Marx, hello Feng Shui'. The *China Times* newspaper (27 June 2015) reports that a survey conducted a few years earlier by the Chinese Academy of Governance estimated that half of all county-level party officials indulged in feng shui and related practices such as astrology and dream interpretations. The still frowned-on and illegal feng shui consultants helpfully provided Party functionaries with invoices for other services – gardening, cleaning, and accounting – to disguise their recourse to one of the supposedly discredited 'Olds'.

As an example of the new normalcy of feng shui in 'Communist' China, the China Architectural Culture Center held the first summit on architectural feng shui in the Great Hall of the People in September 2004. Chen Kuiyuan, President of Chinese Academy of Social Sciences, sent a message of congratulations to the seminar and offered his strong support – a wonder that Mao's image did not fall off the outside wall or at least begin weeping. In 2005 Central China Normal University offered a 23-day feng shui course that attracted participants from 12 countries (Paton 2007). The same year, Nanjing University offered university-credit feng shui courses. An unnamed feng shui practitioner was quoted by Xinhua news agency as saying at least 70% of modern real-estate projects in Nanjing, the capital of Jiangsu province with a population of eight million, had been evaluated by feng shui masters before construction began. This translates as mega-millions of dollars in construction value and a percentage of the millions in feng shui consultancy fees, all of which are ultimately paid for by the hapless citizens.

On the other hand, there have been calls from some academics to have the Nanjing University programme shut down, as Xinhua reported on its English website.[15] Chen Zhihua, an architect and professor at prestigious Tsinghua University, repeated almost exactly Matteo Ricci's words when he said: 'Feng shui is no science. It only fills the wallets of some charlatans'. Tsinghua University is a world leader in Microelectronics, with hundreds of researchers in the field, using state-of-the-art technology and theory, but none of whom have yet isolated any chi energy flow nor developed any instrument to measure such flow. There are instruments for measuring micro-effects on the surface of Mars, but none for measuring, or even registering, the supposed all-pervasive movement of chi. This says a lot. Understandably Tsinghua faculty are both dumbfounded and offended to see modern universities setting up Feng Shui Chairs and departments.

This debate parallels the comparable debate in India where under the former BJP government (1998–2004), universities were allowed to create, and have funded, Vedic Science and Astrology departments (Nanda 2005, Chap. 3). This was also opposed by Indian astronomers and others for the same reasons as university-endorsed feng shui has been opposed in China: the practice is not science; it is a pseudoscience taking precious resources from genuine science; it is fraudulent and

---

[15] See www.chinaview.cn (17 August 2005).

carried on by charlatans preying on the credulity of large segments of the population.

It is a separate, and political question, as to why the Communist Party is tolerant of feng shui and qigong but so intolerant of falun gong a parallel modern qigong belief and therapy system which is likewise predicated upon chi 'science' although lacking the astrological and divination dimensions of the former (Palmer 2007). In 2000 the CCP established the Chinese Health Qigong Association to regulate qigong practice and to stop qigong enthusiasms metamorphizing into oppositional spiritualist or political movements (Karchmer 2002).

Falun gong was articulated in the 1970s by Li Hongzhi and is aimed at using slow movement and stretching to 'open up' internal meridians to facilitate the flow of good chi and to allow the discharge of bad or 'black' chi (Li 1999). Li's writings have been translated into 40 languages; he was feted by different Chinese embassies; there were 20+ million practitioners in China and many millions throughout the world (Ownby 2008; Penny 2012).

But in July 1999, falun gong became Public Enemy #1. On the orders of General Secretary Jian Zemin, the Communist Party set up the 610 Office to oversee the elimination of falun gong in China. The Office arranged the execution of numerous falun gong practitioners; in 2009 it was reported that 2000 incarcerated prisoners had been tortured to death; thousands were imprisoned in 'reform through labour' camps; thousands of public servants were dismissed from jobs on account of practicing falun gong; an even greater number of citizens were forced to flee the country (Lemish 2008). To merely report any of these meant dismissal from a government job, a fine, or a prison term. Outside China, public supporters of falun gong are placed on a 'no entry' visa list. When in 2008 the University of Technology in Sydney placed a falun gong artwork on its website, the CPA closed the university's website in China. Efforts to have falun gong categorized as one of the State-recognized religions and, consequently, tolerated and have failed.

Educational Responses

An engaging question for students, especially where there is coordination between science, social science, and history classes, is what is so wrong with falun gong? Given that China has formally embraced feng shui's chi-based architecture and chi-based Traditional Chinese Medicine (TCM) with its *qigong* and acupuncture, the once given 'false metaphysics' grounds for banning falun gong can no longer be advanced.

These 'public sphere' debates can be an occasion for science teachers to encourage discussion and examination of the basic issues in philosophy of science that the debates are predicated upon: What is science? What is scientific method? What counts as evidence for a theory? What is the difference between proactively and reactively seeking evidence for a theory? What is the legitimate and the illegitimate role of metaphysics in science? Can pseudoscience be identified and demarcated from science?

These well-publicised debates, and court cases where feng shui practitioners are fined and jailed for providing 'poor' advice, provide questions that can easily and legitimately be taken up by science teachers: What constitutes poor feng shui advice? Can there be good advice yet poor outcomes? Is feng shui (or astrology) scientific? Is it just poor science? Is it pseudoscience? Students can reasonably expect their teachers to have an informed opinion on the questions. And such opinions are strengthened to the degree that teachers have familiarity with the history and philosophy of science.

The same questions and same conclusion about teachers' need for HPS were occasioned by the recent US court cases over the teaching of Creation Science in schools: first, the 1981 Arkansas trial over the state's Act 590 which required Creation Science to be taught alongside Evolutionary Science (Ruse 1988) and, second, the 2004 Dover County, Pennsylvania, requirement that Intelligent Design be taught alongside Darwinian evolutionary theory (Pigliucci 2010, chap. 7; Slack 2007). If something related to science is in the newspaper or on television, then it can legitimately be discussed in science classes; in many cases, not do so is educationally irresponsible. Curricula should not be determined by the newspaper front page, but teaching programmes can be adjusted to selected front pages. This is the 'public interest' basis that is appealed to by proponents of the socio-scientific issues (SSI) approach to science education (Zeidler and Sadler 2008). In Asia, the public interest argument lends strong support to the classroom examination of feng shui.

Conclusion

For at least the past 150 years, modern science has been taught in China, and the country leads the world in many scientific and technical fields. As with everywhere else, the population's engagement with science varies along a spectrum from not very interested to very interested and engaged. At the top end, the scientific knowledge can nevertheless be 'superficial', merely a knowledge of scientific facts, laws, technique, and associated technology. The kind of knowledge examined in PISA tests. Acquiring such knowledge is important and not easy. Yet the engagement in China and all countries can be more serious involving learning something about the methodology, history, and philosophy of science (HPS) and so pursuing to an appropriate degree questions such as:

- What is the method of science?
- What is the legitimate domain of scientific method?
- What is scientism, and should it be embraced or avoided?
- How are scientific hypotheses and theories tested?
- How are competing theoretical traditions appraised?
- What are the hallmarks of good experiments?

- What is the legitimate role of values and politics in scientific decision-making?
- What, if any, ontology and metaphysics is presupposed by science?
- What, if anything, distinguishes science from pseudoscience?
- How can productive science be conducted independently of political, social, and cultural configurations?
- What societal structures are required for science to flourish?
- Can a scientific habit of mind be fostered in a closed, undemocratic, illiberal society?
- Are the claims (truths) of science universal or are they local and culturally bound?
- Are there other equally good or legitimate other ways of knowing about the world and its mechanisms?
- Why was China's embrace of modern science so delayed?

All school science programmes provide opportunities for naturally asking these questions. The educational task is to identify when, where, and why some of the above listed HPS questions might be introduced in school programmes and how students might best be encouraged to engage with them.

Part IV
Feng Shui: Considerations from Philosophy of Science

Chapter 10
Joseph Needham on Feng Shui and Traditional Chinese Science

Having discussed the historical narrative and associated philosophical questions that arise from Ricci's sixteenth-century (Chapter 6), Eitel's nineteenth-century (Chapter 7), and Chinese early- (Chapter 8) and late twentieth-century (Chapter 9) accounts of feng shui and Chinese science, it is instructive to turn to Joseph Needham's account of traditional Chinese science and culture. Needham was the greatest and most influential twentieth-century student of premodern Chinese science. He exhaustively documented the interplay of science, technology, philosophy, metaphysics, and Chinese culture and alerts the reader that:

> Superstitious practices flourished in China just as strongly as in all other ancient cultures. Divination of the future, astrology, geomancy, physiognomy, the choice of lucky and unlucky days, and the lore of spirits and demons, were part of the common background of all Chinese thinkers, both ancient and medieval. (Needham and Ling 1956, p. 346)

By 'all thinkers' he means all thinkers – scholars, natural philosophers, and court advisors. This is just how it was in Europe. And Needham goes on:

> The historian of science cannot simply dismiss these theories and practices, for they throw much light on ancient conceptions of the universe. (Ibid)

The Needham Question

Joseph Needham (1900–1995) was a Cambridge biochemist, a 'heterodox' Marxist, a Whiteheadian process philosopher, and an historian of science. He devoted the bulk of his life to researching the development of science, technology, and natural philosophy in China where, during the war, he had travelled for two years.[1] His science, his Marxism, and his philosophy all influenced, some might say 'coloured', his account of traditional Chinese science (Blue 1998; Nakayama 1973).

On 15 May 1947 after his return to England from two momentous years of extensive travel, interviews, and document acquisition in China (1943–1945), he wrote a letter to Cambridge University Press proposing a book on *Science and Civilisation in China*. The letter asked:

> Why did their science always remain empirical and restricted to theories of primitive or medieval type? What were the inhibiting factors which prevented the rise of modern science in Asia? (Winchester 2008, p. 171)

This is the first statement of what was later to become known as 'the Needham Question'.[2] The University Press immediately accepted the proposal; it would become a hugely influential multivolume study (Needham and others 1954–2004), with Volume 2 devoted to Chinese science (Needham and Ling 1956).[3]

Needham, and most historians and philosophers of his time, believed that there had been a scientific revolution in Europe in the sixteenth and seventeenth centuries; something dramatic occurred that began with Galileo and culminated with Newton; a new mode of science, the Galilean-Newtonian Paradigm (GNP) was inaugurated. Putting astronomy aside, above all, what characterized the GNP, and made it revolutionary in investigating terrestrial events and processes, was its commitment to experiment and the mathematization of qualities and claims. This was the default scholarly position and was forcefully stated by the French historian and philosopher, Alexandre Koyré (1892–1964):

> … what the founders of modern science, among them Galileo, had to do, was not to criticize and to combat certain faulty theories, and to correct or to replace them by better ones. They had to do something quite different. They had to destroy one world and to replace it by another. They had to reshape the framework of our intellect itself, to restate and to reform its concepts, to evolve a new approach to Being, a new concept of knowledge, a new concept of science – and even to replace a pretty natural approach, that of common sense, by another which is not natural at all. (Koyré 1943/1968, pp. 20–21)

Alfred Rupert Hall (1920–2009), in a multi-edition widely read work, reinforced this as the default position for Anglo-American scholars and students (Hall 1983). Richard Westfall (1924–1996), the American historian and biographer of Newton, defended the 'revolutionary' position against later 'levellers' or 'continuists'

---

[1] Winchester (2008) provides an account of Needham's life and work and the crucial contributions of his assistant Lu Gwei-djen, a Chinese biochemist and historian. A 'condensed' Needham is available in Ronan (1978).

[2] Needham's arguments and resulting literature are canvassed by H. Floris Cohen (1994, pp. 439–488). Among many appraisals of Needham's work, see contributions to Nakayama and Sivin (1973) and Jin et al. (1996).

[3] Needham studied premodern Chinese science and technology; his research did not extend much beyond the 1644 end of the Ming dynasty. His argument is presented in single-book length in Needham (1969) and in chapter length in Needham (1963, 1964).

(Westfall 2000). He regarded experiment as one of the novel and defining elements in the New Science:

> Method was also an aspect of the Scientific Revolution, which increasingly built itself on experimental procedure. On this subject we must not overstate discontinuity, for experiment as such was not new with the seventeenth century. We do have to look carefully, however, to find experiments before the seventeenth century. Experiment had not yest been considered the distinctive procedure of natural philosophy; by the end of the century, it was so recognised. (Westfall 2000, pp. 48–49)

Joseph Needham held the same view:

> Controlled experimentation is surely the greatest methodological discovery of the scientific revolution of the Renaissance, and it has never been convincingly shown that any earlier group of Westerners fully understood it. I do not propose to claim this honour for the medieval Chinese either, but they came just as near it theoretically, and in practice often went beyond European achievements. (Needham 1963, pp. 147–48)

In his *Science and Civilisation*, he maintained that Chinese natural philosophy lacked entirely a tradition of formulating theories verifiable by experiment. There was no attempt to formulate 'mature hypotheses couched in mathematical terms and experimentally verifiable' (Needham and Ling 1956, p. 346). And so traditional Chinese thinkers were never able to pass beyond the relatively primitive and unquantifiable theories of the five elements and the two yin-yang forces.

Given this, it was then natural to ask: Why did modern science appear in Western Europe in the seventeenth century and not earlier and not elsewhere? And equally naturally for historians, the Needham Question followed.

Historians of science can be grouped into two 'camps' in answering the 'Why the Scientific Revolution?' question. *Internalist* historians of science have concentrated upon the development of scientific concepts, understandings, and methodologies wholly from within the world of scientific ideas including philosophical and methodological ones and largely ignoring the social and cultural milieu in which the ideas develop. The internalists assuredly deny any causative role to social factors on the development of scientific ideas – Newton's inverse square law would not have been an inverse cube law had he lived in a different society. Representative internalist studies are Alastair Crombie (1952, 1959/1990), Rupert Hall (1970, 1983), Steven Weinberg (2015), and David Wootton (2015). *Externalist* historians of science endeavour to causally connect the growth of science to its social circumstances. These can be either the immediate personal and work circumstances of scientists – proximal externalists and sociologists of science such as Steven Shapin (1982) – or the more general economic and political circumstances of the scientists' times, distal externalists such as John Desmond Bernal (1939), and historians in the Marxist tradition, such as Boris Hessen (1931).

The same pattern of internalist/externalist approaches can be found in answers to the Needham Question. A compromise is to see internal factors as necessary, but not sufficient to explain scientific revolutions, or lack thereof; ideas are needed but so also is freedom, incentive, and technology.

There are, of course, historians who reject the very question because they deny that there ever was a Scientific Revolution in Europe; it was 'Fake News', in current political terminology. Preeminent among these is the French historian-philosopher Pierre Duhem (1861–1916), of whom another historian-philosopher and Catholic priest wrote:

> Singlehanded he destroyed the legend of the 'scientific night of the Middle Ages'. Before him, the phrase was a hallowed shibboleth of a self-styled Enlightenment. After him it has become the sign of an inexcusable ignorance which unfortunately lingers on. (Jaki 1969a, p. xvii)

Others who have argued for a non-revolutionary, not-so-new continuity interpretation of the 'New Science' of Galileo and Newton include William Wallace (1981) and James Weisheipl (1985), two modern Thomists who also were Catholic priests.

Ernest Moody's 'compromise' position is sensible. In numerous studies of medieval science (Moody 1975), he finds elements of the new Galilean-Newtonian Paradigm but maintains that the elements were not put together prior to the Herculean intellectual achievements of both scientists.

> [Galileo] did not create his mechanics (nor that of Newton as some seem to suppose) out of thin air, and to this extent Duhem is certainly right. But he conceived the *kind* of science that became classical mechanics, using the materials available to him and building his new science with the foundations in the right place. This his medieval predecessors did not do, and did not even try to do. In this achievement Galileo had no precursors among his medieval predecessors. (Moody 1975, p. 408)

It should be no surprise that scholars take comparable positions on the Needham Question as can be found on the Scientific Revolution Question.

Chinese Technology

China certainly had technology and had it in abundance. China is rightly lauded for the discovery of gunpowder (twelfth century AD), book printing (seventh century AD), paper-making (300 BC), and the navigation compass (early twelfth century AD) – the big four. And these were just the tip of an iceberg of technical and social accomplishments laid bare by Joseph Needham's labours (Needham and others 1954–2004). To these can be added countless other Chinese 'firsts': the making of exquisite porcelain (third century BC), creating a seismograph (132 AD), designing and constructing the Great Wall of China (third century BC), laying out and creating the 1800 km Grand Canal linking the Yellow and Yangtze rivers (fifth century BC to 600 AD), making and using rockets (1360 AD), ball bearings (second century BC), cast and wrought iron (fourth century BC), iron-chain bridges (sixth century AD), and much else.[4] China had industry in the medieval period. One estimate is that

---

[4] Simon Winchester provides a list of 300 such 'firsts' documented by Needham (Winchester 2008, pp. 267–277).

150,000 tons of iron was produced by the late eleventh century. In the thirteenth century, China had water-powered reeling machines for the spinning of hemp thread, this being 400 years before such machines appeared in Europe. In comparison, at the time, the West was agrarian, poor, and underdeveloped (Lin 1995, p. 270).

Marco Polo's fourteenth century *Travels* paints envious pictures of sophisticated Chinese cities. Of Khan-balik (King's city), he writes:

> … it is built in the form of a square with all its sides of equal length and a total circumference of twenty-four miles …I assure you that the streets are so broad and straight that from the top of the wall above one gate you can see along the whole length of the road to the gate opposite. The city is full of fine mansions, inns, and dwelling houses. …All the building sites throughout the city are square and measured by the rule. … Every site or block is surrounded by good public roads; and in this way the whole interior of the city is laid out in squares like a chess-board with such masterly precision that no description can do justice to it. … There is a suburb outside every gate, such that each one touches the neighbouring suburbs on each side. … And in each suburb or ward, at about a mile's distance from the city, there are many fine hostels which provide lodgings for merchants coming from different parts. (Polo 1958, pp. 128–29)

Needham correctly identifies all of these laudable achievements as *technological*, not *scientific*. For Needham there was no question that from the immediate BC centuries through to the fifteenth century, China far surpassed Europe in technological innovation and achievement: 'Chinese civilisation was much *more* efficient than occidental in applying human natural knowledge to practical human needs' (Needham 1969, p. 190). But the enormous array of Chinese discoveries and inventions were disconnected, seldom refined, and little connected to the development of science. Crucially, there was no independent 'research' culture or infrastructure to coordinate, disseminate and exploit the technology. The Confucian-infused imperial system that brought envied uniformity and stability to China over millennia also brought conformity, central control, and fear of free-thinking and intellectual autonomy. That is, fear of the very things that enable science to flourish in Europe.[5] Many scholars have observed that since the establishment of imperial rule under the Ch'in dynasty (221–207 B.C.E.), China has remained a civilization specifically founded upon and sustained by policies aimed at eliminating any diversity of creative and critical thought within the individual and throughout the empire.[6]

Consider the great contrast between China and Europe in the utilization of new technology such as thermometers, barometers, time-keepers, vacuum pumps, steam engines, movable type, and the like. In China, compasses were made, but voyages of discovery did not take place in anything like the European scale; rockets were fired, but there was no concerted effort to describe their trajectory; books were printed, but there was no upwelling of communication, debate and argument as early books occasioned in Europe. China lacked the universities that began to appear in Europe from the thirteenth century.

---

[5] On these matters see Bodde (1991) and Huff (1993, Chap. 7).

[6] In 2016 Fang Lizhi, the dissident Chinese astrophysicist, lamented this characteristic of Chinese culture: 'We move only from "Confucius says …" to "Chairman Mao teaches us …"' (Fang 2016, p. 15).

The obvious contrast between the abundance of sophisticated technology in China, yet the absence of modern science, does make it clear that science does not grow out of technology. The widespread view that common sense plus experience plus technology give rise to modern science cannot be sustained. Since ever, sophisticated technologists, miners, and irrigators realized that a water pump could not draw water more than about 30 ft. For everyone this was a technical problem to be addressed by improved technology (which did not work) and then by more of the same (staging the pumps at different levels). Galileo saw it as a theoretical, that is, scientific problem: why just 30 ft? Aristotle's 'nature abhors a vacuum' theory could not explain the phenomena. Galileo was led to a whole new scientific theory: air has weight, the idea of atmospheric pressure, and the barometer as an instrument to measure these suppositions.[7]

It is useful in considering the long history of technological innovation in China, and elsewhere to distinguish *technics* from *technology*. The former term picks out, or designates, trial-and-error artisan creation that is not driven by, or dependent upon, scientific theorizing, even of an immature or protoscientific variety. As Bunge remarks:

> Most of the inventions proposed until the beginning of the Modern period [17th century] owe hardly anything to science: recall the domestication of most plants, animals, fungi, and bacteria; the plow and metallurgy, architecture and coastal navigation. (Bunge 1988/2001, p. 347)

The pendulum clock, the Watt governor, the barometer, and so on are all technological inventions, not technical ones, because they all followed upon moderately developed theory, and they were utilized in order to refine the theory. It can be illuminating to apply the technics/technology categorization to the cornucopia of Chinese inventions so well documented by Needham and his associates.

Internal and External Impediments to Modern Science in China

Needham documented at great length the *internal* philosophical, intellectual, and cultural factors inhibiting the appearance and growth of Western-style science in China. He picked out four for special mention. First, the narcissistic preoccupation of Confucian thought. For him, Confucians had no 'curiosity about Nature outside and surrounding man' (Needham and Ling 1956, p. 544). Second, the Chinese lacked altogether the Judaeo-Christian-Islamic notion of a celestial and supreme rational creator deity that had 'laid down a series of laws which were to be obeyed by minerals, crystals, plants, animals and the stars in their courses' (Needham and Ling 1956, p. 518). Needham follows this with verses from an eighteenth-century Newton-inspired Anglican hymn:

---

[7] The example is elaborated in Grove (1989, p. 37).

> Praise the Lord, for he hath spoken,
> Worlds his mighty voice obeyed;
> Laws, which never shall be broken,
> For their guidance he hath made. (Needham and Ling 1956, p. 518)

It was this conviction that there was a divine law-giver that animated *all* early modern European natural philosophers; to chart God's handiwork was to do him honour. This motivation was not there for Chinese natural philosophers; natural laws were not looked for. Third, the idealism and lack of systematicity of Chinese metaphysics thwarted scientific development. Fourth, the lack of a serious, science-related, experimental tradition.

Notwithstanding his elaboration of the internal, or intellectual, barriers to the development of modern science in China, his Marxist convictions led him to identify *external* social factors as being largely responsible for science's 'failure to thrive' (Needham 1963). In his volume on physics in traditional China, he documents the studies of magnetism and electrical phenomena, saying that they were far ahead of European studies at the time, and comments that:

> … if the social conditions had been favourable for the development of modern science, the Chinese might have pushed ahead first in the study of magnetism and electricity, passing to field physics without going through the stage of 'billiard-ball' physics. Had the Renaissance been Chinese and not European, the whole sequence of discoveries would probably have been different. (Needham and Ling 1962, p. 1)

In a 1946 essay (reproduced in *The Great Titration*), written while he was still in China, Needham writes:

> So we come to the fundamental question, why did modern science not arise in China? The key probably lies in the four factors: geographical, hydrological, social and economic. All explanations in terms of the dominance of Confucian philosophy, for instance, may be ruled out at the start, for they only invite the further question, why was Chinese civilization such that Confucian philosophy did dominate. (Needham 1969, p. 150)

These social conditions have been long identified: a stratified society where there was a clear separation of manual work and technological experience, from intellectual to philosophical speculation; the imperial anointing, for over 2000 years, of Confucianism as the state religion or official ideology[8]; the primacy given in the Confucian tradition to 'inner' knowledge, ethics, and personal well-being as part of a family and a stable hierarchical community; its denigration of manual and technical work; the rigid court control of higher learning and its dissemination; and other factors.

---

[8] China's formal embrace of Confucianism began with the Han emperor Wu Di's decree of 134 BC making it the official ideology of the Han dynasty. It remained the state ideology (religion, worldview) till the 1911 Revolution. Neo-Confucianism arose in the Song dynasty (960–1279) and became central to Chinese and much of East Asian culture and philosophy through the following eight centuries, during which time it absorbed elements of Buddhism and Daoism (Angle and Tiwald 2017; Berthrong 1998).

Three hundred years ago, the Scottish philosopher David Hume (1711–1776) in his essay 'On the Rise and Progress of the Arts and Sciences' anticipated the Needham Question and pointed to Chinese politics as its answer:

> In China, there seems to be a pretty considerable stock of politeness and science, which, in the course of so many centuries, might naturally be expected to ripen into something more perfect and finished, than what has yet arisen from them. But China is a vast empire, speaking one language, governed by one law, sympathizing in the same manners. The authority of any teacher, such as Confucius, was propagated easily from one corner of the empire to the other. None had courage to resist the torrent of popular opinion. And posterity was not bold enough to dispute what had been universally received by their ancestors. This seems to be one natural reason, why the sciences have made so slow a progress in that mighty empire (Hume 1826, p. 136).

Hume proceeded to opine that 'the only proper nursery of these noble plants [science] is a free state' (ibid). This is, of course, the opinion held in modern times by committed, liberal scientists in all authoritarian states, including China (see Chap. 9).

Most recognize that the internal/external distinction is fuzzy and hard to draw even for Western science – is the availability and standard of mathematics an internal or external factor? Is metaphysics and worldview internal or external? It is a much harder distinction to make for premodern Chinese 'science' where natural philosophy, cosmology, technics, and cultural tradition are intertwined. Nevertheless, some more clearly external factors can be identified – for instance, the stultifying, conservative verging on reactionary impact of the imperial examination system.

Derk Bodde (1909–2003), an American Sinologist, in his extensive study of 'The Intellectual and Social Background of Science and Technology in Pre-modern China' ranges over many internal and external factors impinging on the development or otherwise of Chinese science, leading him to the conclusion that:

> Chinese pre-modern science not only did not, but could not, evolve by itself into anything comparable to what is today called modern science. (Bodde 1991, p. 368).

That transformation required 'the impact of powerful outside stimuli' (ibid).

Bodde had lived in China, first in 1919, then 1931–1937, and 1947–1948. He was a linguist and expert on Chinese language and script. He worked for some years with Needham on a projected joint Volume 7 of the *Science and Civilisation* book series, but in the end, he withdrew as there was an irreducible gulf between both scholars over the influence of Chinese language on the 'retardation' of Chinese science (Bodde 1991, sect. II). Bodde, in a summary statement of his thesis, identifies the core language problems as the lack of punctuation in premodern script, the lack of capitalisation which makes difficult the identification of proper names, the lack of a system of alphabetisation, the lack of continuous pagination, and the inherent ambiguity built into a language having a basic 7000 ideogram characters with 100,000 variants (Bodde 1991, pp. 88–92).[9] These features of Chinese language, which remained unchanged till the twentieth century, made 'innovative' hypothesizing and communication about phenomena difficult. Needham was alert to the

---

[9] Bodde's account of Chinese language is detailed in Huff (1993, pp. 290–96).

impact of Chinese language on Chinese natural philosophy, but understood it in a much more positive light saying:

> the structure of the Chinese language itself encouraged these ancient thinkers to develop an approach, not only to the type of thing usually called Hegelian, or approximating to that of Whitehead, but even more fundamentally and exactly, to what is now being investigated under the head of combinatory logic. (Needham and Ling 1956, p. 77)

In his discussion of the medieval five-element theorists, he comments: 'Notable here once again is the inability of these thinkers to coin new technical terms' (Needham and Ling 1956, p. 260).

The normal difficulties involved with translation into and out of any language became many times more difficult and problematic for premodern Chinese. So apart from a general cultural reluctance to recognize worth in outside 'barbarian' civilisations, there was simply a great difficulty in the competent and accurate translation of outside script, be it Indian, Arabic, Latin, or any European language into Chinese text. Thus, the oft-remarked fact that Indian mathematics texts, complete with the zero numeral, were in circulation but not translated, Euclid was in circulation but not translated, Ptolemy and Copernicus were in circulation but not translated – all support both the difficulty of translation claim and the cultural 'self-satisfaction' thesis.

In support of Bodde's identification of language as a restrictive factor, recall that in Chap. 5 attention was drawn to scholarly disagreement about the proper translation of the fourteenth-century astronomy book *Ge xiang xin shu* by the Daoist priest/astronomer Zhao Youqin. Needham renders it *New Elucidation of the Heavenly Bodies*, while Alexeï Volkov says that this omits the title's all-important allusion to the *I Ching*, with *Ge* denoting the *I Ching* hexagram #49 'Alteration'. So, for him, the title should be *New Writing on the Image of the Alteration* (Volkov 1996a, p. 39). There is variation about the English rendering of hexagram #49, with 'skinning', 'revolution', 'moulting', 'transformation', and 'reform' - all finding champions. In turn there are hundreds, if not thousands, of websites that interpret the particular *gé* hexagram. For instance:

> The elements of this hexagram are fire under water. Fire evaporates water, and water puts out fire. Similarly, change often causes conflict, and conflict brings about change. This hexagram refers to a time in the cycle of human affairs when things are stirring up, and when the hint of dramatic change is in the air. (https://divination.com/iching/lookup/49-2/)

Clearly achieving a 'correct' translation is going to be difficult; maybe just 'adequate' translation suffices. But for scientific understanding and for the contemporary understanding of older or other sciences, exactness is of the essence. There should be differences in the translation into any language of the scalar 'speed' and the vector 'velocity', so too with translation of older Chinese natural philosophy.

Arai Shinji discusses another translation dispute over a sentence in the same book that talks of Zhao cosmology. The sentence reads *Tian biao li you shui* (of course this is already a translation of the original Chinese characters) and Needham renders it 'Inside the lower part of the heavens there is water' (Needham and Ling 1959, p. 217). Shinji cannot accept this translation because it 'makes no reference to the existence of water outside the heavens' (Shinji 1996, p. 62). He refers to half-a-dozen other contributors to the debate over the correct rendering of the sentence, and whether the water referred to is rainwater, river water, ocean water, and so on.

These translation disagreements, among the very best scholars, about one fourteenth-century book simply underline the point argued by Bodde: translations between any languages are fraught and difficult; but it is in the nature of the script, that Chinese translations into and out of other languages are especially fraught and contentious. This constrains the introduction of outside ideas and the formulation of novel indigenous ideas in natural philosophy.

Others pointed to different 'external' factors that inhibited the growth of natural philosophy in traditional China. Toby Huff identifies the overarching, centralized imperial system with its daunting and conformity-inducing national examination system, as having significant responsibility for the thwarting of Chinese science (Huff 1993, Chap. 7). Whereas, by the fourteenth century, Europe had about 100 fully functioning universities and hundreds of colleges where disputation and investigation were encouraged and indeed written into the curriculum, China had no such independent institutions. Huff writes:

> ... the rigidity of the educational content of the examination system – virtually unchanged from early Ming (ca. 1368) to the twentieth century – and its absolute uniformity, bordering on political indoctrination, make it a colossal failure insofar as science, innovation, and creativity are concerned. ... From an early age, young boys were taught to memorize the Confucian classics (initially without even knowing the meaning of what they were memorizing). (Huff 1993, pp. 277, 279)

The same observation was made as early as 1735 by the Jesuit Jean-Baptiste du Halde in his *General History of China* (from which Fig. 6.2 has been taken):

> The chief and only way that leads to Riches, Honours, and Offices, is the Study of the Canonical Books, History, the laws and Morality, it is to learn to write in a polite manner, in Terms suitable to the Subject treated upon; by this means the Degree of Doctor is obtained, and when that is over they are possessed of such Honour and Credit that the Conveniences of Life follow soon after, because then they are sure to have a Government [Post] in a short time; even those who wait for this Post, when they return into their Provinces, are greatly respected by the mandarin of the Place, their Family is protected from vexatious Molestations, and they there enjoy a great many privileges.

> But as there is nothing like this to hope for by those who apply themselves to the speculative Sciences, and as the Study of them is not the Road to Affluence and Honours, it is no wonder that these sort of abstracted Sciences should be neglected by the Chinese. (in Bodde 1991, p. 366)

Needham did not disagree with this assessment, saying in a 1953 essay that the examination system was 'entirely based on literary and cultural subjects, and did not include subjects that could, in any sense, be called scientific' (Needham 1969, p. 179). The examination system was inherently conservative: the syllabus was classic Confucian texts, for which memory and an unquestioning and uncritical disposition was the chief, and only, requirement for success. Children began memorizing the classics at 5 years of age. Several million presented themselves for the first, provincial-level, exam. In 1850, the pass rate was 0.05% (Zarrow 2005, p. 20). These imperial exams persisted through to the brief reforms in the Ch'ing (Qing) dynasty of 1905 when they were abolished. Not everyone cheered. An ambitious village school teacher wrote: 'I woke at first light with my heart like dead ashes. I saw that all was vanity and there was nothing eternal … no one knows what will become of customs and morals' (Zarrow 2005, p. 29). This was a real concern. For centuries the exam system had unified belief and ideology across the vast geographical and ethnic expanse of the Middle Kingdom; post-1905, there was no such unifying vehicle.

Sometimes the Needham Question is posed in a manner that invites an externalist answer. Consider, for instance, the opening sentence of Justin Yifu Lin's paper on the subject:

> One of the most intriguing issues for students of Chinese history and comparative economic history is: Why did the Industrial Revolution not occur in China in the fourteenth century? At that time, almost every element that economists and historians usually considered to be a major contributing factor to the Industrial Revolution in late eighteenth-century England also existed in China. (Lin 1995, p. 269)

Unless readers are attentive, the slide from Needham's 'Scientific Revolution' to Lin's 'Industrial Revolution' is overlooked. These are two separate questions with two separate answers. There are varieties of externalist answers to the second, and historians of science and economic historians debate the merits of the proffered answers.

Neo-Confucianism, the Organicism Worldview, and Chinese Science

Needham identifies an 'organicism' worldview in traditional China that he sees as comparable with the worldview of twentieth-century science as elaborated by his Cambridge colleague Alfred North Whitehead (Needham 1925). Needham follows Eitel in saying that it was not Confucius but the rise, 1000 years later, of neo-Confucian thought in the late Song Dynasty (960–1280) that gave a 'modern' naturalistic, scientific direction to Chinese thinking about nature and led to studies in pharmacy, geography, magnetism, and mathematics.[10] China had the intellectual groundwork, opportunity, and technologies for modern science, but it did not mature (Needham and Ling 1956, pp. 293–296). The Daoists got close (Harper 2017). Needham affirms that the term 'Tao' meant the 'Order of Nature' (Needham and Ling 1956, p. 45), and further:

---

[10] Stephen Angle and Justin Tiwald have published a comprehensive book on neo-Confucian philosophy (Angle and Tiwald 2017), and John Makeham has edited a large anthology on the same subject (Makeham 2010). It is noteworthy that 'experiment' does not occur in the index of either book. Given that experiment is the defining feature of post-Galilean modern terrestrial science, it says something about the neo-Confucian tradition that it ignores experiment.

> If there was one idea which the Taoist philosophers stressed more than any other it was the unity of Nature ... which is the basic assumption of natural science ... [for Taoists] nothing is outside the domain of scientific inquiry, no matter how repulsive, disagreeable or apparently trivial it may be ...Thus the anatomy of an ox and the skill of an anatomist are no less part of the Order of Nature than the movements of the Stars. (Needham and Ling 1956, pp. 46–47)

This was in contrast to Confucians who had no interest 'in seemingly worthless minerals, wild plants and animal and human parts and products' (Needham and Ling 1956, p. 47). Further, the independence of nature meant that human ethical judgement was taken out of Daoist natural philosophy; all natural processes were equally good, and this independently of their relation to humans; a flowing river bringing fish and water for crops and a flooding river bringing death and destruction were equally 'good'; they just were, independently of human interest. Needham surmises:

> It will be remembered that the Taoist thinkers, profound and inspired though they were, failed, perhaps because of their intense mistrust of the powers of reason and logic, to develop anything resembling the idea of laws of Nature. With their appreciation of relativism and the subtlety and immensity of the universe, they were groping after an Einsteinian world-picture, without having laid the foundations for a Newtonian one. By that path science could not develop. (Needham and Ling 1956, p. 543)

The Confucian tradition gave birth to a neo-Confucian stream, which settled in for the duration of modern times. For this group:

> The Neo Confucians arrived at what was essentially an organic view of the universe. Composed of matter-energy (*Ch'i*) and ordered by the universal principle of organization (*Li*), it was a universe which, though neither created nor governed by any personal deity, was entirely real, and possessed the property of manifesting the highest human values (love, righteousness, sacrifice, etc.) when beings of an integrative level sufficiently high to allow of their appearance, had come into existence. (Needham and Ling 1956, p. 412)

Against the atomistic, colliding billiard-ball, mechanical worldview that in the West came to replace Aristotelian teleological self-actualising natures:

> The Chinese world-view depended upon a totally different line of thought. The harmonious cooperation of all beings arose, not from the orders of a superior authority external to themselves, but from the fact that they were all parts in a hierarchy of wholes forming a cosmic pattern, and what they obeyed were internal dictates of their own natures. Modern science and the philosophy of organism, with its integrative levels, have come back to this wisdom, fortified by new understanding of cosmic, biological and social evolution. (Needham and Ling 1956, p. 582)

Needham then raises the tantalizing counter-factual question: 'Yet who shall say that the Newtonian phase was not an essential one?' (Needham and Ling 1956, p. 582). And earlier he had affirmed that 'Greek atomism and mathematics are doubtless rightly regarded as the foundations of the Cartesian-Newtonian science of the European 17th century' (Needham and Ling 1956, p. 339). And Daoists, as cited above, 'were groping after an Einsteinian world-picture, without having laid the foundations for a Newtonian one'.

Needham's 'process' rendering of neo-Confucian philosophy was prepared by his Cambridge undergraduate absorption in the process philosophy of Whitehead and Bernal.[11] One commentator observed:

> All these studies in philosophy proved invaluable to him later on when he came to investigate mediaeval Chinese philosophy and found that so much of it could only be understood in the light of the organicism which Europeans had had to discover for themselves. Indeed the perennial philosophy of China, and the Neo-Confucianism of the +12th century, can only be interpreted in this way; it was a remarkably prophetic and prescient school, which built up the whole universe from *Li* (organisational pattern at all levels), and *Ch'i* (what we might call matter-energy), together with a considerable understanding of inorganic, biological and social evolution. (Lu 1982, p. 26)

Other studies make the same point: features of Needham's Cambridge milieu prepared him for a sympathetic attention to the 'organic' dimension of Chinese worldviews and philosophy (Nakayama 1973).

Needham maintains that the Chinese simply could never embrace, even provisionally for the sake of doing science, the mechanical worldview that underwrote Europe's Scientific Revolution:

> The key word in Chinese thought is *Order* and above all *Pattern* (and, if I may whisper it for the first time, *Organism*). … Things behaved in particular ways not necessarily because of prior actions or impulsions of other things, but because their position in the ever-moving cyclical universe was such that they were endowed with intrinsic natures which made that behaviour inevitable for them. If they did not behave in those particular ways they would lose their relational positions in the whole (which made them what they were), and turn into something other than themselves. They were thus parts in existential dependence upon the whole world-organism. And they reacted upon one another not so much by mechanical impulsion or causation as by a kind of mysterious resonance. (Needham and Ling 1956, p. 281)

And later:

> We conclude, therefore, that 'law' was understood in a Whiteheadian organismic sense by the Neo-Confucian School. One could almost say that 'law' in the Newtonian sense was completely absent from the minds of Chu Hsi and the Neo-Confucians in the definition of *Li*; in any case it played a very minor part, for the main component was 'pattern', including pattern living and dynamic to the highest extent, and therefore 'organism'. In this philosophy of organism all things in the universe were included; Heaven, Earth and man have the same Li. (Needham and Ling 1956, p. 568)

This can be seen as a Chinese version of the Aristotelian worldview of self-actualizing natures that was abandoned as the Scientific Revolution became embedded in Europe.

A problem for Needham, and all advocates of an organicism worldview, is to separate such a view from primitive animism which is simply incompatible with modern science and prevents its appearance. In most traditional societies, including

---

[11] Needham's own eclectic mix of socialism, Christianity, and process philosophy can be read in his *Science, Religion and Reality* (Needham 1925) and *Order and Life: The Terry Lectures* (Needham 1936/1968). For a sophisticated endorsement of Whitehead's process metaphysics by a leading quantum physicist and philosopher, see Abner Shimony (1965).

Chinese, the world is 'alive': all manner and means of beings live everywhere; the division between animate and inanimate objects is barely made; spirits and demons are active throughout nature. This does not inspire, or support, a search for stable, natural, and repeatable causes, for Kepler's laws of planetary motion or Newton's laws of nature.

Needham is saying that China did not have the mechanical worldview which, as a matter of fact, characterized the New Science of the European Scientific Revolution; it did have the suggestion, or hint, of the 'organicism', evolutionary worldview which he, following Whitehead, says characterizes the post-mechanical worldview of contemporary science. But although the 'billiard-ball' materialist model may have gone with nineteenth-century developments in electricity and magnetism, the mechanical worldview was not abandoned. Newtonian attraction is still mechanical; magnetic fields act on iron filings just as mechanically as billiard balls act in collisions. Nothing is gained, and much is lost if these, interactions are thought of as 'mysterious resonance'. Needham's expression might well resonate with New Age mysteriums, but it only clouds scientific comprehension.

Adopting the mechanical worldview was a necessary step in the appearance and early growth of science,[12] and China did not take that step. Despite some Daoist writings, that step was not available to it. China had a worldview formed by, and insulated within, philosophy; not a worldview tempered and formed by experimental science, as there was none. Needham makes a counterfactual historical claim: If China had had a mechanical metaphysics, then given all the other prerequisites that were in place, it could have had modern science at or before the time it developed in Europe.

Needham is correct in pointing out that modern science has a metaphysical component; whether it is the Whiteheadean one he identified with is another question. All cultures and traditions have metaphysical convictions – Western, Eastern, Middle-Eastern, and all others – which frame and enter into the sciences of all cultures. Western science, from its pre-Socratic origins to the present, has developed in conjunction with metaphysics; no one denies this, and all attempts to rid science of metaphysics have failed.[13] But there is a difference between productive or warranted metaphysics and speculative or unwarranted metaphysics. What has typified the former has been its genuine engagement with the best science of its time. Serious metaphysics learns from science or protoscience and adjusts its ontology and epistemology in light of the maturation of science (Bunge 1959; Shimony 1965, 1989, 1993a, b).

As mentioned in Chap. 9, the protoscience (natural philosophy) of the classical and Hellenic materialists and atomists – Thales, Anaximander, Leucippus, Democritus, Epicurus, Anaxagoras, and others – was in constant struggle with the dualist, finalist, teleological, purposeful worldviews of Platonists and Aristotelians. For two millennia, the latter triumphed. In the West, things changed when ancient

---

[12] See at least Dijksterhuis (1961/1986) and Harré (1964).

[13] See at least Agassi (1964), Amsterdamski (1975), Bunge (1977, 1998, 2009, 2010), Burtt (1932), Dilworth (1996/2006), Popper (1963), and Wartofsky (1968).

materialism reappeared in the seventeenth century with Galileo's writings and other contributors to the Scientific Revolution.[14]

Any contemporary metaphysics that does not engage with science is lazy; it is mere hand-waving, or more literally book-waving, or chatter. What is talk of cosmology without reference to science? Talk of the mind without reference to neuroscience? Talk of ontology without reference to physics or chemistry? Talk of human origins without reference to evolutionary science? And so on. On these grounds feng shui metaphysics is condemned. The mixed Daoist, Buddhist, and Confucian metaphysics of feng shui had little, if any, engagement with modern, post-seventeenth-century, science and, consequently, shows little, if any, evidence of *intellectual* or *cognitive* growth despite its enormous cultural and commercial growth.

For Needham 'elemental' metaphysics was an obstacle to the development of any form of modern science in China. He comments that: 'Literature on the Five-Element theory from just before and during the Han time [200 BC–200 AD] is very large (and also tedious, fanciful and repetitive)' (Needham and Ling 1956, p. 248). By the medieval Sui dynasty, the late sixth century AD, 'the five-element theories had become a universal commonplace of Chinese thought' (Needham and Ling 1956, p. 253), with the principal text being Hsiao Chi's *Main Principles of the Five Elements*. However, by then the theory 'became more and more bound up with pseudo-sciences such as fate-calculation' (Needham and Ling 1956, p. 253).

The five-element theory and yin-yang cosmology were unified by Tsou Yen (305–240 BC) traditionally regarded as the father of Chinese naturalistic thought (Bodde 1991, pp. 100–103). The 'ontology' of basic elements is found outside of Daoism and its associated Chinese traditions; it was a staple of pre-Socratic Greek thinkers and was most highly developed by Aristotle who in his *De Caelo* identified as the four fundamental elements *water, earth, fire,* and *air.* Following his analysis of natural and unnatural motion, he was moved to admit a fifth element. This was not the Daoist *metal* but rather his own postulated *ether* which he saw as a necessary constituent of primary matter.

The five elements of feng shui, the *Wuxing*, were elaborated in early Daoist writing; they are not the elements of Hellenistic philosophy, much less the elements of the modern periodic table. In the period of Mao's domination of China, there was some inclination to see *Wuxing* as the same elements found in the Western tradition as this showed the materialism of traditional Chinese thinking. The categorical difference between the elements of either the early Hellenistic, or the later scientific traditions, and those of feng shui, can be seen in a 2014 exposition of the latter:

> Two Confucian classics, the *Shangshu* [*The Book of Documents*] and the *Liji* [*The Book of Rites*], lucidly articulate the Five Elements and their cyclical progression. In tandem with natural phenomena and seasonal changes, the Five Elements, Earth, Wood, Metal, Fire, and Water, form two parallel cycles: the cycle of creation (sheng) and the cycle of conquest (ke).

---

[14] On the philosophy of the Scientific Revolution as a resurrection of ancient materialism, see Bunge (2001, Chap. 3), Pigliucci (2010, Chaps. 8, 9), Popper (1963, Chap. 5), and Vitzthum (1995, Chap. 2).

In the cycle of creation, one element generates its successor by following a prescribed pattern of succession that accords with natural interactions. For example, water nourishes the trees (wood), and then burning wood makes fire. By contrast, in the cycle of conquest, one element destroys its predecessor. For example, water extinguishes fire, and then fire melts metal. (Chen 2014, p. 327)

Graham Parkes discusses how the elements are understood in the feng shui tradition:

The major background assumption that needs to be highlighted is the understanding of the world as a dynamic play of forces, or energies, rather than an aggregate of material things, with a corresponding emphasis on 'becoming' over 'being'. (Parkes 2003, p. 191)

And adds:

As the seasons proceed, the cosmic qi alternates between yang and yin: yang qi is in the ascendant from the beginning of spring and reaches its highest point in midsummer; it then diminishes as the yin qi begins to increase, which peaks at the winter solstice. (Parkes 2003, p. 197)

Elements are here in a different 'ball game' than they are in the Hellenic and the scientific traditions. Parkes affirms feng shui's 'process' metaphysics against the Cartesian-Newtonian 'dead matter in motion' metaphysics that he mistakenly ascribes to modern science (Parkes 2003, p. 185) and to which he, in part, attributes the environmental degradation of the developed, science-dominated world. In this he repeats a common opinion shared by many environmentalists, feminists, postmodernists, multiculturalists, romantics, and other critics of modern science. But 'dead matter' theory plays no part in modern science. It played a diminishing role in eighteenth-century science and was basically eliminated from science after the consolidation of electromagnetic theory following James Clerk Maxwell's (1831–1879) 1873 *Treatise on Electricity and Magnetism* (Maxwell 1873). There is environmental degradation aplenty in the developed science-dominated world, but to attribute this to 'dead matter in motion' metaphysics is implausible and misleading; it is almost 200 years since 'dead' matter featured in any scientific theory or worldview (Schofield 1970; Yolton 1983).

Importantly, the elements, whatever they might be, are not merely background colour, or stage props for feng shui practice. Rather, it is supposed that they are localized at any site, causally efficacious and essential to the practice. The more surprising then it is that they have not registered on any scientific instrument and that they seem unbound by any prosaic consideration such as the conservation of energy requirement. When Lavoisier electrically separated water into its components of oxygen and hydrogen, that was the scientific end of water as a fundamental element. But removed from any engagement with modern science element talk still lingers on in feng shui. Needham sensibly said of traditional five-element thinking that:

The only trouble about the Chinese five-element theories was that they went on too long. What was quite advanced for the $+1^{st}$ century was tolerable in the $+11^{th}$ century, and did not become scandalous until the $+18^{th}$. The question returns once again to the fact that Europe had a Renaissance, a Reformation, and great concomitant economic changes, while China did not. (Needham and Ling 1956, p. 294)

Appraisal of Needham

Everyone acknowledges the immense debt that all scholars, especially historians, have to Needham for initiating the serious study of Chinese, and more broadly Asian, science. But with the passage of half-a-century, philosophical criticism of the Needham project has emerged. This is nicely stated by Wen-yuan Qian, a Chinese theoretical physicist, turned historian of science[15]:

> Needham must be appreciated as a remarkable encyclopaedist, but not as a great philosophical historian; someone who raised many more important questions than he could satisfactorily answer. (Qian 1985, p. 131)

The dominant philosophical criticism is that Needham assumes, but does not defend, a 'positivist', 'universalist', 'linear', or simply unsophisticated 'ecumenical' understanding of science.

An early such critic was Angus Graham (1919–1991) who said that the very idea of asking the Needham Question – Why did modern science not emerge in traditional China? – is a mistake because all we know is what led to modern science in seventeenth-century Europe, and it is altogether hypothetical whether those conditions might have led to science anywhere else (A. C. Graham 1973). Yung Sik Kim articulated similar misgivings saying that scholars should first 'understand Chinese natural knowledge in its own intellectual and social context without any inappropriate assumptions, before embarking on comparisons with the conditions under which modern science emerged in the West' (Kim 1982, p. 103). Morris Low writes:

> In Needham's scheme, the significance of local indigenous knowledge has tended to be measured according to how much it contributed to the formation of what we now know as science. … The history of science is, happily, moving toward a more inclusive historiography that values non-Western forms of knowledge. (Low 1998, pp. 4, 6)

Patricia Fara criticizes Needham for viewing 'scientific knowledge as absolute and universal', writing:

> Needham's image of flowing rivers implies that the great flood of scientific discoveries is leading inevitably towards the Ocean of Truth. Or to switch metaphors, he concluded that China reached the level of Leonardo da Vinci but never made it as far as Galileo. (Fara 2009, p. 64)

Fara aptly condenses prevalent, constructivist, anti-realist epistemology when she writes:

> Knowledge about the world can appear in various forms, developed in different places for different purposes – there is no unique route towards Truth. (Fara 2009, p. 64)

---

[15] Qian was born in 1936, and in 1959 he completed a Soviet-type education in theoretical physics at Peking University then taught physics at Zhejiang University, having as hundreds of thousands of other interlectuals did a 10-year interlude of 'bourgeois correction' during Mao's Cultural Revolution. In 1980 he went to the History Department at University of Michigan, completed a PhD, and depressed by Chinese realities and convinced that an open society would never eventuate, stayed in the USA.

This charge is endemic among historians. For the overwhelming majority of historians, every culture's or subculture's collection of ideas about the working of the world is labelled 'knowledge' or 'science'. For Barry Barnes and David Bloor, knowledge simply is 'any collectively accepted system of belief' (Barnes and Bloor 1982). Needham can be defended against this charge. Fara does not recognize that her 'no unique route towards truth' is quite consistent with there being a truth, even a universal one. The uncontroversial 'no unique route' claim is often understood as the different but controversial and contested 'no unique truth' claim. This does seem to be Fara's position. Realist philosophers say of Fara's assertion that if 'knowledge' is replaced by 'beliefs', 'views', 'opinions', 'understandings', or any other such subjective term, then it becomes an utterly uncontroversial anthropological, sociological, or psychological claim, whose details can, with benefit, be examined by the appropriate discipline. But Fara's assertion is a philosophical one for which no argument is advanced and against which there have been substantial realist objections.[16]

Fara's 'no unique route' claim also puts aside, or rules out, the task of comparing and evaluating routes. Even if one asserts (without evidence) that there are other routes to knowledge of the world than the modern scientific route, then the alternatives can nevertheless be judged, appraised, and ranked. Wandering around in the desert for 40 days before finally locating the spring is not as efficient as using star guidance to get there in 10 days, compass guidance to get there in 5 days, or GPS guidance to get there in 3 days. There are individual, and in a larger scale social, costs for each route and method – perhaps more apparent when the goal is cure of smallpox rather than finding a spring. Methodological relativism puts aside such ranking exercises.

It is well known that traditional societies make use of numerous plants to alleviate and cure countless ailments and illnesses (Lewis and Elvin-Lewis 1977). Dioscorides, in his first-century, five-volume *De materia medica* listed the medicinal properties of more than 600 plants. The classic case is the widespread practice, in traditional societies, of chewing the bark of myrtle and willow trees to reduce fever and relieve pain. In 1763 a contribution to the *Proceedings of the Royal Society* noted that this practice was recommended in an Egyptian medical text of 1543 BC. In the late nineteenth century, chemists at the Friedrich Bayer Company in Germany had isolated the active ingredient of the bark as salicylic acid and found ways to synthesize acetyl-salicylic acid in commercial quantities. For half-a-century Bayer's aspirin dominated the anti-inflammatory and pain-relief market. With coming of the Spanish flu pandemic in Europe in 1918, multiple millions of tablets were sold. In the 1960s, after other drugs took over its primary role, it was reborn as a blood-thinning, anti-clotting agent (Jeffreys 2008). Some cultures had traditional knowledge of the anti-inflammatory effect of the bark, but they did not have scientific knowledge of it; nor could the first 'grow' into the second without all the requisite

---

[16] See Matthews (2015a, Chap. 9) and contributions to Agazzi (2017) and the literature cited therein.

theoretical, methodological, instrumental, cooperative, and educational scaffolding or 'furniture' of modern science.

Another case is the award of the 2015 Nobel Prize for Medicine to the Chinese researcher Youyou Tu at Peking University. The antifebrile effects of the herb *Artemisia annua* (sweet wormwood) had long been known in Traditional Chinese Medicine (TCM). It was mentioned in the fourth-century medical text of Ge Hong. Tu, using modern analytic chemistry, refined technique, and sophisticated equipment, was able to identify and extract the herb's active component, artemisinin and determine its composition and structure. Chinese companies were then able to synthesize, produce, and market this as an effective anti-malarial drug having minimal side effects. Again, TCM had traditional knowledge, but it did not have scientific knowledge of plant biochemistry that now constitutes the scientific discipline of pharmacognosy that allows efficient and largescale production of medicine for the effective treatment of countless illnesses and diseases (Tyler 1985). To claim Tu's identification and synthesizing of artemisinin as an outcome of TCM is a step too far; TCM contributed something, but the 'heavy lifting' was done by orthodox science and modern technology.

Needham did hold a philosophical position, namely, that there is one universal science. All cultures can have it in different degrees, but European-born, modern science is its yardstick. As is apparent in multicultural science education debate, the universalist theory of science is at once philosophical and political; and is commonly decried on political grounds. For realists, the former should ultimately dictate to the latter; for constructivists, the politics can dictate the philosophy. But there are acute dangers in the constructivist and multiculturalist position: witness what happened in Stalin's Russia, Hitler's Germany, and Mao's China where universalist science was rejected in favour of the science of the party.

The lesson Joseph Needham draws from his life-long comparative study of the histories of European and Chinese science and technology is one that should inform, and give direction to, all multicultural education discussion:

> Let us take pride enough in the undeniable historical fact that *modern* science was born in Europe and only in Europe, but let us not claim thereby a perpetual patent thereon. For what was born in the time of Galileo was a universal palladium, the salutary enlightenment of all men without distinction of race, colour, faith or homeland, wherein all can qualify, and all participate. Modern universal science, yes; Western science, no! (Needham 1963, p. 149)

Conclusion

Although modern science arose in Europe – the so-called Western world – it is a geographically universal, knowledge-seeking enterprise; all countries and people contribute to its growth. However, because it began in the West, and there was nothing like it in China or the 'East', then there was no scientific structure or ethos which could be used as a yardstick against which traditional feng shui could be measured. This despite individual efforts by the likes of Ricci, Eitel, and Chinese modernizers, feng shui existed and expanded in what was, for centuries, a non-science environment, and so it was the default worldview of China and Southeast Asia. Needham's life-long study of science and technology in traditional China demonstrated that it was really technology, not science that characterized Chinese progress. In the 'West', pursuits like alchemy, magic, and astrology were all eventually moved into the pseudoscience basket because of the development of modern science, but Needham showed that there was no such Eastern measuring stick whereby feng shui could be similarly sidelined. But now there is, and feng shui should take its proper place on the pseudoscience sideline.

Chapter 11
The Science and Teaching of Energy

The central feng shui notion of *chi* has been for centuries translated into English as 'energy', this being the term used in the first English translation of Ricci's *China Journals* (Ricci 1615/1953). Clearly attention needs to be paid to the changing meaning of the word in English or more accurately to the progressive differentiation of the scientific meaning of 'energy' from the multifarious, everyday meanings of the term. When *chi* is translated as 'energy', there is a tendency to think that because scientists and feng shui practitioners are using the same word, they are talking about the same thing. They are neither talking about the same thing nor talking about it in the same way.

Energy and Metaphysics

Energy is the central conceptual component of modern science – all modern science, not just physics (Coopersmith 2010; Feekes 1986; Sherman 2018). Mario Bunge observed: 'All the sciences that study concrete or material things, from physics to biology to social science, use one or more concepts of energy' (Bunge 2000, p. 458). There are as many types of energy as there are types or classes of natural processes: kinetic, potential, magnetic, nuclear, chemical, and so on. Energy is a property of all existent things; hence it is universal. But it does not exist apart from things of which it is a property. Einstein's 'most famous equation in the world' unites energy with mass and energy, with something; there is no 'free-floating' energy (Fernflores 2012; Lange 2001). For Bunge:

> Because it is ubiquitous, the concept of energy must be philosophical and, in particular, metaphysical (ontological). (Bunge 2000, p. 458)

But being metaphysical does not license obscurantism, incoherence, or contradiction. Richard Feynman in his famous *Lectures on Physics* well stated the centrality of energy in modern science:

> There is a fact, or if you wish, a *law*, governing all natural phenomena that are known to date. There is no known exception to this law — it is exact so far as we know. The law is called the *conservation of energy*. It states that there is a certain quantity, which we call energy, that does not change in the manifold changes which nature undergoes. That is a most abstract idea, because it is a mathematical principle; it says that there is a numerical quantity which does not change when something happens. (Feynman 1963, Lect. 4.1–4.2)

And that:

> It is important to realize that in physics today, we have no knowledge of what energy *is*. We do not have a picture that energy comes in little blobs of a definite amount. It is not that way. However, there are formulas for calculating some numerical quantity, and when we add it all together it gives "28"—always the same number. It is an abstract thing in that it does not tell us the mechanism or the *reasons* for the various formulas. (Feynman 1963, Lect. 4, sect. 4.1, p. 2)

Although he modestly says that he does not know what energy is, he is prepared to defend the claim that whatever it is, it is conserved. And far from Feynman's lecture legitimating all alternatives to orthodox scientific understanding of energy, he is prepared to say that psychoanalysis is not a science and that witchdoctors' recourse to spirits as the cause of disease likewise marks them as non-scientific (Feynman 1963, Lect. 3, sect. 3.6). So long as chi remains in the spiritual heavens, then its violation, or otherwise, of energy conservation cannot be ascertained. But as soon as chi putatively engages with worldly processes and enters terrestrial energy chains, such as the supposed 21 bodily meridians intercepted in acupuncture, then it enters the domain of science and can be shown to violate the conservation law. So there is a stark choice: maintain the cornerstone of science or maintain chi; having both is not possible.

History and Conceptual Refinement

Some history is also important because it shows fundamental differences between modern (post seventeenth century) scientific approaches to understanding energy and approaches in the ancient and contemporary feng shui tradition. The scientific term 'energy' was introduced by Thomas Young (1773–1829) in his Royal Society lectures of 1802 and given currency when these were published in 1807. For Young:

> The term 'energy' may be applied, with great propriety, to the product of mass or weight of a body, into the square of the number expressing its velocity. (Young 1807, p. 52)

He is saying that 'energy' is the word to designate the quantity mv^2 which was Huygens, Leibniz, and Laplace's measure of *vis viva* (vital force). Eventually half of this quantity, $\frac{1}{2} mv^2$, became the measure of one of the many manifestations of energy, namely, kinetic energy, that is the energy of a moving body.

Young did not invent the word; the English word 'energy' had a heritage of 2000–3000 years stretching back to the pre-Socratic philosophers. Heraclitus (c.535–450BC) uses the Greek word ενέργειας to refer to the source of movement. The term was Latinized as *enérgeia*, a compound of *en* (to) and *ergon* (work). For Heraclitus:

> En-ergon is the father of everything, king of all things and, out of it, all forms of contrast originate. Since 'en-ergon' is common to everything, it is vital for life itself. (Feekes 1986, p. 1)

This pre-Socratic conceptual connection of work with energy was thereafter cemented.

Aristotle's use of *enérgeia*, as with all of his philosophical concepts, cast a long shadow in premodern Western philosophy and natural philosophy. Aristotle took up and to some degree refined extant meanings of the term, but still it has different meanings in his own writings. In his *Ethics* and his *Metaphysics*, it refers to actual activity in contrast to the potential or disposition for activity which is Latinized as *dýnamis*; change (the process of moving from potential to actual) cannot happen without involvement of *enérgeia* (Lindsay 1974, chaps. 1, 2).

Although movement, especially 'natural motion', was important for Aristotle's physics, there is little attempt to quantify it. And the vital distinction between work (effort) being done (expended) over *time* and effort being expended over *distance* was not made. In the late-medieval period, better measurement, conceptualization, and experimentation on what Aristotle labelled 'local motion' (what is now called simply movement), and the formulation of the conservation of momentum law, was the prelude to modern understanding of energy. This was an ambiguity which centuries later caused significant, but needless, controversy between Descartes and Leibniz over the conservation, or otherwise, of force.

As with so much else in classical mechanics, it was Galileo's pendulum investigations that were central to ideas of the conservation of momentum.[1] When a pendulum bob at rest was released, it swung down gathering speed and rose to exactly the same height; if the swing was interfered with by a nail in a board or wall, then the bob still rose to the same height even if it travelled a shorter path. In all cases, ideally, the 'motion' gained at the nadir of the swing was precisely that needed for it to regain its initial height. Galileo knew that in reality, due to impediments and accidents, the bob did not regain its exact initial height. If the bob collided with a suspended identical stationary bob, the latter rose to the original bob's height. Galileo in his *Dialogues Concerning Two New Sciences* writes (Fig. 11.1):

> But all these momenta which cause a rise through the arcs BD, BG, and BI are equal, since they are produced by the same momentum, gained by fall through CB, as experiment shows. (Galileo 1638/1954, p. 172)

Galileo's idea or 'law' of conservation of movement (*momenti*) which was a multiple of mass and speed (mv) enabled him and others to do many things in early modern mechanics, including to answer the seemingly incontrovertible argument of

---
[1] See the 32 contributions to Matthews et al. (2005).

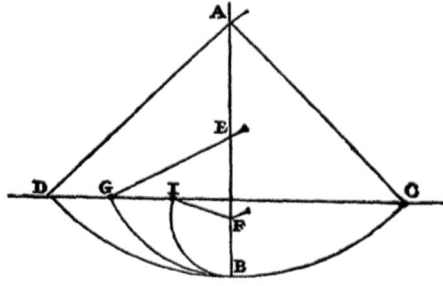

Fig. 11.1 Galileo's Pendulum (*Two New Sciences*, 1638)

the stationary earth proponents that bodies dropped from the top of a tower on a spinning earth should *not* fall to its base, but fall a considerable distance to the west (Finocchiaro 2010, pp. 124–29). Descartes in his *Principles of Philosophy* famously took up the conservation idea giving it a philosophical and theological interpretation:

> … it seems to me evident that it is God alone, who in His omnipotence created that matter with motion and rest, and who now preserves the universe in its ordinary concourse as much motion and rest as He put into it at creation. For although motion is nothing but a mode of moved matter, it exists in a certain quantity which never increases or diminishes, although there is sometimes more and sometimes less in certain of its parts. (Hall 1970, p. 268)

Neither Galileo nor Descartes clearly separated force from momentum, and they did not conceptually separate motion as a *scalar* quantity (modern day 'speed') which had no direction, from motion as a *vector* quantity tied to direction (modern day 'velocity'). What will become the Newtonian and modern 'conservation of momentum' law applies only to momentum (**p**) as a direction-dependent vector quantity: **p** = m**v**. It is only for vector quantities that Descartes collision laws held and then only for elastic collisions. In these cases, momentum before a collision equals momentum after the collision, and given that mass is known, velocities after a collision can be determined.[2] The problem was by-passed by Huygens who used mass multiplied by the square of velocity as his measure of momentum (force); this was then a scalar, that is, direction-independent quantity. It became known as a measure of a body's *vis viva* (vital force).

This was fundamentally a conceptual, and ultimately a philosophical, not an empirical problem, at the birth of modern physics. The concepts under discussion included force, quantity of motion, momentum, quantity of progress, vis mortua (dead force), and vis viva (living force). The issue 'raged' between followers of Newton (1643–1727), Descartes (1596–1650), and Leibniz (1646–1716) over about a 50-year span following publication of Leibniz's 1686 critical essay 'A Brief Demonstration of a Notable Error of Descartes and Others Concerning a Natural Law' (Leibniz 1686/1956). The debate is known as the *Vis Viva Controversy* (Iltis 1971). The subtitle of Leibniz's essay is revealing: 'According to Which God is Said Always to Conserve the Same Quantity of Motion'.

---

[2] This is the textbook equation: $m_1 v_1 + m_2 v_2 = m_1 v_1' + m_2 v_2'$.

All the foremost Western scientists of the period were comfortable with invoking the deity as justification both for their research and as foundation for their axioms and principles. Modern science assuredly was not godless – Leibniz, for instance, wrote a defence of the Roman Catholic doctrine of Eucharistic Transubstantiation (Leibniz 1668/1956) – but God did not figure in Newton's, Descartes', or Leibniz's scientific systems. God does not appear in any equation or explanation of an experimental outcome. Where in the latter cases, this might appear to happen, God is there as a placeholder until a naturalistic explanation is uncovered. In no case in modern science is it sufficient to say: 'God did it' and then walk away. "How God did it?" and "Does He always do it?" are the inevitable questions to be asked. This is in stark contrast to comparable yin-yang and five-element explanations in the fung chi tradition.

The *vis viva* controversy was finally clarified using the mathematical derivation which begins with the definition of uniformly accelerated motion and is now readily found in standard physics textbooks. The standard expression for force times displacement is:

$$Fs = \frac{1}{2}mv_2^2 - \frac{1}{2}mv_1^2$$

and it shows that, when a resultant force is considered as acting over a *displacement*, then the familiar terms for kinetic energy appear, with force times displacement being the change in kinetic energy.

Similarly, the expression for force times time is:

$$Ft = mv_2 - mv_1$$

and, in turn, this shows that, when a force is considered as acting over a period of *time*, then the familiar terms for momentum appear, with force times time being the change in momentum. It was Jean d'Alembert (1717–1783) in 1743 who saw that both measures (force times distance and force times time) were just different conventions for measuring the manifestation or effect of the one force.

Newton – by 'standing on the shoulders of giants' as he says of his dependence on the achievements of Galileo, Brahe, and Kepler – came to believe that nature was united by the operation of a very limited number of universal forces:

> For the basic problem of philosophy seems to be to discover the forces of nature from the phenomena of motions and then to demonstrate the other phenomena from these forces … we derive from celestial phenomena the gravitational forces by which bodies tend toward the sun and toward the individual planets. Then the motions of the planets, the comets, the moon, and the sea are deduced from these forces by propositions that are also mathematical. If only we could derive the other phenomena of nature from mechanical principles by the same kind of reasoning. (Newton 1726/1999, p. 382)

And he adds:

> For many things lead me to have a suspicion that all phenomena may depend on certain forces by which the particles of bodies, by causes not yet known, either are impelled toward one another and cohere in regular figures, or are repelled from one another and recede.

Since these forces are unknown, philosophers have hitherto made trial of nature in vain. But I hope that the principles set down here will shed some light on either this mode of philosophizing or some truer one. (Newton 1726/1999, pp. 382–383)

His other 'phenomena of Nature' were to include optical, magnetic, chemical, thermal, and sundry other effects. Newtonians attempted to bring all of these, along with electrical phenomena, into one explanatory force-based, atomistic system which became known as the 'mechanical worldview' or 'corpuscularianism' (Dijksterhuis 1961–1986; Schofield 1970; Westfall 1971). There was an ambiguity in this programme between 'mechanical' meaning 'by contact', the Cartesian view, and 'mechanical' meaning 'contact and deterministic force acting at a distance', the Newtonian view. The common denominator was uniformity and determinism; things happened because of external, regular, 'knowable' causes (Westfall 1971). The mechanical worldview was opposed to all chaotic, spirit-moved, undetermined views of the world. This 'naturalism' became a presupposition for the conduct of modern science. As Fang Lizhi observes, it was in contrast to the situation in traditional Chinese natural philosophy:

> A major difference between ancient Chinese astronomy and its Western counterpart was that the latter was always concerned with *explaining* the apparent irregularities and anomalies in the motions of the planets, while the former was not. … Chinese astronomers were untroubled by phenomena that departed from the norm, and in fact were quite taken with them. (Fang 1989/1992, p. 34)

The whole Newtonian idea of recourse to unseen force, or forces, as the explanation of visible events was criticized by Ernst Mach (1838–1916) from his staunchly positivist perspective. He maintained that force (or any other such theoretical construct) was unnecessary and uninformative; the idea of force added nothing to what was already seen, and it had only a 'book-keeping' function (Mach 1893/1974, pp. 298–305).[3] Nevertheless the Newtonian programme as laid out in his *Principia* (Newton 1726/1999) prepared the ground for the modern physics of energy which began appearing in the middle nineteenth century (Harman 1982).

Of course, the extant concepts that were the object of Mach's empiricist critique did have a creditable book-keeping function because they were rigorously defined and connected to empirical states of affairs. Chi lacks any reference; nor can it perform even a book-keeping role as it is persistently ill-defined or not defined at all.

Animal Magnetism

Natural philosophers had done many quantitative and experimental investigations of different energy phenomena – heat, magnetism, light, electricity, etc. Through the eighteenth and into the nineteenth centuries, they were each thought to be independently the result of the workings of separate fluids or substances – animal

---

[3] Many philosophers have discussed this issue. See at least Ellis (1976) and Hunt and Suchting (1969).

magnetism, phlogiston, caloric, igneous fluid, magnetic fluid, electrical fluid, luminiferous ether, and so on. They were referred to as 'imponderable' or 'subtle' fluids.[4] They are as close as Western science has come to entertaining anything comparable to universal chi energy. Attention to how mainstream science examined and made judgement on these conjectured subtle fluids provides lessons for the diligent appraisal of chi claims, lessons for both the West and the East. Such lessons highlight an important difference between science and pseudoscience: the former applies a critical and sceptical analysis to its theories and hypotheses, dismissing or modifying them when they don't measure up, whereas the latter has no such internal evaluative mechanisms; conjectures and assertions are not checked, and they are taken on authority or self-interest.

A prominent example of the development of a chi-like 'science' in Europe was that promoted by the German physician Franz Anton Mesmer (1734–1815). He was well educated in philosophy, theology, and law and then studied medicine at the University of Vienna, where in 1766 he completed a doctoral thesis 'Physical-Medical Treatise on the Influence of the Planets' (Mesmer 1766/1980). The title reflects his conscious, or otherwise, connection to the Hermetic tradition in European thought. He took as a given the influence of the moon on bodily function, repeating the commonplace observation that: 'The symptoms of epileptics tend to reappear at the new moon and especially at the full moon, resulting in their being called *lunatics*' (Mesmer 1766/1980, p. 15).

Mesmer's early Viennese studies and clinical experiences led him to postulate an all-pervasive, superfine, invisible, cosmic 'fluid' that held the solar system together and enabled Newton's gravitational forces to operate at a distance. But beyond astronomical duty, this fluid also seeped into human and animal bodies affecting their constitution and behaviour. So far very chi like. His 1766 thesis launched 'animal magnetism' on to the Western scientific and medical stage:

> There is a force which is the cause of universal gravitation and which is, very probably, the foundation of all corporal properties; a force which actually strains, relaxes and agitates the cohesion, elasticity, irritability, magnetism, and electricity in the smallest fluid and solid particles of our machine, a force which can, in this report, be called *animal gravity*. (Mesmer 1766/1980, p. 14)

By 1775, in a letter to Dr. A.M. Unzer, he had changed his terminology to *animal magnetism* and reported on many properties of the fluid and on its clinical manipulation by natural and artificial magnets: 'Menstrual periods and hemorrhoids were restored to their normal condition … I cured all kinds of hypochondriac, convulsive, and hysterical irregularities' (Mesmer 1775/1980, p. 28). Animal magnetic fluid could be manipulated by magnets because of its 'extreme subtlety and similarity to nervous fluid' (Mesmer 1775/1980, p. 29). And the fluid could be transferred from doctor to patient over distances:

> Without any direct communication and from a distance of eight to ten feet, moreover being hidden behind a man or a wall, I roused jolts in another part of the patient that I wanted to, and with a pain as ardent as if one had hit her with a bar of iron. (Mesmer 1775/1980, p. 28)

---

[4] Among many studies, see Alexander (2015) and Schofield (1970, chap. 8).

In his 1779 Dissertation, he quotes another 1775 letter to another 'Foreign Physician' in which he gave an 'exact idea of my theory':

> I set forth the nature and action of ANIMAL MAGNETISM and the analogy between its properties and those of the magnet and electricity. … all bodies were like the magnet, capable of communicating this magnetic principle; that this fluid penetrated everything; that it could be stored up and concentrated, like the electric fluid; that it acted at a distance … Finally, I accounted for the various sensations and backed these assertions with experiments which enable me to put them forward. (Mesmer 1779/1980, p. 51)

Elsewhere he emphasizes the 'alternating "flow" of streams – coming and going – of a subtle fluid which fills the space between two bodies. … it is impossible to cause a displacement without a corresponding replacement' (Mesmer 1779/1980, p. 107). This morphs into the supposition of both positive and negative animal magnetism or positive and negative fluids. He encapsulated his theory in a series of 27 propositions, including (Fig. 11.2):

> 1. There exists a natural influence between Heavenly Bodies, the Earth and Animate Bodies.
> 2. A universally distributed and continuous fluid, which is quite without vacuum and of an incomparably rarefied nature, and which by its nature is capable of receiving, propagating and communicating all the impressions of movement, is the means of this influence.
> 3. This reciprocal action is subordinated to mechanical laws that are hitherto unknown.
> 4. This action results in alternate effects which may be regarded as an Ebb and Flow. (Mesmer 1779/1980, pp. 67)
> 27…In conclusion, this doctrine will enable the physician to determine the state of each individual's health and safeguard him from the maladies to which he might otherwise be subject. The art of healing will thus reach its final stage of perfection. (Mesmer 1779/1980, p. 70).

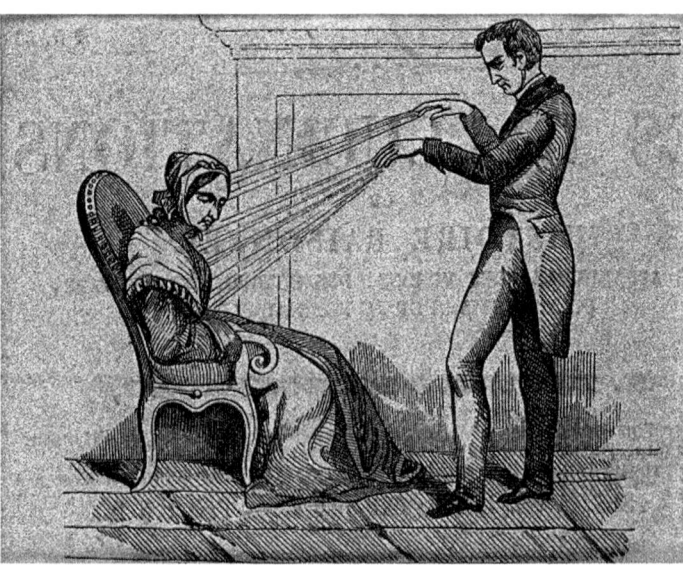

Fig. 11.2 Animal magnetism therapy

Mesmer describes curing many patients in Vienna, Swabia, and Switzerland. Overwhelmingly they were women suffering from a range of maladies including colic, apoplexy, hemiplegia, ophthalmia, and blindness. He used natural and artificial magnets and touch to add, subtract, or move a patient's animal magnetism. His fame spread; he was admitted to the Scientific Academy of München; he wrote folk 'flocked to my house'. Between 1774 and 1775, he was caught up in a dispute over the curing practices of an Austrian Catholic priest: half of Vienna regarded him as a charlatan; half saw him as the agent of God's miraculous interventions. Mesmer thought neither: the priest had unknowingly become a 'tool of Nature' by stumbling onto the manipulation of animal magnetism (Mesmer 1779/1980, p. 57). But popular fame was accompanied by professional disappointment:

> However, being wearied by my labors extending over twelve consecutive years and still more so by the continued animosity of my adversaries …I felt I had done my duty by my fellow citizens. (Mesmer 1779/1980, pp. 64–65)

In 1778 Mesmer left Vienna and set up a large medical and healing practice in pre-Revolutionary Paris. Newspapers of the time had stories headed: 'Everyone is occupied with mesmerism', 'The great subject of all conversations in the capital is still animal magnetism', and 'Men, women, children, everyone is involved, everyone mesmerizes' (Darnton 1968, p. 40). In most major cities, 'Societies of Harmony' were founded to promote magnetic therapy. But there was professional opposition. By 1781 Mesmer had 'insulated' his theory by claiming that animal magnetism could only be detected by a hitherto unrecognized sixth 'artificial' sense, something he had and could only be enabled by training: 'it must be transmitted by experience. Experience alone can render my theory intelligible' (Mesmer 1781/1980, p. 135).

Enough Parisian physicians and natural philosophers regarded him as a charlatan, or perhaps simply as an out-of-town competitor, that in 1784 they succeeded in having Louis XVI establish a Royal Commission, chaired by Benjamin Franklin and including Antoine Lavoisier, to investigate Mesmer's claims.[5] It issued, just five years before the beginning of the Revolution, 'The Franklin Report' (Franklin et al. 1784/2002, 1784/2014).[6]

The terms used nearly 250 years ago to justify the Commission have a contemporary ring:

> The government could not remain indifferent to a question of this kind that affected the health and the life of its citizens, and, because according to the system of M. Mesmer and his disciples any individual could simply by practising magnetism, cure anyone, the whole science of medicine would become useless; it would be necessary to close the medical schools, to change the system of instruction, to destroy the corpus of works believed until the present to be depositories of medical knowledge and to change everything to the study of magnetism. In an affair of this importance the government had to be on its guard against both too facile belief and too unbridled incredulity. It was necessary to acquire information

---

[5] The Commission actually examined a surrogate for Mesmer, another magnetic physician, M. Deslon.

[6] The context, composition, and functioning of the Commission are described in Darnton (1968, pp. 62–64).

before coming to an opinion, it was essential at least to avoid the reproach of precipitate prejudice. Such are the motives that have given rise to the setting up of the Commission whose report we present. (Franklin et al. 1784/2014, p. 30)

These are all the very same concerns raised in 'Alternative Medicine' debates, especially when matters of government funding, health insurance, university credentials, and treatment efficacy are on the agenda.

The Commission gave an honest digest of Mesmer's theory, saying that for him, animal magnetism was:

> An agent universally distributed throughout the whole of Nature; it is the mechanism of a mutual influence between the celestial bodies, between the earth and animate bodies; it is combined in such a way as not to allow any vacuum; its subtlety is beyond comparison; it is capable of receiving and propagating all the impressions of movement; it is susceptible to flux and reflux. The animal body demonstrates the effects of this agent and it is by insinuating itself into the nerves that it affects them immediately. One recognizes, in the human body in particular, properties similar to those of the magnet; in it diverse and opposite poles can be distinguished. The action and the power of animal magnetism can be communicated from one body to other bodies, animate and inanimate. This action takes place at considerable distance without the intervention of any intermediate body; it is augmented [when] reflected by mirrors, and communicated, propagated and increased by sound; this power can be accumulated, concentrated and transported. Although this fluid is universal, all bodies are not equally susceptible to receiving and transmitting it; there are even some, though a small number, which have a property so opposed [to it] that their mere presence destroys all the effects of the fluid in other bodies. (Franklin et al. 1784/2014, p. 30)

The commissioners recognized, in the way that most current feng shui and alternative medicine advocates do not, that merely testing the effectiveness of treatment did not advance the main question because many illnesses cure themselves, placebo effects are rampant, and cures have ultimately to be linked to specific mechanisms in order to have scientific sanction. Not surprisingly, given Franklin's and Lavoisier's involvement, the Commission was sensitive to methodological issues:

> The art of drawing conclusions from experiments and observations consists in evaluating the probabilities and in judging whether they are large enough, or numerous enough, to amount to proof. This type of calculation is more complicated and more difficult than one thinks; it demands great sagacity and is, in general, beyond the powers of most men. It is upon their errors in this type of calculation that is founded the success of charlatans, sorcerers and alchemists; and, in other times, of magicians, enchanters and all those who deceive themselves and attempt to pretty on public credulity. (Franklin et al. 1784/2014, p. 26)

And:

> After having seen the effects at the public treatment it was necessary to disentangle their causes and to look for evidence of the existence and the utility of Animal Magnetism. Animal Magnetism can well exist without being useful, but it cannot be useful if it does not exist. (Franklin et al. 1784/2014, p. 43)

They observed that:

> Touchings, imagination, imitation, these are the real causes of the effects attributed to this new agent advertised under the name *animal magnetism*. The practice of magnetism consists of the art of increasing the imagination by degrees; gaze, pressure, touching seem to act as a preparation, the nerves begin to be agitated, imitations communicates and expands impressions. (Franklin et al. 1784/2014, p. 36)

And: 'Magnetism seemed ineffective on those who submitted to it with some scepticism' (Franklin et al. 1784/2014, p. 52). They conducted many 'cures' and concluded that 'imagination necessarily accounts in great part for the effects attributed to magnetism' (Franklin et al. 1784/2014, p. 55). After much testing and examination, their conclusion was comprehensive and damning:

> The Commissioners, having recognized that this animal magnetic fluid cannot be perceived by any of our senses, that it has no effect, either on them or on the patients who are subjected to it; being assured that the pressures and touching occasion changes which are rarely favourable to the animal economy, and disturbances that are always disturbing to the imagination, having finally shown by decisive experiments that the imagination, with magnetism, produces convulsions, and that magnetism without imagination produces nothing, have concluded unanimously about the question of the existence and utility of magnetism, that nothing proves the existence of animal magnetism; that this non-existent fluid is consequently useless. (Franklin et al. 1784/2014, p. 37)

And they ruled out the possibility of any refinement of Mesmer's theory because its ontological presuppositions were simply false; they did not accord with how the world was:

> If M. Mesmer announces a larger theory, it will only be more absurd; celestial influences are an old chimera whose falsity has been long recognized; all this theory can be judged in advance by that which magnetism necessarily as for its base, and it can have no reality because the animal fluid does not exist. (Franklin et al. 1784/2014, p. 70).

They could have, but did not, point to the 'get out of jail' card that Mesmer had dealt himself by adding 'negative' animal magnetism to his ontology. Among his 27 Summary Propositions is:

> 19. This opposing property also penetrates all bodies; it may likewise be communicated, propagated, stored, concentrated and transported, reflected by mirrors and propagated by sound; this constitutes not merely the absence of magnetism, but a positive opposing property. (Mesmer 1779/1980, pp. 67–68)

So the failure of a test does not count against the animal magnetism theory; rather it supports the theory by indicating the presence of negative magnetism. Despite the scientific penumbra around mesmerism, this in-built, no-failure clause renders it non-scientific on standard Popperian and falsificationist criteria. Given that mesmerism wears a scientific cloak, this built-in no-failure clause makes it a pseudoscience. Such clauses are indicative of pseudosciences.

With Franklin and Lavoisier among the commissioners, the Commission's procedures and methods reflected Enlightenment understanding of scientific method. This makes the animal magnetism episode instructive for scientists, philosophers, and students. The commissioners were not simple positivists. They were not of the 'If we cannot see it, it is not there' school. They recognized that while scientific entities and realities were legitimately inferred from observation, not all such inferences were warranted. Different Mesmerists were riding on the back of gravitational attraction, magnetic fields, caloric, and electric phenomena to claim existence for their particular 'unseen', 'weightless', 'colourless' magnetic fluid. The commissioners were conscious of needing to rule out the latter in a way that did not rule out the former; to rule out Mesmer but not Newton.[7]

---

[7] On this see Donaldson (2005) and Weyant (1980).

The Commission's findings dealt a blow, but not a fatal one, to Mesmerism. Many leaders and supporters of the Revolution embraced mesmerism on account of it being opposed by the cultural, academic, and medical elite of the *Ancien Régime* (Darnton 1968, chap. 3). Some opined that the Commission's tests and their methodology were 'stacked against' a version of science that was not adequately captured by the tests: 'the commissioners were mistaken in treating it as a science by performing controlled empirical tests' (Weyant 1980, p. 102). Two hundred years later, this is the same complaint made by feng shui practitioners who likewise are averse to controlled empirical tests of their craft.

Magnetisers crossed the channel and operated in England. Dr. Mainauduc lectured to packed halls and founded the Hygeian Society. Dr. Benjamin Perkins effected cures by use of a metallic tractor (hand-held instrument) which powerfully directed the patient's magnetic fluid. This was impressive; but became decidedly less so when Dr. John Haygarth cured his patients by using a 'metallic' tractor made of wood! He read a paper on the matter to the Literary and Philosophical Society of Bath and in 1800 published a pamphlet 'Of the Imagination as a Cause and Cure of Disorders, exemplified by Fictitious Traction' (Haygarth 1800). For the same group of five patients, he used both wooden and metal tractors; with patients thinking they were the same instrument, the same cures were affected by both instruments. This might be the first research published on the placebo effect.[8]

The commissioners rejected Mesmer's magnetism as the cause of the convulsions and cures that they observed in magnetic therapy, but their favoured cause – 'imagination' – was not without its own problems: What did imagination have that magnetism did not? Charitably, it can be seen as a 'placeholder', a belief that mental or psychological states can affect bodily states and that the exact mechanisms and ontological status of the mental are yet to be uncovered. This is a legitimate and standard scientific procedure. Positively it directs research within the individual; negatively, it says do not bother looking for extra-individual or trans-individual forces or entities. But the positive work had to be done; otherwise the Commission's position is just hand waving. A century later, the psychologist Alfred Binet (1857–1911) and physician Charles Fèrè (1852–1907) in their book on *Animal Magnetism* said of the putative 'imagination' cause that:

> In our day, this would appear to be an insufficient explanation. We might as well say that hysteria is due to the imagination. (Binet and Fèrè 1890, p. 17)

The subsequent history of scientific studies of psychosomatic illness can be seen as the 'cashing in' of the Commission's 'imagination' promissory note. Whether the

---

[8] The whole episode is described in Gratzer (2000, chap. 5). The term comes from the Latin rendering of Psalm 114:9, *Placebo Domino*, 'I will please the Lord'. Haygarth said that doctors were giving medicine more to please the patient than cure the patient. Two centuries later, this has not changed. Studies routinely show that doctors (a quarter in Canada, a half in the USA) regularly give patients medicine, or prescribe treatment routines, in order to please not cure the patient; in some cases, just to 'get them out of the surgery' (Alcock 2018, p. 336). It is a mistake to speak of placebo 'effects' as the placebo (water, sugar, soothing words, white coat) has no effect. For an individual in a particular circumstance, it has no effect but generates a response. So the expression should be 'placebo response'.

Freudian medical tradition is part of the cashing in has been a matter of century-long debate about its scientific or otherwise status.

German physicians, scientists, and philosophers did persist with Mesmer's ideas; the second volume of Mesmer's collected studies was published in Berlin in 1815. Karl von Reichenbach (1788–1869), the prominent German chemist, engineer, industrialist, and esteemed member of the Prussian Academy of Science, claimed to have discovered the hitherto unknown Odic force which cannot be detected by any instrument, but can only be seen and felt for special individuals who he labelled 'sensitives' of which he was one, others being some young girls in his circle. All substances were either Odic positive, negative, or neutral. The positive substances were warm and disagreeable and emitted a red light; the negative substances were cool and agreeable and emitted a blue light. Odic force extended through the entire universe. Reichenbach thought that his Odic force could explain many hitherto puzzling phenomena, so the frequent appearance of ghosts in graveyards is just Odic light emanating from decaying corpses. While experimenting he:

> Maintained a strict regimen of rest and diet and refrained from touching metals all day. Then at evening he held the hand of his sensitive while she reported hour by hour the variations of odic force transmitted through the Baron's hand; all of which is charted in curves, as professionally as the record of an experiment in thermodynamics. (Jastrow 1935/1962, p. 347)

Parallels with feng shui practice are obvious, but Reichenbach did not have the whole superstructure of chi cosmology to support his lonely speculation.

Though mesmerism, Odic forces, and magnetic therapy had been rejected by the scientific and medical establishments by the early nineteenth century, it lived on in 'alternative' circles. And it emerged full-blown and unembarrassed in the late twentieth century. On 11 August 1997, the American *ABC World News Health Report* had a programme describing 'biomagnetic therapy', a multimillion-dollar business sweeping across international sports medicine, and beyond locker rooms. The therapeutic magnets, costing $89, were being sold in a thousand golf pro shops. On the programme a plastic surgeon explains how 'magnets provide a static or magnetic force that allows changes in the tissue'. A physical therapist informs viewers that 'magnets are another form of electric energy that we now think has a powerful effect on bodies'. The CEO of the main magnet firm explains that 'All humans are magnetic, every cell has a positive and negative side to it'. A graphic shows blood being drawn to damaged tissue by the magnet. Viewers probably think that it is OK as everyone knows that blood contains iron. But the iron in haemoglobin is in a chemical state that is not ferromagnetic; there is absolutely zero scientific basis for the claim that blood is influenced by magnetic fields.[9] Although a multimillion-dollar business, it is really a multimillion-dollar racket. It is simply another incarnation of the kind of fraud and racketeering that Matteo Ricci identified being perpetrated by feng shui practitioners in sixteenth-century China. Five centuries later, little has changed.

---

[9] This business, and more on contemporary magnetic therapy, is detailed in Park (2000, pp. 58–62).

Versions of the whole mesmeric picture can be found in ancient and modern feng shui manuals. For example, a 2002 publication claims:

> Chi is the cosmic breath of the life force. It is the vital breath and maintains physical, emotional, and environmental balance, and links man to his environment. It connects knowledge and substance. It creates a process of rapid flow and cessation, flowering and wilting, increase and decrease, and so on. Chi is vitality. (Taylor 2002, p. 14)

Some feng shui practitioners point to the affinity with mesmerism as a demonstration of the universality of feng shui cosmology and of how its basic principles have been recognized, under other names, by different cultures in different times (Fenton 1996). Given the history of Mesmerism, this is a doubtful tactic.

Imponderable Fluids

The findings of the French Royal Commission may have dinted enthusiasm for animal magnetism, but not the idea of imponderable fluids. By the middle of the eighteenth century, it was apparent that the scientific mechanical, corpuscularian worldview that underpinned the new science of Galileo, Boyle, and Newton was incapable of coherently explaining the documented and diverse range of magnetic, electrical, optical, and thermal phenomena with which natural philosophers were engaged. Different 'imponderable fluids' were resorted to in order to explain some or all of the mechanical-resisting phenomena (Schofield 1970, chap. 8). As early as 1745, Cadwallader Golden (1688–1776) – an Irish natural philosopher, Governor of the New York colony, friend of Franklin, and 'corrector' of some mistakes in Newton – thought that investigations of electricity would illuminate the 'nature of that subtle, elastic and aethereal medium, which Sir Isaac Newton queries on, at the end of his *Opticks*' (Schofield 1970, p. 159). Benjamin Franklin, Mesmer's *bête noire*, postulated electrical fluid to explain the phenomena he was documenting (Cohen 1956). He thought that such fluid was diffused through all matter and that the earth was its inexhaustible reservoir.

In 1839 Golding Bird (1814–1854), the English physician and chemist, in his *Elements of Natural Philosophy* claimed that:

> The subtle and invisible forms of ethereal matter, when caused to assume a vibratory or undulatory movement with sufficient rapidity, produce a peculiar set of phenomena, whose effects are known by the terms of light and heat; effects of vast importance, for without them nature would be dead to us, its beauties no longer apparent, and this world a cheerless place. (Alexander 2015, p. 2)

A decade later, Henry Morley wrote:

> We are in the present day upon the trace of a great many important facts relating to the imponderable agencies employed in nature. Light, heat, and electricity are no longer the simple matters, or effects of matter, that they have aforetime seemed to be. New wonders point to more beyond. (Alexander 2015, p. 1)

Herman von Helmholtz (1821–1894) in his 1862 lecture 'On the Conservation of Force' said that the extensive investigations of thermal phenomena by eighteenth- and early-nineteenth-century natural philosophers documented processes that 'could at that time be explained in no other manner than that heat is a substance [caloric]' (Helmholtz 1862/1995, p. 114). That is, heat is a fluid; just how imponderable was to be investigated.

The new scientific era of exact measurement, sophisticated instruments, and precision is well demonstrated in nineteenth-century investigations of the putative imponderable fluids. Careful attempts were made to ascertain the weights of the conjectured fluids, after all if they were fluids, they should have weight, but no such weight was ever registered, at least none that was not accounted for by standard measurement deviations. Using ever more sensitive instruments in ever more controlled circumstances, it was found that when iron was heated, it did not gain weight and when pith balls were electrostatically charged, they did not gain weight. But as natural philosophy uncovered many strange things, some thought that a weightless fluid could just be added to the list of remarkable things; such a supposition was not prima facie ruled out of court nor out of science.

These imponderable fluids are perhaps the closest that modern European science has come to the universal chi of Asian feng shui. But the former were carefully and experimentally investigated in the way that the latter never was, and eventually the whole programme of imponderables was abandoned by Western scientists. William Grove (1811–1896), in his Presidential Address to the 1866 Nottingham meeting of the British Association for the Advancement of Science, could say:

> … when we pass to the consideration of those other attributes of matter which were at one time supposed to be peculiar kinds of matter itself, or, as they were called, 'imponderables', but which are now generally, if not universally, recognized as forces or modes of motion, we find the evidence of continuity still stronger. (Grove 1866/1970, p. 90)

Grove was convinced, as also was Helmholtz, by the works of Humphry Davy who melted ice by sliding one block on another within a cooled and insulated container; there was no outside source of heat or imponderable fluid, yet ice moving on ice seemingly created heat sufficient for melting.

Benjamin Thompson (Count Rumford 1753–1814) began doubting the caloric fluid theory when he weighed water then froze it and could not detect any loss of weight as the departure of caloric might predict; even with scales sensitive to one part in a million, he could detect no loss of weight. More famously Rumford in his 1798 *Philosophical Transactions* paper – 'An Inquiry Concerning the Source of the Heat which is Excited by Friction' (Thompson 1798) – had noted the immense heat generated in boring brass cannons in the Munich Arsenal where the drill bit was powered by two horses tethered to either end of a rotatable bar.[10]

Beyond noting the heat, in the characteristic manner of modern science, he quantified it by having the brass bored in a weighed water bath and measuring its rise in

---

[10] He further noted that the measurable heat generated by the boring was roughly the same as the heat generated by burning the hay eaten by the working horses! This observation and measurement might be seen as the beginning of biophysics.

temperature. On the imponderable fluid account, the caloric was supposedly released from the brass rod when the filings were opened and separated; but he found that these discarded filings had the very same specific heat as the original brass from which they had been bored. If they had given up their igneous fluid, they should have carried less heat than untouched strips of new brass. He heated both the filings and brass strips from the cannon and put both into the same quantity of water at the same temperature, and both waters rose exactly to the same temperature. He wrote:

> The heat of the mixture, at the end of 1 minute, was just 63^0 as before. ...Each of the above experiments was repeated 3 times, and always with nearly the same results. (Thompson 1798, p. 82)

After other experiments, all described in detail so that they could be reproduced by anyone reading his Royal Society account, he concluded:

> the capacity for heat, of the metal of which great guns are cast, *is not sensibly changed* by being reduced to the form of metallic chips, in the operation of boring cannon. (Thompson 1798, p. 87)

He noted 'that the source of the heat generated by friction, in these experiments, appeared evidently to be *inexhaustible*' (Thompson 1798, p. 99) and added:

> any thing which any *insulated* body, or system of bodies, can continue to furnish *without limitation*, cannot possibly be a *material substance*. (Thompson 1798, p. 99)

After other controlled experiments, Rumford proposed that heat was the consequence of motion not of the transfer of caloric or igneous fluid. He says that he could not form any idea of how:

> the heat was excited and communicated in these experiments, except it be MOTION. (Thompson 1798, p. 99)

Rumford refined this claim by ingeniously showing that heat was only the consequence of *resisted* motion. He had a set-up of wheels turning wheels, turning other wheels, with the last turning atop a piece of phosphorus; it was only when the last was braked did the phosphorus below it explode (Grove 1866/1970, p. 91).

He was conscious of Newton's methodological advice against 'feigning hypotheses',[11] writing:

> But, although the mechanism of heat should, in fact, be one of those mysteries of nature which are beyond the reach of human intelligence, this ought by no means to discourage us, or even lessen our ardour, in our attempts to investigate the laws of its operations. (Thompson 1798, p. 100)

---

[11] In the General Scholium of the *Principia*, Newton wrote:

> I have not as yet been able to deduce from phenomena the reason for these properties of gravity, and I do not 'feign' hypotheses. For whatever is not deduced from the phenomena must be called a hypothesis; and hypotheses, whether metaphysical or physical, or based on occult qualities, or mechanical, have no place in experimental philosophy. (Newton 1726/1999, p. 943)

On this contentious subject, see at least Harper (2011, pp. 343–346) and Wootton (2015, pp. 388–391).

Helmholtz in his 1862 lecture, delivered 4 years before Grove's BAAS address, surveyed the theoretical and experimental field and concluded:

> These facts no longer permit us to regard heat as a substance, for its quantity is not unchangeable. It can be produced anew from the *vis viva* of motion destroyed; it can be destroyed, and then produces motion. We must rather conclude from this that heat itself is a motion, an internal invisible motion of the smallest elementary particles of bodies. (Helmholtz 1862/1995, p. 118)

The idea of 'imponderable fluids' was abandoned in physics, though some qigong promoters are resuscitating the notion under the banner of 'subtle energy' and saying that this is the physicist's name for chi energy (Wozniak et al. 2001, p. 130). This is nonsense. But in the absence of decent science education, or even of a sceptical disposition, who is to know?

Spiritualist Science

In the mid-nineteenth century, beginning in the USA, there was an outbreak of enthusiasm for the influence of spirits in the world, of communications with the dead, and of engagement with 'the other side'. This constituted the spiritualist movement, which with some ups and downs is still extant, now mostly taking the form of parapsychology.[12] Spiritualism began in 1848 with the young Fox sisters from Hydesville New York and their travelling road show of séances and ghostly interventions causing tables to move, bowls to shake, pointers to swing, Ouija boards to reconfigure, and so on. Soon mediums were setting up business everywhere on both sides of the Atlantic offering to put audiences in contact with their departed loved ones (Brandon 1983). The words 'telepathy' and 'medium' made their appearance in print. In 1855, the US chemist Robert Hare wrote of his experience at one séance where he was convinced the spirit of his long-dead father was in the room and communicating via a moving pointer (Hines 2003, p. 49). His scientific reputation and testimony were great boosts to the spirit business. The movement known as *spiritualism* spread and resonated way beyond its Anglo-American roots.

In June 1853 the *Illustrated London News* was disappointed and surprised that:

> Railroads, steam, and electricity, and the indubitable wonders which they have wrought, have not proved powerful enough to supersede and destroy that strong innate love of the supernatural which seems implanted in the human mind. Thousands of people in Europe and America are turning tables, and obstinately refusing to believe that physical and mechanical means are in any way connected in the process. (Noakes 2004, pp. 25–26)

For a period in the late-nineteenth century, a community of British, American, and European scientists entertained the notion of spirits as legitimate scientific enti-

---

[12] See Brandon (1983, chap. 3), Ferguson (2012), Lamont (2013), Noakes (2004), Oppenheim (1985), Shermer (2001, chap. 8), and Webb (1971, chap. 1).

ties in order to explain psychic phenomena. These included top-rank scientists such as Charles Richet, Robert Hare, Oliver Lodge, Alfred Russel Wallace, William Crookes, Gustav Fechner, and J.J. Thompson. In 1866 Alfred Wallace (1823–1913) wrote a defence of spiritualism, saying:

> A little enquiry into the literature of the subject, which is already very extensive, reveals the startling fact, that this revival of the so-called supernaturalism is not confined to the ignorant or superstitious, or to the lower classes of society. On the contrary, it is rather among the middle and upper classes what the larger proportion of its adherents are to be found …. And among those are numbers of literary, scientific, and professional men. (Shermer 2001, p. 190)

In 1885 the Austrian Archduke Johann was alarmed at the spread of spiritualism writing:

> That this modern superstition flourishes not only among the weavers of the Braunauer country, or among the workmen and peasants in Reichenberg, but it has also fixed its abode in numerous palaces and residences of our nobility, so that in many cities of the monarch, and especially in Vienna and Buda-Pesth, entire spiritualistic societies exist, carrying on their obscure nuisance without any interference. (Webb 1971, p. 16)

In a 1785 autobiographical essay, Wallace wrote of his journey to spiritualism with which many feng shui adherents would identify:

> Up to the time when I first became acquainted with the facts of spiritualism, I was a confirmed philosophical sceptic, rejoicing in the works of Voltaire, Strauss, and Carl Vogt, and an ardent admirer (as I am still) of Herbert Spencer. I was so thorough and confirmed a materialist that I could not at that time find a place in my mind for the conception of spiritual existence, or for any other agencies in the universe than matter and force. Facts, however, are stubborn things. My curiosity was at first excited by some slight but inexplicable phenomena occurring in a friends's family, and my desire for knowledge and love of truth forced me to continue the inquiry. The facts became more and more assured, more and more varied, more and more removed from anything that modern science taught, or modern philosophy speculated on. The facts beat me. (Shermer 2001, p. 197)

But what were the facts that drove Wallace's journey? In 1882 the Society for Psychical Research was established to scientifically investigate the alleged, and widely believed, medium-induced occurrences and phenomena. Its journal has been published quarterly and continuously since 1884. Initial areas of research were thought transference, mesmerism, mediumship, Reichenbach phenomena (also known as Odic force), apparitions, haunted houses, and séances.

Oliver Lodge (1851–1940), the knighted, physicist, socialist, Christian, and pioneer of radio theory and technology, was one of the most prominent spiritualists (McCabe 1914), Lodge did significant original work on electromagnetic radiation that convinced him of the reality of the ether, and he came to believe that this was the realm of real, active, causally empowered spirits, including that of his loved son Raymond who had been killed in the Great War and with whom he still regularly communicated (Lodge 1909, 1916, 1930). Many, including Lodge, saw a scientifically approved spiritualism as a counter to the ungodly materialism associated with nineteenth-century science. Among other things, it reconciled theistic belief in the human soul, shared by nearly all religions, with Darwinian accounts of the evolution of *homo sapiens* which did not provide for the onset of souls. The First Spiritual Temple website says of Lodge that:

> Sir Oliver sought to bring together the transcendental world with the physical universe. He affirmed, with great conviction, that life is the supreme, enduring essence in the universe; that it fills the vast interstellar spaces; and the matter of which the physical world is composed is a particular condensation of ether for the purpose of manifesting life into a conscious, individual form.[13]

Given the repute and calibre of the 'spiritualist' scientists, it is clear that the intellectual heart of the late-nineteenth-century debate – carried on in books, journals, conferences, newspapers, clubs, fairs, and meeting halls – was not between scientists and pseudoscientists or between science and religion; it was about the proper understanding of science; and it was, at core, an internal debate among scientists about philosophy of science. The spiritualists were scientists and understood themselves as advancing, or opening up, a scientific research field.

Clearly not all believers saw spiritualism as endorsement of religious, specifically Christian, faith and ontology; many were wary of being associated with something functioning outside of official channels and likely to blow up. Michael Faraday (1791–1867), the founder of electromagnetic theory, was a lifelong devout Christian of the Sandermanian sect who in 1853 dedicated serious experimental effort investigating spiritualist phenomena. His conclusion was there is no such specifically *psychic* phenomenon; the visible movements, once deliberate fraud was discounted, were the consequence of unconscious muscular movement of the psychic (Hyman 1985). Psychologists have now recognized this as an outcome of the medium being in a *dissociative state*. This state can, and has been, scientifically investigated without recourse to non-natural entities or forces.

However, apart from all other considerations, Einstein's relativity theory rendered Lodge's ether redundant (Einstein 1922). The ether followed crystalline spheres, *impetus, elan vital*, phlogiston, and caloric out of the vocabulary and practice of science. Spirits were homeless. By the 1940s, scientific spiritualism had disappeared. Although some maintained their beliefs and even established the supposed discipline of parapsychology (Grove 1985), most of the spirit-scientists came back to the 'fold' or drifted out of science. For some scientists, the ether remained, but it had stricter occupancy rules: 'no spirits here'.

It is yet another testament to the credulity of people that spiritualist enthusiasm was barely dinted by early revelations that the Fox sisters were charlatans who rigged all of the 'psychic' phenomena dramatically seen at their séances. And predictably, a good many other mediums were exposed as frauds (Kurtz 1985).

The contrast of spiritualism with feng shui is instructive. They both posited immaterial causally effective entities (spirits and chi), to account for phenomena; and they asserted parallel material and nonmaterial universes. But to their credit, most of the spiritualists took science seriously. They engaged with it, tried to quantify and measure their posited entities, and, in the end, were prepared to walk away

---

[13] See http://www.fst.org/lodge.htm

from the spiritualist edifice when the accumulated scientific evidence failed to support it. There is no parallel in feng shui: it is not conducted by scientists and does not engage with science. It is a pseudoscience.

Conservation of Energy

The numerous different 'forces' were found to be convertible: chemical force into electricity (Volta), electricity into magnetism (Oersted), magnetism into electricity (Faraday), heat into electricity (Ritter and Scheele), mechanical energy into heat (Joule), and work into heat (Rumford). Not only were they convertible, but by diligent and carefully controlled experiment, they were shown to be *precisely, uniformly,* and *always* convertible. For instance, Helmholtz cited Joule's experiments on conversion of work to heat and his formulation of a *unit* of heat as 'the heat necessary to raise one gramme of water through one degree centigrade' (one calorie) saying that it is equal to 'the work which a gramme would perform in falling through a height of 425 metres'; Helmholtz cited other experimenters who found equivalences of 426.3 m, 425.3 m, and 424.9 m (Helmholtz 1862/1995, p. 117). The mounting evidence for conversion undermined the explanatory power of the imponderable fluids; they were spectators, left sitting idly and doing no work. At the same time, the nineteenth-century conversion experiments prepared the ground for postulation of a common something – energy – which was conserved in the conversions.

For Helmholtz, and others, all of this pointed to a foundational conservation law[14]:

> The law in question asserts, that the *quantity of force which can be brought into action in the whole of Nature is unchangeable,* and can neither be increased nor diminished. (Helmholtz 1862/1995, p. 98)

He explains that 'quantity of force' is 'more popularly expressed … as *amount of work* in the mechanical sense of the word' (ibid).

The energy-focused conservation and unification movement of the nineteenth century culminated in the great German chemist, Nobel Laureate, and monist, Wilhelm Ostwald's (1853–1932) *Energetics* programme (Hakfoort 1992; Ostwald 1901). Convinced by Mach's empiricist critique of atomism, Ostwald held that energy was a substance, indeed it was the *only* substance, with everything else being its manifestation. It was an alternative to both materialism and idealism; it could be seen as a move from Western to Eastern science. But energetics as a scientific programme collapsed; what became entrenched was the understanding that energy is a

---

[14] Thomas Kuhn names Helmholtz and 11 other natural philosophers who between 1842 and 1847 independently enunciated a conservation of force law by which a single quantity or feature was conserved across its numerous manifestations – dynamic, thermal, electrical, chemical, etc. He said: 'The history of science offers no more striking instance of the phenomenon of simultaneous discovery' (Kuhn 1959/1977, pp. 68–69).

property of something, not a process, much less the universal process of Ostwald. He was correct in seeing energy as universal, but this is because it is a universal property of all things that exist; there can be no energy without things; there is no 'free-floating' energy. Fragments of the energetics programme can still be read in different popularizes of feng shui and Eastern mysticism.

It was not just better technical solutions or more efficient industrial processes that emerged from the multitude of natural philosophers conducting all this exacting and public measurement, observation, experiment, and hypothesis testing. A scientific outlook on the world, a scientific or naturalistic worldview, emerged. Helmholtz painted it in moving terms in the conclusion of his 1862 lecture:

> What I have to-day mentioned as to the origin of the moving forces which are at our disposal, directs us to something beyond the narrow confines of our laboratories and our manufactories, to the great operations at work in the life of the earth and of the universe. The force of falling water can only flow down from the hills when rain and snow bring it to them. To furnish these, we must have aqueous vapour in the atmosphere, which can only be effected by the air of heat, and this heat comes from the sun. The steam-engine needs the fuel which the vegetable life yields, whether it be the still active life of the surrounding vegetation, or the extinct life which has produced the immense coal deposits in the depths of the earth. The forces of man and animals must be restored by nourishment; all nourishment comes ultimately from the vegetable kingdom, and leads us back to the same source.
>
> You see then that when we inquire into the origin of the moving forces which we take into our service, we are thrown back upon the meteorological processes in the earth's atmosphere, on the life of plants in general, and on the sun. (Helmholtz 1862/1995, p. 126)

Here is the germ of the we-are-all-in-it-together, environmentalist, naturalistic worldview which 150 years later underpins most global governmental, scientific, cultural, and educational climate change and environmental education efforts.

Many accounts of feng shui theory or principles seem, superficially, to echo this mid-nineteenth-century depiction of the modern naturalistic worldview. Consider, for instance, one such statement:

> Feng shui doctrine is encapsulated in a saying from the *Book of Burial* [Pu 300 BC]: 'Ch'I blows as wind, ascends to the clouds, falls as rain, nourishes the soil and gives birth to everything on earth'. Since energy is the universe, the goal of feng shui is not to obstruct or harness it. (Rolnick 2004, p. 31)

Others talk of yin and yang as forces of attraction and of repulsion. Angle and Tiwald in their *Neo-Confucianism* book translate *qi* as 'vital stuff'. Such commonly repeated statements can give an appearance of scientificity to feng shui; but really such statements are simulacrums: they are deceptive substitutes for serious scientific claims or analysis. Scientific words are sprinkled over pages without attention to their scientific meaning or their coherence in the page. Feng shui's energy-related claims have been largely stable for 3000 years; they are little affected by experiment nor informed by the critical spirit which is so evident in the history of elaboration, refinement, and correction of notions of force and energy in orthodox science.

Feng shui theory is not informed by development in other sciences, something which is a characteristic of all scientific disciplines. Utilizing John Donne (1572–

1631), a contemporary of Galileo: 'No science is an island entire of itself; every science is a piece of the continent'. Indeed, cross-disciplinary borrowing and information-sharing is so normal that 'mixed disciplines' are a standard marker of the growth of science: geochemistry, biochemistry, electrochemistry, biophysics, geophysics, paleobotany, paleoanthropology, and so on. Such hybrids as there are in feng shui – romantic feng shui, monetary feng shui, and lucky feng shui – are merely 'come here with your money' banners for the gullible; they are not mixed sciences.

Teaching About Feng Shui in the Science Programme

One obvious and non-controversial way to raise issues about feng shui in science classes is to do so when teaching about energy. The saturation of everyday conversation, TV, newspapers, and social media with concerns about new forms and sustainability of energy provides an appropriate base for such teaching. But the very same saturation gives rise to multiple and well documented misunderstandings and barriers to learning the science of energy: conceptual and ontological confusion abounds for both teachers and students. Such confusion is predictable as the term 'energy' has had a documented history of at least 2500 years and it is only in the past 150 years that the modern and coherent scientific meaning of the concept has emerged while leaving a forest of other everyday and technical meanings intact.

Most of the everyday talk – for example, of 'energy creation' – are simply inconsistent with scientific understanding. Separating out the scientific meaning of 'energy' from the 'buzzing confusion' of everyday multiple meanings is a prime task of science education. In the foregoing case, all talk of energy 'creation' should be instead of energy 'conversion'. A simple matter, but a whole worldview, pivots on it. And this clarifying task also then involves separating out the scientific understanding of energy from a host of scientifically related concepts with which it is oft confused – power, force, momentum, work, impulse, entropy, and so on.[15] 'Energy' is a polysemic word.

From the earliest age, children hear about and are taught the need for 'responsible use of energy'. Later they are taught about the global energy crisis; about the deleterious effects of fossil fuel as an energy source; about the importance of solar, wind, and other 'green' or renewable energy sources; about the dangers associated with domestic use of nuclear energy and the catastrophic impact of its military use; about being environmentally responsible and utilizing low-energy appliances and transport; and about many other energy-related matters. Energy is a recurrent concept in the marketing of drinks, foods, clothing, and 'supplements' or potions. Seemingly everything good for a person either saves, boosts, or gives them energy;

---

[15] Among hundreds of substantial studies on the teaching and learning of energy, see at least Arons (1999), Chen et al. (2014), Coelho (2009), Duit (1986, 1987), Goldring and Osborne (1994), Papadouris and Constantinou (2011), William and Reeves (2003), and contributions to Bevilacqua (2014a). A comprehensive review of the field is provided by Bächtold and Guedj (2014).

just as cars are high or low in petrol, so people are high or low in energy. Chinese-influenced Asia has all of the foregoing and much more because feng shui discussion and deliberation is a commonplace part of life for so many; it is part of the woof and warp of much Asian culture.

Throughout the world at all educational levels – elementary, secondary, and college – energy is a ubiquitous topic in science programmes simply because energy is ubiquitous in nature. The study of energy is one of the core requirements in the English school science programme where students are expected to detail different sources of energy and know how different cultures utilize these sources. Energy is also one of the core 'cross cutting concepts' in the USA *Next Generation Science Standards* (NRC 2013).[16]

Energy Literacy

In the USA, the Department of Energy (DoE) produced an educational guide that specifies an 'energy literate' person has:

> An understanding of the nature and role of energy in the universe and in our lives. Energy literacy is also the ability to apply this understanding to answer questions and solve problems. (Department of Energy 2012, p. 4)

The US Department further elaborates that such a person 'Can assess the credibility of information about energy'.

Given the centrality of 'energy' in feng shui discourse and practice, and given the wide influence and impact of feng shui in Asian and now Western culture, these DoE objectives can assuredly be advanced by explicit treatment of feng shui in science classes. The DoE guide can reasonably be taken to mean that students be able to assess the credibility of feng shui energy claims. If daily feng shui forecasts are on the morning news, then asking a class to assess their credibility is hardly an outside imposition on the class.

All feng shui claims make reference to energy. As previously noted, Elliot Jay Tanzer, a modern astrologer and feng shui consultant, says of chi that: '*Qi* is the Chinese word for "energy." Everything animate and inanimate, real or conceptual, has *qi*'.[17] Many thousands of feng shui websites attest to the centrality of this primeval, all-pervasive, encompassing 'peculiar' energy that has such a multitude of amazing properties. One site authoritatively announces:

> The dynamic movement of energy changes with the passing of time — in the year, month, day, and hour. Flying Star Feng Shui shows you how these changes and movements of energy can affect space and people. That's why Flying Star is considered one of the most dynamic and effective techniques in Classical Feng Shui. It is used not only to assess the

---

[16] Papers presented at an international conference to link research to the teaching of energy in the NGSS can be seen in Chen et al. (2014).

[17] See http://abodetao.com/feng-shui-guidelines-to-energy-flow-analysis-what-is-*qi*-and-how-*qi*-flows/

current condition (i.e., the energetic quality) of the home, but it can also be used as a valuable forecasting tool.[18]

HPS-Informed Teaching

The confusions about energy are mirrored in teaching about force. In the popular mind, force is connected to energy and was an important stepping stone – via the metaphysical idea of *vis viva* (vital force) and the law of inertia – in development of the science of energy. Max Jammer in his book on *Concepts of Force* writes:

> It is the purpose of this book to clarify the role of the concept of force in present-day physics. It is pointed out how ancient thought, with its animistic and spiritual interpretations of physical reality, laid the foundations for the development of the concept and how in pre-classical science the concept of force became invested with a multitude of extra-scientific connotations that greatly influenced the interpretation of the concept until very recent times. (Jammer 1957, p. viii)

He justifies his book because:

> … in our present system of academic instruction a thorough and critical discussion of fundamental and apparently simple concepts in science is consciously omitted. … A historico-critical analysis of the basic conceptions in science is therefore of paramount importance, not merely for the professional philosopher or historian of science. (Jammer 1957, p. vii)

Such teaching is also of paramount importance for science teachers. But even a moderately sophisticated understanding of energy requires a number of things: first, some historical awareness of how the concept emerged from investigations in different domains of physics and then chemistry and biology; second, some epistemological appreciation of how scientific claims about energy, in particular the conservation law, are justified partially by experiment and partially by fiat; and third, some ontological awareness about what energy is – is it the state of some thing (or multiples of things) or is it itself something?

Such HPS-informed teaching of energy results in better and deeper understanding of the science of energy and also a richer understanding of the nature of science (NOS), something which is increasingly a separate curriculum objective. Two Cypriot researchers have written:

> We propose a teaching approach, for students in the age range 11–14, that introduces energy as an entity in a theoretical framework that is invented and gradually elaborated in an attempt to analyze the behavior of diverse physical systems, and especially the various changes they undergo, using a coherent perspective. This theoretical framework provides an epistemologically appropriate context that lends meaning to energy and its various features (i.e., transfer, form conversion, conservation and degradation). (Papadouris and Constantinou 2011, p. 961)

Commitment to a 'coherent perspective' is simply a precondition for a scientific treatment of energy. It should be obvious that the entry of chi into the picture, or equation, makes the treatment incoherent.

---

[18] http://www.fengshuitoday.com/shop/advanced-flying-star-feng-shui-gm-dr-stephen-skinner/

Photosynthesis is rightly a universal component of science programmes. It is a topic that well illustrates the contribution of HPS to good science teaching and also shows the superfluousness, vacuity and irrelevance of the chi concept. Proper teaching of photosynthesis exposes students to a scientifically revealed marvel of nature; this should diminish the need to seek marvel or awe outside the workings of the world that science has illuminated. All animate and inanimate processes on earth depend on energy whose ultimate source is solar radiation from the sun, which in turn derives from nuclear fusion inside the sun. All life on earth is predicated on the property of the chlorophyll molecules of green plants to utilize sunlight to convert carbon dioxide and water into organic compounds (starch) for animal and human food and oxygen for animal and human breathing. Without this action of chlorophyll, called photosynthesis, all life on earth would slowly come to an end and indeed would not have started. The complementary processes are:

Photosynthesis:

$$\text{Carbon dioxide} + \text{Water} + \text{Energy}(\text{light}) \rightarrow \text{Organic compounds}(\text{starch}) + \text{Oxygen}$$

Respiration:

$$\text{Organic compounds} + \text{Oxygen} \rightarrow \text{Carbon dioxide} + \text{Water} + \text{Energy}$$

The first process represents both 'carbon capture' and the restoration of air by oxygen. The second process shows that when buried coal seams are dug up or forests cut down and burnt, there is massive carbon dioxide release. The two processes are major components of the earth's 'carbon cycle'. Clearly it is important for students to learn about these processes and their wider social, economic, cultural, and ethical dimensions and impacts. This is part of responsible citizenship. But additionally there is great value in learning about how these fundamental processes came to be discovered and understood; such learning provides appreciation and understanding of the nature of science and the scientific enterprise. And it provides enough marvel and awe to make yin-yang and chi talk less captivating.

The English polymath Joseph Priestley (1733–1804) and the Dutch medico-scientist Jan IngenHousz (1730–1798) were the separate discoverers of photosynthesis. Priestley was a most remarkable man.[19] Frederic Harrison in his introduction to a nineteenth-century edition of Priestley's *Scientific Correspondence* said of him:

> If we choose one man as a type of the intellectual energy of the eighteenth century, we could hardly find a better than Joseph Priestley, though his was not the greatest mind of the century. His versatility, eagerness, activity, and humanity; the immense range of his curiosity in all things, physical, moral, or social; his place in science, in theology, in philosophy, and in politics; his peculiar relation to the Revolution, and the pathetic story of his unmerited sufferings, may make him the hero of the eighteenth century. (Bolton 1892, Introduction)

---

[19] The definitive biographical study of Priestley is Robert Schofield's two-volume work (Schofield 1997, 2004). See also Matthews (2015a, chap. 7) and literature cited therein.

Priestley did not begin serious chemical studies until his early 30s, during his ministry at the Leeds Presbyterian Chapel (1767–1773). In quick succession, by utilizing a new method of collecting airs over water and mercury, and by utilizing a new and massive burning lens as a source of heat, he created, isolated, and listed properties of a dozen or more of the major 'airs'. The experiments and investigations of airs, conducted in Leeds by Priestley, were announced to the scholarly world in a series of talks he delivered to the Royal Society in London in March 1772. The talks subsequently were published as his famous 118-page paper in the Society's *Philosophical Transactions* of the same year – 'Observations on Different Kinds of Air' (Priestley 1772). This paper was translated into many European languages; it was widely read, including by Lavoisier; and in 1773 it was awarded the coveted Copley Medal of the Royal Society – the eighteenth-century equivalent of the Nobel Prize.

In the Royal Society address, Priestley noted:

> The quantity of air which even a small flame requires to keep it burning is prodigious. It is generally said, that an ordinary candle consumes, as it is called, about a gallon in a minute. Considering this amazing consumption of air, by fires of all kinds, volcanoes, etc. it becomes a great object of philosophical inquiry, to ascertain what change is made in the constitution of the air by flame, and to discover what provision there is in nature for remedying the injury which the atmosphere receives by this means. Some of the following experiments will, perhaps, be thought to throw a little light upon the subject. (Priestley 1772, p. 162)

Priestley's Christian worldview motivated this quest: with centuries of animal and human respiration, plus volcanoes and natural fires, the atmosphere should be progressively rendered unfit for human life; but there were theological reasons why this could not happen. A beneficent all-powerful creator would not design such a world; God must have made some provision for the natural restoration of air. Priestley's first thought, or hypothesis, was the commonsensical one: as air is necessary both for animal and vegetable life, then both animals and plants must process air in the same manner. But experiment led him to reject this idea. As he wrote:

> One might have imagined that, since common air is necessary to vegetable, as well as to animal life, both plants and animals had affected it in the same manner, and I own that I had that expectation when I first put a sprig of mint into a glass jar standing inverted in a vessel of water; but when it had continued growing there for some months, I found that the air would neither extinguish a candle, nor was it at all inconvenient to a mouse, which I put into it. (Priestley 1772, p. 162)

He continued:

> This observation led me to conclude, that plants, instead of affecting the air in the same manner with animal respiration, reverse the effects of breathing, and tend to keep the atmosphere sweet and wholesome, which it had become noxious, in consequence of animals living and breathing, or dying and putrefying in it. (Priestley 1772, p. 166)

Priestley did suggest a mechanism for the beneficent effect: 'this restoration of vitiated air is affected by plants imbibing the phlogistic matter with which it is overloaded by the burning of inflammable bodies' (Priestley 1775–1777, vol. 1, p. 49), but in keeping with his strict epistemological principle of only giving cautious or provisional status to conjectured unseen mechanisms, he added 'whether there be any foundation for this conjecture or not, the fact is, I think indisputable' (ibid). His distinction between observational facts upon which there could and should be agreement and unseen, putative mechanisms was a fundamental one for Priestley.

He was prepared to think further about the source of the pure air he saw released by green matter and plants in his phials. Initially he thought it came *from* the green matter or leaves, but he was able to devise a nice experimental test of this hypothesis. In September 1779 he wrote to his good friend Benjamin Franklin (1706–1790) relating that:

> Though you are so much engaged in affairs of more consequence [drafting the Declaration of American Independence], I know it will give you some pleasure to be informed that I have been exceedingly successful in the prosecution of my experiments since the publication of my last volume [his *Experiments and Observations on Different Kinds of Airs*].
>
> I have confirmed, explained, and extended my former observations on the purification of the atmosphere by means of vegetation; having first discovered that the *green matter* I treat of in my last volume is a vegetable substance, and then that other plants that grow wholly in water have the same property, all of them without exception imbibing impure air, and emitting it, as *excrementitious* to them, in a dephlogisticated state. (Schofield 1966, pp. 178–79)

Other experiments confirmed that the green matter, along with aquatic green leaves, only produced pure air in the presence of sunlight; heat was no substitute for light. Thus the vegetable hypothesis was restored and indeed extended: not only did vegetation restore atmospheric air depleted by fires and animal respiration; it also restored water that had dissolved unhealthy air and whose dissolved air was being rendered noxious by respiration of fish. Priestley's research on the restoration of air basically finished at this point. Most of the outlines of what would in the late nineteenth century come to be called 'photosynthesis' were in place.[20]

Forty years after Priestley's death, Julius Robert von Mayer (1814–1878), the German chemist, physicist, and founder of thermodynamics, was able to write straightforwardly and without argument in 1845 that:

> Nature set herself the task of capturing the light flooding toward the earth, and of storing this, the most elusive of all forces, by converting it into an immobile force. To achieve this, she has covered the earth's surface with organisms which while living take up the sunlight and use its force to add continuously to a sum of chemical difference. The plants take in a force, the light, and bring forth another force, the chemical difference. (Sherman 2018, p. 240)

Subsequently the exact atomic process whereby light 'ejected' electrons off chlorophyll molecules thus enabling breakdown of carbohydrate molecules has become known (Sherman 2018, pp. 240–245). Chi does not appear anywhere in this truly marvellous process of energy-conserving and life-giving photosynthesis; here was a chance for it to appear, but it did not; it remains scientifically idle.

---

[20] For the history of photosynthesis, see Nash (1948); for the contribution of Jan IngenHousz, see Magiels (2010).

Compare the foregoing accounts of the detailed, prolonged scientific investigation and measurement of energy transformation with what, as mentioned in Chap. 4, the International Feng Shui Society (UK) says of the putative omnipresent reality of chi:

> Chi, or *qi*, is the oriental word for the vital intangible natural energy that emanates from everything in our universe, a combination of both real and abstract forces: energy from the earth's magnetic field, sunlight, cosmic influences, colour vibrations, the nature of our thoughts and emotions, the form of objects, the quality of the air around us. Depending on whether it flows harmoniously or not, chi influences how a place feels and how we feel in it.[21]

This may or may not be a nice word picture, but it is fundamentally nonsense and dangerous nonsense when house and office construction, health, therapy, and planning policies are predicated upon it. It opens a wide door for charlatans of every kind to walk in and exploit people.

A coordinated, cross-disciplinary approach to the inclusion of feng shui in a curriculum could look like Fig. 11.3. Where the columns contain topics in each of the subjects in the curriculum.

Such cross-disciplinary coordination is characteristic of Liberal Education (Carson 1997; Schwab 1949/1978); it is a feature of all proposals for General Education (Bernal 1939/1949; Conant 1945); and it typifies the current international calls for integrated STEM-Humanities, or STEAM (science-technology-engineering-arts-mathematics) education programmes (Kim 2014; Rosicka 2016). This cross-disciplinary pattern has been fleshed out for a more connected teaching of photosynthesis in biology (Matthews 2009c, 2015a, p. 293) and pendulum motion in physics (Matthews 2014b, p. 51). Other authors have done so for other topics.

Otto Blüh in the Preface to his Ernst Mach-inspired 1950s physics textbook indicated one way in which this can be approached:

> This book further offers the student the opportunity of becoming acquainted with the historical and cultural relations of physics, in the belief that the education of a scientist can be advanced most effectively by giving a significant place to the philosophical, social, and moral implications of physical science within the physics curriculum rather than through supplementary so-called humanistic and social studies. (Blüh and Elder 1955, p. vii)

The different school disciplines are connected. This is a matter of intellectual and historical fact. There is no physics without mathematics or without technology and philosophy; and physics grew while intertwined with theology, music, and social forces. There is no biology without chemistry, mathematics, and microscopes; and biology developed in a philosophical and theological context. Integrated or even coordinated teaching brings this cultural reality into focus for students. Having feng shui as a node or science topic in such an integrative curriculum can shed great light on feng shui, on the nature of science, on Chinese history, on cultural transmission and commercialisation, and on much else.

---

[21] http://www.fengshuisociety.org.uk/

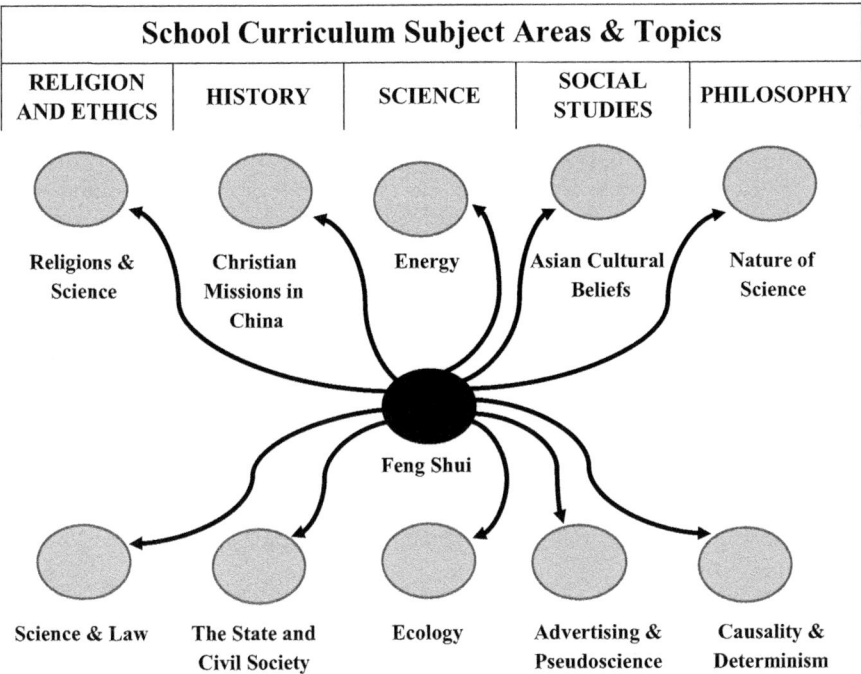

Fig. 11.3 Cross-disciplinary teaching of feng shui

From the beginnings of modern science, 'near enough' was never 'good enough'. For example, the long-standing assumption that the earth was spherical was rejected by Huygens on account of his pendulum clock having to be shortened by 3 mm in order to beat seconds in equatorial regions.[22] By the nineteenth century, precision became the hallmark and norm for all sciences; it was an Age of Precision.[23] It was typified by Sir William Thomson's (Lord Kelvin 1824–1907) oft-quoted opening assertion when giving his 1883 address to engineers on the need for units of measurement in electricity:

> In physical science a first essential step in the direction of learning any subject is to find principles of numerical reckoning and methods for practicably measuring some quality connected with it. I often say that when you can measure what you are speaking about, and express it in numbers, you know something about it; but when you cannot measure it, when you cannot express it in numbers, your knowledge is of a meagre and unsatisfactory kind; it may be the beginning of knowledge, but you have scarcely, in your thoughts, advanced to the stage of *science*, whatever the matter may be. (Thomson 1891–1894, vol. 1, p. 73)

The implication of Thomson's position for chi-based feng shui thinking is clear: it is not science. And given that thinking, or at least talking, about chi has been going on for three millennia, it is unlikely anytime soon to become science.

---

[22] The episode is discussed in Matthews (2015a, pp. 232).
[23] See Winchester (2018) and contributions to Wise (1995).

Multicultural Considerations

All of the foregoing is predicated on the premise that modern science is the most reliable and vindicated source of knowledge about the natural and social worlds that we have and so should constitute the content of science programmes. But in the past half century, this unremarkable premise has been challenged by multicultural educators appealing on the one hand to constructivist epistemology wherein no epistemological distinction or ranking can be made between different systems of cogent belief about the natural and social worlds and on the other hand to ethical considerations fuelled by awareness of the bleak and bloodied history of Western colonial and imperialist traditions (Matthews 1994, chap. 9). Multiculturalists affirm that feng shui is an integral part of Chinese, and more broadly Asian, culture and so needs to be considered and appraised from a multicultural perspective; specifically, its truth claims cannot be challenged from the standpoint of 'alien' modern or Western science.

There is a spectrum of alternatives that have been adopted by science teachers when teaching science in multicultural situations where fundamental cultural beliefs clash with science:

Imperialist, where traditional understandings of nature and phenomena are ignored and modern science is taught as it is in the metropolitian centres: PSSC Physics in Polynesia, Nuffield Science in Newfoundland, CBA Chemistry in Colombia, NGSS in New Guinea, and so on. Traditional beliefs and systems are only attended to in order to prepare the ground for new knowledge which will supplant the old.

Integrationist, where alternative understandings and ways of thinking about nature are recognized, respected, and made use of, but in the last resort only as a more effective means of having students learn about modern science. Ethnoscience is dealt with in an anthropological, non-judgemental way: what other cultures believe and the reasons for their beliefs are pointed out, and students can make their own decisions.

Non-interventionist, where ethnic or traditional science is recognized as an intellectually legitimate alternative to modern science and cultivated in its own terms along with varying degrees of modern science and technique. In some places both traditions are fully taught, and a 'best of both worlds' approach is taken; in other places, just traditional sciences are taught.

The non-interventionist view is partly supported by ethical and political considerations that say that existing belief systems and cultures need to be respected and that only internally instigated changes ought be sanctioned. Australian aboriginal culture is saturated with spirit belief; beliefs about landforms, lakes, and rivers being formed by movements of the Dingo Spirit or other primeval spirit sources. Breaking traditions or taboos invites the attention of bad spirits and much more. Engaging with this 'spirit world' is a serious and pressing educational issue. As one African educator has said:

> The purpose of education in Africa is not to destroy its own civilization or its own culture, in order to replace it with something that is conceived to be 'better'. To proceed in that direction or with that implicit attitude is to create unnecessary difficulties in science education in Africa. (Urevbu 1987, p. 8)

And another maintains that it is:

> Unfair if Western science is forced upon students who do not share its values, meanings, or practices. (Lee 1999)

It needs to be acknowledged that non-intervention is an easier option when serious practical matters of life and death are not in play; leave traditional beliefs and worldviews alone provided no significant damage is being done. There is a powerful ethical argument for respect and acknowledgement and in many countries such as New Zealand, a legal argument. But non-intervention is severely tested when, for example, in New Zealand some decade ago Maori parents were criminally charged for gouging out the eyes of their young son because they thought his recurring epilepsy was due to spirit possession and it was only by taking out his eyes could the evil spirit be released. Their defence of traditional belief was not upheld. Overwhelmingly Pakeha and Maori were in agreement.

The same sets of argument would be mounted for the maintenance of those parts of Asian culture for which feng shui is integral.

It is not intended here to appraise these complex cultural, ethical, and political arguments for non-intervention; instead only the philosophical arguments proffered for the position will be examined. The core epistemological argument for non-intervention is the rejection of universalism as a theory of knowledge. Universalism is rejected in favour of some form of relativism which says that different knowledge systems are equally valid, and so there is no good cognitive reason to introduce modern science to traditional cultures. One can find many statements of this in the literature on multicultural science education. The following are a sample:

> Science is a way of knowing and generating reliable knowledge about natural phenomena. Other cultures have generated reliable knowledge about natural phenomena, therefore reason invites exploration of the possibility that other cultures may have different sciences. But science teachers wanting to celebrate this diversity have been so indoctrinated in the Western cultural tradition of science that they lack a methodology enabling examination of the science of other cultures with little more than tokenism. (Pomeroy 1992, p. 257)

> There is a need to struggle to assert the equal validity of Maori knowledge and frameworks and conversely to critically engage ideologies which reify Western knowledge (science) as being superior, more scientific, and therefore more legitimate. (Smith 1992, p. 7)

> For the developing countries of Africa dominated and governed by non-western sociocultural factors, western science means an imposition of one culture over another. It means the replacement of the anthropomorphic worldview with a mechanistic one. (Jegede 1997, p. 1) ... From this can be seen the need to design science education that satisfactorily meets with the needs of Africa in such a way that the African view of nature, sociocultural factors, and the logical dialectical reasoning embedded in African metaphysics are catered for within a changing global community. (ibid. p. 15)

These observations are sufficient to illustrate the argument of this book, namely, that some core epistemological and ontological assumptions of modern science are in objective conflict with core assumptions of some traditional, including feng shui, belief systems.

The matters of ontological contention are:

1. Is the world constituted in such a way as to serve human interests?
2. Are processes in the world teleological? That is, do events and behaviours occur in order to bring about some fitting end state?
3. Are inanimate and nonhuman animate processes activated and controlled by spiritual influences?

The modern scientific tradition, which began in the West with Galileo and Newton in the sixteenth and seventeenth centuries, answers 'no' to each of the above questions, while many traditional belief systems, feng shui included, affirm some or all of the propositions. The basic matters of epistemological contention are the following:

1. Does knowledge come from the observation of things as they are in their natural states?
2. Are knowledge claims validated by successful predictions?
3. Do particular classes or authority figures define knowledge or become the custodians of knowledge?
4. Is knowledge a fixed and unchanging system?

Modern science, after centuries of debate, answers 'no' to each of these questions, while many traditional societies, and the core feng shui tradition, affirm some or all of them. Of course, there is some debate about these questions of natural states, prediction, the institutionalization of knowledge, and accretion versus revolutions in knowledge development. But these are in-house or family debates; they do not affect the general incompatibility between modern science and most traditional worldviews or indigenous knowledge systems. Robin Horton in his classic 1971 study drew attention to a fundamental difference between traditional and modern sciences:

> The key difference is a very simple one. It is that in traditional cultures there is no developed awareness of alternatives to the established body of theoretical tenets; whereas in scientifically orientated cultures, such an awareness is highly developed. It is this difference we refer to when we say that traditional cultures are 'closed' and scientifically oriented cultures are 'open'. (Horton 1971, p. 153)

As will be documented in Chap. 13, despite modern China's unequalled material and technological advances, it has the disturbing elements of a 'traditional closed culture'; alternatives to the established body of CCP-dictated theoretical tenets cannot be entertained; doing so is prohibited, illegal and punished by fines, loss of employment or imprisonment. Science is easy to find; scientific attitudes are harder to find.

Conclusion

One can recognize among the pre-Socratic philosophers the slow, awkward attempts to distance their thought about the world from the mythical worldviews that were characterized by anthropomorphic, animistic, and teleological dimensions (Sambursky 1975). There have been the same naturalistic efforts in the Chinese tradition. In traditional China, Needham identifies it with some of the Daoists and Mohists of the same period as the pre-Socratics. In the twentieth century, the thread continued with the New Thought movement and later with Marxist Materialism. The CCP did enormous damage to Chinese science, scientists, and culture with its enthronement of 'Marxism-Leninism-Mao Zedong Thought' (the phrase Deng Xiaoping introduced at the Twelfth Party Congress in 1982) as the judge and arbiter of all economic, social, and cultural questions, including, disastrously, fundamental scientific questions. This should have been seen as insane when first introduced in 1949; its insanity became more obvious with each passing decade. With the upper echelons of CCP losing intellectual and philosophical credibility, it is important that ordinary philosophers, scientists, educators, and teachers themselves grapple with the compound philosophical, scientific, cultural, and educational issues presented by the omnipresence of feng shui belief and practice in Chinese society. The same conclusion of course applies for philosophers, scientists, educators, and teachers in all societies where feng shui belief and practice, or other pseudosciences have taken hold.

Chapter 12
Scientific Testing of Chi (*Qi*) Claims

It is noteworthy that feng shui, such an ancient and still widespread belief about something so all pervasive, powerful, and supposedly a part of nature, is never examined in science programmes. The previous chapter showed that this foregoes an opportunity for students to learn about science, its methodology, its scope, its cultural import, and its utility. It is also noteworthy that despite the centrality of chi claims through all of feng shui theory and practice, including in all variants of Traditional Chinese Medicine (TCM), there has been precious little effort expended in scientifically testing these claims. This chapter will survey such efforts as there have been and will draw conclusions about the scientificity of the chi construct.

Remarkable Qigong Claims

There have been some embryonic steps towards scientific validation of chi claims. Of special interest were those made by Dr Yan Xin, a former TCM practitioner who had worked in or visited different Chinese and US universities. He was a celebrity super qigongist with a national reputation for healing thousands of patients at a distance by generating and casting his own *qi* over them; he made and sold personalized qi-ized healing water for them to drink; he was able to increase the alcohol content of wine by *qi* power and was reported in the papers doing many other astonishing things (Lin et al. 2000, p. 113ff). Some of his lectures were attended by tens of thousands. His followers have founded the International Yan Xin Qigong Association. In 1986 he was attached to the Qigong Cooperative Research Group at the prestigious Tsinghua Technical University in Beijing, and while connected there with established Professors Lu Zuyin and Li Shjengping, he put *qi* claims to a scientific test. Yan's claims were not modest. He said that:

the mind power or *Qi* emitted by a trained Qigong master can influence or change the molecular structure of many test samples, including those of DNA and RNA, even if these test samples are 6 to 2,000 kilometers away from the master. *Qi* can also affect the half-life of radioactive isotopes and the polarization plane of a beam of light as emitted from a Helium-Neon laser.[1]

This is certainly a stunning claim, and, as Karl Popper would say, Yan should be commended for putting his theory's neck on the experimental block. To his credit he conducted these chi-confirming and qigong master-demonstrating experiments. He, with colleagues, published papers purportedly showing how the external *qi* (chi) that he generated could travel distances of several kilometres and bring about phase changes in liquids and change infrared absorption spectra in biological media. Ten of his 'scientific' papers are reproduced in the Yan Xin Qigong Association handbook (Wozniak et al. 2001, Chap. 3), while a number are reproduced and discussed in Lu (1997). One study, for example, was 'The Study of Qigong Effect on Bacteria Strain Improvement' which showed, predictably, that the 'high-yield strain produced by this method showed promising potential for industry' (Wozniak et al. 2001, p. 123).

One famous paper was 'Experimental Research on the External Qigong Effect on Substances Over a Distance of 2,000 Kilometers'. This made the front page of Chinese newspapers and into TV news bulletins. Not surprisingly Yan became even more of a sensation lecturing to packed auditoriums throughout China and in 1990 to the USA where his qigong lectures were, for their enthusiasm and credulity, matched only by those of auditorium-filling fundamentalist preachers.

Yan's US tour was a series of highs: in San Francisco Yan lectured to 'a thousand admirers' at the Pasadena Hilton, one scheduled 25-min presentation was extended to 3.5 hours; he made a presentation at the Massachusetts Institute of Technology (MIT) on how he used his *qi* power to change the molecular structure of water, and in this presentation, he was supported by Dr Lu, a prominent Chinese nuclear physicist[2]; in Washington the Chinese ambassador attended a lecture; at a San Francisco banquet, Yan projected his *qi* power to make two wheelchair-bound people walk for the first time in many years, an event so dramatic that a businessman took out his credit card and paid for the whole banquet; President George H.W. Bush invited Yan to a formal dinner, described him as 'a contemporary sage' and praised his 'research on the scientific principles behind qigong healing'; and he was photographed with President Clinton.[3]

A biography of Yan was written, in which it was mused that the spreading of Yan's powers among all citizens would put China on top of the world: 'everyone

---

[1] See www.item-bioenergy.com/infocenter/chinesechiresearch.doc

[2] Lu Zuyin was a renowned professor of nuclear physics at the Institute of High Energy Physics, Chinese Academy of Sciences, as well as a physics professor at Tsinghua University. He was responsible for the comprehensive planning of nuclear physics parameter measurements during China's nuclear tests, and he led the effort of measuring the forces, neutron fields, and gamma fields of atomic bombs in China.

[3] All of these, and more, extraordinary things are related by Vicente Ongtenco, a council member of the World Medical Qigong Association, and detailed in Wozniak et al. (2001, pp. 127–132).

could do things by thought, move objects, control the cosmos with our minds' (Lin et al. 2000, p. 103). The consequences of any of this even being remotely true are staggering. So staggering that scientific testing of the claims should be not just a Chinese priority but an international one.

Yan's followers straight-facedly maintain that: 'His discoveries are changing the way modern science is viewed and challenging many of its assumptions'.[4] For some gullible people, with minimal grasp of science, such claims might, sadly, change *their* view of modern science. What is less understandable, but of significance, is that many of his followers have sophisticated scientific and technical backgrounds, including PhD degrees in science. But his astonishing *qi* claims need to be proved, and there was ample evidence that they were false.

Professor Chao Nanming, Head of the Biology Department at Tsinghua University who had agreed to the 1986 studies being conducted, has made troubling comments about the Qigong Cooperative Research Group, saying: the Chi Research Group had no formal connection to the university but were merely using its name without authority; there were just three experiments conducted, and Yan would not repeat any of them; the distance from Yan to the samples was 100 m not 2000 km; a Raman laser machine was supposedly used, but Yan did not know how to operate it; and water was used from different sources, so its molecular composition was not controlled. Chao's doctoral student, who had a role in the experiments, reported that all standard experimental care and control were neglected. Finally, the initial papers were published with Chao's name among the authors, but not with his permission. Professor Chao thought the studies were worthless (Lin et al. 2000, pp. 62–65). This did not affect Yan's Chinese, then international, stardom; devotees flocked to learn more. As most people would, given the extraordinary claims being made.

On 7 March 1990, the *Xinmin Evening Paper* reported that at the 18,000 seat Shanghai auditorium:

> The great super-Qigongist, Yan Xin, was delivering a six-hour Qigong lecture in one session. By means of the microphone and 48 loudspeakers, his voice resounded through the whole conference hall. …He talked slowly, telling the meaning of Qigong, and mentioned some diseases that can be cured. … Less than five minutes into the lecture, some in the audience began to shout, laugh, cry, and swing to and fro as if they were drunk. (Lin et al. 2000, pp. 56–57)

Dr Yan's 'research' career continued. In a 30-page 2002 paper in *The Journal of Scientific Exploration*, co-authored with 10 Chinese and US scientists,[5] the claim is made that:

> This paper reviews a portion of the data generated via the external *qi* emitted by Dr. Yan Xin. Included here are (1) strong responses developed in LiF thermoluminescent dosimeters, (2) strong responses in aqueous solution structure as probed with laser Raman spectroscopy and (3) alterations in the half-life of 241Am as probed with both ray spectroscopy

---

[4] See www.item-bioenergy.com/infocenter/chinesechiresearch.doc

[5] The authors' institutional affiliations included Harvard University, Mass General Hospital, University of Oklahoma, MIT, and the Institute of High Energy Physics, Chinese Academy of Sciences, Beijing.

and a solid-state nuclear track detector. According to the different circumstances, external *qi* of Dr. Yan Xin can display different attributes such as being distance transcending, bi-directional, reversible or targeting. Although external *qi* of Yan Xin Life Science Technology has not been identified with any of the four known and accepted fundamental physical forces, its influence on physical reality is robustly confirmed. (Yan et al. 2002, p. 381)

All of this is very impressive. The study reports a 3-h *qi*-emitting lecture by Dr Yan to a packed Chinese Academy of Sciences (CAS) auditorium in Beijing where sophisticated monitors were set up everywhere to record the audience's positive changes after being radiated by Yan's *qi* and to measure the increase of auditorium 'high-energy' *qi*. Different locations recorded a five- to tenfold increase in the latter, comparable to the impact of 'gamma rays and neutrons'. Similar results are reported from an 11-h *qi*-emitting lecture of Dr Yan at the Red Flag Avenue Auditorium in Beijing.

The study includes page after page of highly technical detail and mathematical tables. It has all the appearance of sophisticated science. But among the pages is the above-reported 1986 'study' done at Tsinghua University where the head of biology's name was included among the authors though he thought the study was worthless. Of this discredited study, the new 2002 study says:

The results [1986 study] on liquid water provided the first direct evidence that Yan Xin Life Science Technology healing is physical and external *qi* from Dr. Yan may cause physical adjustments in the human body since water makes up about 65% of human body weight. (Yan et al. 2002, p. 389)

After still more pages of molecular formulae, atomic weights, spectrometer specifications, Raman laser spectra readings, and much more, we are told that:

The *qi*-effects on the structure and properties of liquid water were also observed using a different technique later in 1991. Changes were repeatedly observed in the ultraviolet (UV) absorption of de-ionized water treated by external qi emitted by Dr. Yan from the US to Beijing, China. (Yan et al. 2002, p. 392)

So, now Yan's self-generated *qi* affects molecular structure and behaviour at a distance of 10,000 km, across a continent and across the Pacific Ocean. Remarkably scientists from half-a-dozen reputable universities signed off on this, and it was published in a supposedly scientific journal. The International Yan Xin Qigong Association was formed. The authors were confident enough to conclude that: 'Yan Xin Life Science Technology has already emerged as an important scientific discipline deserving more substantive exploration' (Yan et al. 2002, p. 408). This is truly something for scientists and serious sociologists of science to examine: how can such a scandalous state of affairs come to be? Is it a scaled-down version of the USSR's Lysenkoism which also sustained its own journals and supposed experimental culture (Joravsky 1970)?

Yan's *qi* powers interfere not just at the difficult-to-see micro-levels but at the easy-to-see macro levels. He speaks of moving cups of tea by his *qi* power and, 'when friends come, transporting a pot of tea for them', and if a lot of people come, qigong masters can 'convert earth to cups', but he cautions that the last 'demands a lot of *Qi*' and that 'the energy of the human body is limited and should be used ingeniously' (Wozniak et al. 2001, pp. 74–75). He especially does not like to use his precious *Qi* when video equipment is present, and he relates how during a 1986 visit to Japan someone, against his instructions, tried to videotape his use of transport powers but 'their video camera stopped working' (Wozniak et al. 2001, p. 73).

These, and other such papers by Yan and colleagues, certainly warrant 'more substantive exploration' by established biophysical and biochemical researchers. But the claims are so outrageous and so contrary to all established scientific knowledge that obvious questions should be asked: why have these exceptional, revolutionary claims not been published in any mainstream, established, peer-reviewed journal? And, are we here examining good science, bad science, or pseudoscience?

In support of the pseudoscience answer, it is worth noting that during Dr Yan's triumphant US tour in 1990, a tour during which he constantly asked the scientific establishment to seriously investigate qigong claims, the British Columbia Skeptics Group invited him to demonstrate his *qi*-healing and molecular-changing powers. The invitation was turned down by his assistant, a Chinese professor of agriculture, who wrote:

> Dr. Yan Xin and I are not interested in the very low-level test which was very popular in China ten or fifteen years ago. He is busy on some cooperating research subjects with several important organizations in U.S. (Beyerstein and Sampson 1996b)

One of the 'cooperating research subjects' was the study of 'subtle' energy that might be the physicists' chi. It can be assumed that a 'low-level test' is one of those objective, controlled, and repeatable tests preferred by bothersome scientists.[6]

Dr Hui Lin, of the Chinese Chi Research Centre, and co-author with Yan of the above study, offers the following striking example of chi power:

> Consider a simple experiment on Qigong potential. In this experiment people used their qi to shake pills out of a sealed bottle. However, the intermediate process was undetectable by any available means. The pills passed through the bottle (analogous to conducted experiments in which a person passes through a solid wall), even though the bottle is completely sealed and intact, without any possibility of tampering.[7]

Accepting at face value the results, he concludes that:

> This demonstrates the probable existence of a form of energy associated with qi which transcends the three or four [gravitational, electromagnetic, strong and weak interaction] fundamental forces. (ibid)

This all sounds very scientific and certainly would cause a revision in our understanding of science and of the world picture that science has given us. But in Hui Lin's 'experiment', no independent witness to such 'transportation' is noted; and no replication study is reported. Independent observation and replication should be the

---

[6] Unfortunately Dr Yan seems unable to take further part in such testing. At some time around 2000, he was made a 'national treasure' by the Chinese government, and his travel and appearances have been constrained. How much this was a result of doing bad science and how much it was the party's fear of a mass qigong movement, are unknown.
See https://www.thedaobums.com/topic/30754-what-became-of-qigong-master-yan-xin/

[7] See www.item-bioenergy.com/infocenter/chinesechiresearch.doc

beginning of any effort to bring these 'truly remarkable results' into the scientific fold.[8] They should be among the first things that any scientifically literate student or adult asks of the remarkable experiment. Their absence is a powerful indicator that the whole feng shui practice is pseudoscientific.

Zhang Xiangyu was another scientist cum qigongist who shot into national and international fame on the back of her 'scientific' proof of external *qi* power. She graduated from the Tongji Medical University and persuaded Professor Bi of its biology department to allow her to test the claim that her self-generated chi energy could affect the phagocytic function of macrophages in the abdominal cavity fluid of mice, seemingly a thoroughly scientific hypothesis. Her experimental mouse, the one she radiated with her self-generated *qi*, was reported to live longer than those in the control group, and the relevant molecular structure in its cavity fluid did change. The result confirmed the claim, but the experiment was not repeated. She wrote a report, sent it to the Qigong Science Institute of China, titled herself a 'supernatural being', and enjoyed the celebrity, fame, and money that followed (Lin et al. 2000, pp. 65–67).

But as with Dr Yan, things were not quite as they appeared. Zhang's paper was published as being from the Tongji Biology Department despite Professor Bi rejecting the entire methodology, pointing out that the surviving mouse was the one in the group that had a heart transplant and that only Zhang's students could see the molecular change. Bi, and others, said that: 'imperfect experimental design' was coupled with 'fraudulent evaluation of results' (Lin et al. 2000, p. 66).

Lu Zuyin, the nuclear physicist collaborator of Yan Xin and one-time senior figure in China's atomic bomb project, went to the USA and subsequently published his own 404-page book *Scientific Qigong Exploration: The Wonders and Mysteries of Chi* (Lu 1997).[9] The book concentrates on the remarkable *external qigong* (EQ) energies created and 'thrown' by Yan Xin and other qigong masters. A review of the book by Kevin Chen, a US professor of psychiatry and enthusiast for qigong, notes that:

> The subjects of the book cover the major findings in physical, chemical and biological studies of EQ phenomenon, and the distance effect of EQ. The discussed experiments of EQ include infrared radiation effects, magnetic effects, electromagnetic wave and infrasonic wave effects, bi-directional effects, multi-functionality and target adaptability, spatial characteristics (distance effect) and automatic targeting capacity, temporal characteristics (two after-effects), as well as the ability to affect matter at a microcosmic scale (such as molecules and atomic nuclei). (Chen 2002, p. 484)

And he observes that: 'In some ways, studying EQ is similar to studying PSI phenomenon since many well-trained qigong masters have both psychokinetic and ESP capabilities' (Chen 2002, p. 486). He also notes that the Chinese Society of Qigong Science has tried to bring order and consistency to the scientific study of qigong by restricting genuine EQ to that produced by 'a well-trained qigong practitioner under

---

[8] The magician (illusionist), James Randi, has rendered a great public service by replicating, exposing, and debunking these sorts of claims (Randi 1987, 1992, 1995).

[9] It should be noted that the publisher is 'a small specialist publisher'.

the qigong state, and so is tied to the intention, *yi*, of the practitioner'. Consequently, as Chen announces: 'effective study of EQ requires new methodology and new scientific framework' (Chen 2002, p. 486). Yan Xin had said the same thing: 'Currently, the essential qualities of qigong and *qi* are difficult to study in a detailed, qualitative, and quantitative manner'.[10]

This is, of course, completely mistaken: the remarkable claims are easy to study in a detailed, qualitative, and quantitative manner. It just takes textbook experimental design and control of variables. The onus is on the qigong practitioners to specify the variables and set measureable outcomes. If the latter cannot be done, then the pretence of scientificity should be abandoned.

Not surprisingly, qigong has spawned a veritable army of fraudsters. Wang Lin a qigong multimillionaire claimed to have spiritual powers enabling him to cure incurable blindness, cancer, and mostly everything else; further, he could kill people at long distance. Clearly he is very marketable. He made his fortune from credulous citizens and lived in a five-storey luxurious mansion in Luxi. Chased out of China, he began teaching tai chi at Stanford University. Li Yi, the deputy head of the Chinese Taoist Association, gained national, and probably international, attention by holding his breath underwater for 2 h and 22 min. This was shown to be, of course, a pantomime, but hundreds of thousands of yuan poured into his account; and, predictably, he was accused of using his statue to sexually exploit numerous young devotees. And on it all goes.[11]

Science is simply incompatible with the truth of the foregoing and other comparable bizarre claims. They would never be reported in any serious science research journal. Pleasingly serious scientific research journals require studies to be done in a 'detailed, qualitative, and quantitative' manner, and so a high bar is set for publication of any research in reputable journals. This includes feng shui 'research'. But inevitably unpublished and unpublishable feng shui and qigong research finds outlets in its own journals, book series, and websites. Dr Qian Xuesen (1911–2009), chairman of the Chinese Association of Science and Technology (CAST), declared Dr Yan's research to be scientific. This was a big boost to the qigong enterprise.

But it needs be noted that Chairman Qian gave 'scientific' support to Mao's disastrous 1958–1962 Great Leap Forward by publishing a supposed 'physics' article in the *People's Daily* 'establishing' that the number of calories of vegetable that can be produced per unit of land is far higher than what the Great Leapers had themselves been trumpeting. Qian held that cabbages weighing 250 kg and sweet potato harvests of 4000 tons per acre were possible provided farmers dug deeper, worked harder, and leapt higher. All of which they did to no effect except hastening the death by famine-created starvation of about 45 million Chinese citizens and the deaths of about 3 million poor souls by beatings in gulags over the four disastrous 'Great Leap' years (Dikotter 2010). Mao's personal physician subsequently wrote:

---

[10] See www.item-bioenergy.com/infocenter/chinesechiresearch.doc

[11] For a report on these fraudsters, see *Chinese Global Times*, 19 August 2013.

> It was not until the Great Leap Forward, when millions of Chinese began dying during the famine, that I became fully aware of how much Mao resembled the ruthless emperors he so admired. Mao knew that people were dying by the millions. He did not care. (Li, Zhisui 1996, p. 125)

Yet Mao's visage still beams down over Tiananmen Square, the site of 4 June 1989 student killings and arrests; and the official CCP verdict is that Mao was only 30% wrong.

The dissident astrophysicist, Fang Lizhi, who was being 'rehabilitated' by farm work at the time – counter-productively digging deeper and deeper furrows for planting – saw that the physics was all wrong and that Qian was merely a scientific gun for hire by deluded CCP ideologues and thugs. Fang in his autobiography makes a telling comment on this episode:

> Freedom is vital to science; science dies without it. When Qian's ridiculous article came out, Chinese physicists could see it for what it was, but no one had the freedom to say so. Not even purely scientific criticisms were possible, because this author was a favorite of Mao Zedong and this article's conclusions supported the Great Leap Forward. But the even more baleful fact is that the dictator of a mammoth political party could be so benighted and reckless as to use obsequious 'science' to make policies that affected nearly a billion people. How can a country that imprisons science expect anything but disaster? (Fang 2016, p. 101)

The foregoing roll-call of fanciful claims, and the enthusiasm with which they are greeted, is reminiscent of what Matteo Ricci wrote of feng shui some 400 years ago and which has been cited in Chap. 5: 'Fraud is so common and new methods of deceiving are of such daily occurrence that a simple and credulous people are easily led into error' (Ricci 1615/1953, pp. 83). The modern cases may not be conscious fraud; it might be that credulity is now so established even among professionals that fraud and trickery are no longer required. The commonplace academic constructivist ontology of 'alternative realities' and 'many worlds', and its associated relativist epistemology, lessens the 'need to deceive'. The more so when any correction can be dismissed as 'realists would say that', or more directly ignored, as with the current president of the USA, as 'fake news'.

After dismissing qigong as superstitious and one of the 'four olds', the CCP has now embraced it as a way of better connecting the party to the Chinese tradition. Over 20,000 health qigong management centres have been established with 1.5 million practitioners. They are committed to the thesis that qigong is scientific not superstitious and that its practice has the widest range of health benefits.

Paucity of Tests

But after 200 years of modern scientific study of energy,[12] with pressing contemporary universal concern with clean energy production and use and with numerous major international energy summits and conferences, it is noteworthy that feng shui energy is yet to be identified and measured in any reputable laboratory. There are many putative such measures, providing 'convincing' evidence for the efficacy of chi, but as with the above cases from the China Chi Research Centre, they do not bear close scrutiny.

What constitutes 'close scrutiny' is something that can be taught in science classes when teachers have a modicum of HPS competence and interest. Students can read the texts; follow the experimental footsteps of Galileo, Newton, Huygens, Priestley, Darwin, Rutherford, and many others; and see and appreciate the difference between on the one hand close scrutiny, careful measurement, and appraisal of alternative hypotheses and, on the other, the lazy uncritical holding of original and perhaps prejudicial opinions.

Feng shui 'research' has the appearance and trappings of science and promotes itself as such. Some practitioners might function in 'research centres'; but nevertheless, feng shui is not science. It is not just mistaken science, in the way that phlogiston accounts of combustion; Lamarckian accounts of evolution or behaviourist accounts of learning were mistaken; feng shui, as will be shown in Chap. 12, is simply not science; it lacks a number of the crucial necessary features of genuine science. The foregoing discussion of the China Chi Research Centre's research does introduce an added layer to the identification of pseudoscience. One mark of the latter is nonengagement with the scientific community and failure to publish research in scientific journals; but once a group starts its own journal, with a scientific name, then the designation 'scientific journal' cannot be settled just by name alone. There is some disciplinary/sociological dimension that needs feed into the naming.

Examination of feng shui, and more generally any other substantial yet mistaken account of nature, in science classes enables other educational goals to be advanced: appreciating the difference between science and pseudoscience, understanding the impact of science on culture, understanding scientific experiment and methodology, and learning features of the nature of science. As with all fields in science, an appreciation of the history of feng shui prepares the way for wider philosophical and cultural understanding of the field and for seeing how disciplinary understanding has been intertwined with philosophy and culture.

Much is rightly made of the underdetermination of scientific theory by evidence – the Duhem-Quine thesis (Harding 1976; Weinert 1995). More specifically there is serious argument, initiated by Ernst Mach in the late nineteenth century, about premature realist inferences to the reality of hypothetical constructs or entities in explanatory scientific theories (Matthews 2015a, Chap. 9). There were, and still

---

[12] See at least Bevilacqua (2014b), Coopersmith (2010), Harman (1982), and Sherman (2018).

are, detailed philosophical and scientific debates about the ontological status of phlogiston, caloric, atoms, genes, fields, electron shells, forces, and so on.[13] Nothing comparable occurs in the uncritical, unreflective feng shui move from rural farming practice or architectural common sense to the putative reality of chi. The mere juxtaposition of debates about realism in science and in feng shui can illustrate important features of science and of pseudoscience.

Philosophical Insulation of Feng Shui

Lillian Too, the earlier mentioned feng shui consultant, identifies the reason for the exponential growth of Western interest in feng shui as the fact that:

> [people are] … beginning to realize that there are alternative ways and methods of viewing the Universe, of understanding the way energy moves and works, and how these energies affect our well-being. (Too 1998, p. 17)

There are of course 'alternative ways and methods of viewing the Universe'; there can be no dispute about this anthropological and sociological claim.[14] But that leaves the obvious question of whether these 'alternative ways and methods' are equally correct, good, or productive of truth about the 'way energy moves and works'. Are the alternative methods just different but equal sciences? Are they non-science? Are they pseudoscience? The persistent question is whether the alternative perceptions of reality, or alternative systems, are true or false; do they connect or not connect to reality?

Persistently, these questions have not been asked in multicultural, or cultural studies, science education debates. Indeed, it is a matter of educational principle that they are not to be asked; and for such educators, it is a matter of philosophical principle that they cannot be asked (McCarthy 2018). Ken Tobin thus 'insulates' his TCM and acupuncture claims against such persistent questions by citing Ted Kaptchuk:

> Chinese medicine is a coherent system of thought that does not require validation by the West as an intellectual construct. Intellectually, the way to approach Chinese concepts is to see whether they are internally logical and consistent, not to disguise them as Western concepts or dismiss them because they do not conform to Western notions. (Kaptchuk 2000, p. 77)

The editor of the journal *Alternative Therapies* had earlier claimed such philosophical insulation, or protection, in these terms:

> New forms of evaluation will have to be developed if alternative therapies are to be fairly assessed. (in Sampson 1996, p. 195)

---

[13] See essays in any anthology on the realism/instrumentalism debate in philosophy of science. For a recent collection, see Agazzi (2017).
[14] See at least the 23 chapters in Selin (2003).

It might be sufficient to ask, as Tobin, Kaptchuk, and the journal editor do, of random systems of concepts that they be 'internally logical and consistent'. That is the least that can be expected. But if the conceptual system relates to medicine, then more than just conceptual consistency is required; some demonstrable connection to the world, to bodily systems, and to improvement in health is required. It is a contradiction to admit that alternative medicine does not conform to Western notions yet at the same time to want them to be regarded as scientific and deserving of the social support (licences and degrees) and government money given to science.

Given that mega-million-dollar state health subsidies and private insurance payments in nearly all countries of the world, hinge on assessment of these putative practices, then they need to be rigorous and removed from both philosophical faddism and economic opportunism. Merely being internally consistent does not cut either the philosophical or economic mustard. Pouring money and resources into unproved or bad medicine is in no one's interest. This should be obvious even in wealthy countries, much more so for poor countries where social resources are limited.

For constructivists, even ones who have 'moved on', there is not even a speed bump on the road to 'alternative ways and methods of viewing the Universe' or to 'alternative medicines' (Tobin 1993, 2000). Indeed, there are no speed bumps at all in the constructivist and postmodernist roads to 'different but equal' viewing platforms; smooth constructivist motorways lead to each.[15] Tobin, speaking for many, insulates his claims by recourse to 'multilogicality':

> In contrast to the mainstream of research in science education, I advocate a multilogical methodology that embraces incommensurability, polysemia, subjectivity, and polyphonia as a means of preserving the integrity and potential of knowledge systems to generate and maintain disparate perspectives, outcomes, and implications for practice. In such a multilogical model, power discourses such as Western medicine carry no greater weight than complementary knowledge systems that may have been marginalized in a social world in which monosemia is dominant. (Tobin 2015, p. 3)

What the first 45-word sentence might possibly mean can be left aside. 'Multilogical methodology' does sound akin to Paul Feyerabend's much-criticized 'no methodology' (Feyerabend 1975). But that modern medicine - that is successfully treating illness and disease, and underwriting heart, lung, knee, kidney transplants, and so much else, and is adopted in all countries that are in a position to do so – is called a 'power discourse', is indicative of the standing and regard for science in the multicultural and cultural studies academic communities. So also is Tobin's sloganizing claim that it is: 'Scientism, crypto-positivism, neoliberalism, and meritocracy, among other factors' that enable such 'power discourse'. Clearly the above 'multilogical' claim is meant to function as philosophical insulation. But not all philosophical insulation works, sometimes criticism gets through, and this particular Kuhnian-inspired incommensurable insulation has been gotten through.

---

[15] For philosophical appraisals of educational constructivism, see Matthews (2000, 2012, 2015a, b, Chap. 8) and contributions to Matthews (1998).

Perhaps in anticipation of the standard truth-related criticisms that are raised against his and other such claims, Tobin avers that:

> The components of a multilogical bricolage are not systems of truth but are ways of seeing the world and making sense of what happens. By heightening awareness in new ways, possibilities for action are expanded and new directions are forged. (Tobin 2015)

But taking LSD also 'heightens awareness and opens expanded possibilities for action'. Science requires something more than just this. Leaving aside what 'multilogical bricolage' might mean, the idea of truth as a putative relation between an assertion (proposition, belief, or statement) and the world is not a monopoly of the West; all cultures are concerned with the separation of correct claims from fantasy. Wang Chong, the esteemed early Chinese naturalist philosopher of the Han Dynasty wrote in his *Discourses Weighed in the Balance* (ca. 80 AD) that: 'no discussion which lacks facts and fails the test of facts can be convincing, no matter how appealingly it is phrased'. He constantly appealed to this principle to dismiss much of the Daoist superstitious and magical thought that had engulfed China in his time.[16] Despite a long line of impatient undertakers, including Massimo Pigliucci (2010, p. 236), the correspondence theory of truth is still alive and far from being buried (Alston 1996; Bunge 2012b; Devitt 1991).

All societies have a real interest in the separation of truth from falsehood and, more fundamentally, the separation of reliable truth-seeking processes (science) from unreliable, imitative, masquerading truth-seeking processes (pseudoscience). The interest of all cultures is ultimately served by knowing whether claims about the natural and social worlds are true or not true, correct or fanciful: is the next village *really* 1 or is it 10 miles distant? Is my brother *really* living here or not? Is this plant *really* poisonous or not? Are crocodiles *really* in the river or not? Sometimes these simple questions are not so easy to answer, but to say that such questions are a Western monopoly is nonsense. Repressive and obscurantist dictatorships of both Western and Eastern strip do not like the distinction between truth and fantasy being made as truth is a persistent enemy of power, but ultimately across all cultures, such questions demand to be asked. To say that they are Western or Eastern is nonsense.

Methodological Naturalism

Assertions made about the kinds, distribution, and powers of chi can be appraised by science. This is an inescapable conclusion of adopting 'methodological naturalism' (MN) in science; and such adoption, at least in pragmatic form, is required for any investigation to be scientific. The minimal claim that is widely accepted by philosophers, scientists, and curriculum writers is that adopting MN is a requirement for any inquiry and explanation to be scientific. In Robert Pennock's words:

---

[16] See, at least, Graham (1989).

> ... science does not have a special rule just to keep out divine interventions, but rather a general rule that it does not handle any supernatural agents or powers since these are taken by definition to be above natural laws. (Pennock 1999, 284)

The National Academy of Sciences in the USA affirms the same position: 'Because science is limited to explaining the natural world by means of natural processes, it cannot use supernatural causation in its explanations' (NAS 1998, appendix C).

Consistent with this historical and sociological reality, Stephen Jay Gould made his much-repeated declaration that science and religion occupy two independent non-overlapping magisteria (NOMA), with the magisteria of science concerned with the workings of the natural world and the magisteria of religion or faith being concerned with the supernatural world and/or the realm of value and life's ultimate meaning (Gould 1997). So, one cannot judge the other; the non-science domain is insulated from scientific appraisal. This became an almost universal dogma in science education debate and research. But it has been cogently criticized.[17] As soon as a supposed 'off limits' body of beliefs makes claims about worldly processes and events, then science can appraise the claim. And chi beliefs have for thousands of years, as they do today, make claims about the natural world.

While methodological naturalism (MN) has been widely supported,[18] some have argued the stronger claim that ontological naturalism (ON) is a requirement of science. For Martin Mahner:

> ... metaphysical naturalism [ON] is a constitutive ontological principle of science in that the general empirical methods of science, such as observation, measurement and experiment, and thus the very production of empirical evidence, presuppose a no-supernature principle. (Mahner 2012, p. 1437)

The simple reading of an instrument assumes that no supernatural entity or process is interfering with the causal chains linking the instrument to the natural process or event it is responding to. So not only does science require MN; it also requires ON.

Other philosophers have argued that it is a mistake to try to identify *any* presuppositions for science. For instance, Yonatan Fishman and Maarten Boudry argue that:

> ... science presupposes neither MN nor ON and that science can indeed investigate supernatural hypotheses via standard methodological approaches used to evaluate any 'non-supernatural' claim. Science, at least ideally, is committed to the pursuit of truth about the nature of reality, whatever it may be, and hence cannot exclude the existence of the supernatural *a priori*, be it on methodological or metaphysical ground, without artificially limiting its scope and power. (Fishman and Boudry 2013, p. 921)

It needs be recognized that methodological naturalism (MN) is neither methodological nor ontological *materialism*. The latter characterized the mechanical worldview of the seventeenth and eighteenth centuries,[19] but with the progress of

---

[17] See at least Boudry et al. (2012), Fishman (2009), Slezak (2012), and Stenger (1990, 2007).

[18] The rich philosophical literature on the methodological and ontological presuppositions, if any, of science is reviewed in Fishman and Boudry (2013).

[19] See at least Dijksterhuis (1961/1986), Harré (1964), Schofield (1970), and Westfall (1971).

nineteenth- and twentieth-century science, where there was recourse to stable, yet nonmaterial, explanatory, and causal entities (non-contact forces, fields, radiation, gravitational waves, etc.), materialist ontology became a hindrance rather than an asset for science, and such novel nonmaterial entities became 'naturalized' and so a part of orthodox science.

The term 'physicalism' was coined for the methodology that allowed nonmaterial but scientifically verified entities to figure in scientific explanations; this term was also used for the ontological position that claimed only such scientifically verified entities actually existed (Stoljar 2010). Physicalism is proposed against vulgar materialism. Unfortunately, physicalism is standardly understood as 'vulgar materialism + extras', where the extras are fields, attractive forces, photons, and so on for all the other legitimate, accepted scientific explanatory entities. This still leaves matter as inert, a contention that needs be separately argued.

Mario Bunge has advanced and defended a detailed *emergent* materialist ontology that has the scientific gains of physicalism without its defects (Bunge 2003, Chap. 1; 2006, Chap. 1; 2009; 2012a, Chap. 2). On the limitations of naturalism as historically conceived, he writes:

> The great merit of naturalism is that it rejects magical thinking, in particular, supernaturalism. But naturalism is limited, for it denies the emergence of qualitative novelty and consequently the qualitative distinctions among levels of organization – physical, biological, and social, among others. In particular, naturalism does not account for the specificities of the social and the technological. ... This alone suggests that naturalism should be expanded to encompass the articicial and the social. (Bunge 2009, pp. 60–61.

In Bunge's amended naturalist ontology, *all* scientific entities have 'emergent properties'; when entities join, the aggregate has properties that the components do not have. No viable natural or social scientific ontology can be reductionist. At every level there are more complicated, but still natural, entities; and at each level – atoms, molecules, cells, organisms, people, populations, and societies – the behaviour of the whole cannot be resolved into the behaviour of the parts. Reductionism is an unscientific programme (Bunge 2001, Chap. 3).

Some maintain that scientific naturalism, of both methodological and ontological kinds, bias the appraisal of chi. Liu JeeLoo argues for a distinctive chi naturalism, writing:

> By emphasizing its *naturalistic* dimension, this chapter aims to show that even though this whole tradition of qi-cosmology falls outside the scope of contemporary natural sciences, it is nonetheless a rational, coherent and respectable view of nature. (Liu 2015, p. 33)

He advocates a more relaxed 'liberal naturalism', and specifically a 'humanistic naturalism', which gives legitimacy to chi talk. Such a naturalism legitimizes chi-ontological claims such as:

> Furthermore, the flow of qi runs freely within and without a person's body, hence one's bodily conditions are constantly affected by changes in the external environment. (Liu 2015, p. 37)

But the argument has never been about legitimacy; all manner and means of views might be legitimate; the issue is whether the chi view of nature is 'rational, coherent, and respectable', to use Liu's words. The argument of this book is that the chi view fails on each ground. And Liu's own proposed humanistic naturalism does not rescue the patient. He says that for humanistic naturalism:

> The world consists of nothing but entities of the natural world and humans are part of this natural world. Furthermore, there can be no supernatural interactions with entities in the natural world. Natural entities are accessible to humans' cognitive capacities, and statements about the existence and nature of natural entities are truth-apt. (Liu 2015, p. 36)

Scientific naturalism concurs with all of this; there is nothing specifically 'humanistic' about the characterization. Mario Bunge, a defender of scientific naturalism, is explicit in saying, against all reductionist programmes, that 'naturalism should be expanded to encompass the artificial and the social' (Bunge 2009, p. 61). The issue with Liu's argument is whether the truth of 'truth-apt' claims about natural entities can be determined outside of science.

Chi as an Intervening Variable

One fallback option for chi theorists who wish to maintain the scientificity of their system and of the chi construct, yet who also acknowledge the reality that chi has never yet been found or measured, is to abandon the referential dimension of chi, to give up even muted realism about chi, and to swing over to seeing chi as an intervening variable not a hypothetical construct.

The distinction between theoretical terms that are hypothetical constructs and those that are intervening variables can be traced to the positivist Ernst Mach. It was developed by the Logical Empiricists and given prominence by the psychologists Kenneth MacCorquodale and Paul Meehl in their much-cited 1948 *Psychological Review* paper 'On a Distinction between Hypothetical Constructs and Intervening Variables' (MacCorquodale and Meehl 1948). For two decades this paper dominated methodological discussion in psychology.[20]

For MacCorquodale and Meehl, to see theoretical terms (atom, field, electron, intelligence, libido, class consciousness, conscience, drive, magnetism, habits, will power, mind, and so on) as having referents, as referring to entities that though unseen nevertheless exist and exert influence, is to see them as 'hypothetical constructs'. This is the standard realist interpretation of theoretical terms. The referents may or may not exist, but if the term is a hypothetical construct, then they are supposed to exist. But there is an honourable scientific alternative, namely, the empiricist interpretation of theoretical terms, whereby the terms are seen not to be making existence claims at all; they are just shorthand for linking together measured variables of interest; they are 'intervening variables'.[21] So it is a mistake to look for intelligence in a person; it will never be found because it is not there. To say of

---

[20] In 1974 it was still required reading in the Sydney University honours psychology programme.
[21] Bas van Fraassen is one well-known defender of this position (Fraassen 2002).

someone that they have intelligence is merely to say that performance on test X is regularly correlated with performance on test Y and maybe with performance on test Z. We take good scores on tests X, Y, and Z to mean that they have intelligence. And all of this is a matter of professional negotiation over tests, cut-offs, preparation, and everything else. Having or not having intelligence is not an issue settled by holding something in a hand or looking through a microscope. On this interpretation, people do not *have* intelligence; they just *do* intelligent things. Similarly, a person does not *have* a high libido; to go looking for libido is a mistake. Rather to say someone has a high libido means just that in certain defined situations they reliably do certain things. Once again, all this is open to social and professional negotiation. Notoriously, libido goes up and down depending on the expectations and culture of the investigator. Despite the name, intervening variables do not intervene in the world; to do so requires existence; rather they intervene or link mathematical variables. They have a conceptual role, not a causal role.

Intervening variables are akin to dispositional properties (Cross 2005). To say a glass is brittle just means that when dropped, it breaks; it does not mean that the glass *has* brittleness that can be pointed to. In advance of the manifesting behaviour, there is nothing to be seen, no matter how closely one looks. To say someone is biased does not mean that they *have* bias; it just means that when put in certain situations they will behave in particular ways.

The conversion of intervening variables, or dispositions, into hypothetical constructs is commonly called *Reification* and is something that needs be done with caution. Realists have learnt to proceed slowly on that front; the history of science is littered with rash and premature reifications of constructs that seemed good and existent at the time.

The intervening variable or disposition option might seem an attractive one to chi theorists. We do not have to believe in chi; we admit there is no chi; there are just certain things that uniformly go together, and chi names the uniformity. Contentment goes along with living by a lake with nice views. To say there is good chi by the lake, and hence comfortable living, adds nothing to what is already known. Doing certain exercises goes along with feeling of mental and physical ease. To say that the exercises have manipulated internal chi adds nothing to what is already known. Chi is just shorthand, in the way that intelligence, habits, and class consciousness are. This is the position of the numerous chi theorists who end up saying that chi talk is just common sense. Here, chi occupies conceptual space but pays no rent.

So, for instance, consider Lau Tzu's fifth-century BC account of Dao (cited in Chap. 3):

> We look at it and do not see it;
> Its name is The Invisible.
> We listen to it and do not hear it;
> Its name is The Inaudible.
> We touch it and do not find it;
> Its name is The Subtle (formless).
> These three cannot be further inquired into,
> And hence merge into one.

or the previously cited claim (Chap. 5) of two chi theorists that:

> Life is defined by *Qi* even though it is impossible to grasp, measure, quantify, see, or isolate. Immaterial yet essential, the material world is formed by it. An invisible force known only by its effects, *Qi* is recognized indirectly by what it fosters, generates, and protects. (Beinfield and Korngold 1991, p. 30)

A chi realist would say that what is being looked for but not found is a hypothetical construct and we just need better and more appropriate instruments. They can maintain the faith and keep looking and eventually perhaps lose their faith. Alternatively, a chi empiricist would say that chi is an intervening variable and cannot, in principle, be touched, smelt, grasped, or heard. But this does not make the construct unscientific. It is exactly in the same situation as intelligence, drive, libido, and mind.

But this empiricist refuge has its own problems.

1. Intervening variables link measured variables, but notoriously chi theorizing is never accompanied by stable, reliable measurement in the way that, over decades, has been achieved by, for instance, intelligence testing. There are no chi-relevant stable measured variables; everything is chaotic and unmeasured. The presence or absence of chi varies with every observer and theorist, and literally for many, on the time of day and the day of the year. This is complete chaos from which stable intervening variables could never be rescued.
2. The measured variables are no longer the effects of chi; they cannot be generated or protected by chi; this can only be said if chi is a hypothetical construct, so there is no reason to use the term 'chi' as the name of any particular correlation. An intervening variable cannot explain anything; it has no existence, so it cannot have any explanatory power or function.
3. To give up realism about chi, that is to say it is not a hypothetical construct, is to abandon the entire three- to four-millennium-long cosmogenic tradition that underpins so much Chinese and Asian culture. Chi supposedly is the ultimate explanation of everything; if it explains nothing, then it has lost its cultural value and purpose.
4. The intervening variable option wipes out the 'mysteriumism' on which so much feng shui promotion is based; intervening variable talk is bad for business. Recall Ernst Johann Eitel's observation (Chap. 7):

> Well, if Feng-shui were no more than what our common sense and natural instincts teach us, Chinese Feng-shui would be no such puzzle to us. But the fact is, the Chinese have made Feng-shui a black art, and those that are proficient in this art and derive their livelihood from it, find it to their advantage to make the same mystery of it, with which European alchemists and astrologers used to surround their vagaries. (Eitel 1873/1987, p. 1)

The intervening variable option is a case of getting out of the frying pan and into the fire. It becomes just the mouthing of words and the waving of hands. Neither of which advance anything except the bank accounts of shysters.

Chi as Metaphor

The final option for chi theorists who wish to retain the concept, while acknowledging that there is no non-inferential, immediate evidence for it, is to say that chi talk is metaphorical not literal. When it is said that chi is a special form of energy, what is being said is that chi is not literally a special form of energy, but rather it is like energy. The chi construct is a metaphor. This is not exceptional; mainstream science is replete with metaphorical constructs as was shown 50 years ago by Max Black (1962) and Mary Hesse (1966).[22] The very ideas of natural selection, current flow, electron layers, light waves, light particles, covalent bonds, the 'invisible hand', and so on, are deeply metaphoric. It is routine in teaching science that students are told that a remote or new scientific idea A is like experiential or known idea B (Duit 1991; Holton 1986).

The metaphor option is standard and routine in theology. The gospel of Matthew (Chap. 2, 1–12) recounts how the three wise men from the east followed a wandering star that settled over Bethlehem, and there in a manger they found, and gave homage to, the new born child. A scientific, literal reading of the gospel would lead, and has led, believing astronomers to search for such a star. Other believers have said that such a search is pointless as the gospel is not a scientific text, but a literary, religious text, and star-talk is metaphoric. The wandering star coming to rest simply needs to be understood in a nonliteral manner. Science is irrelevant to the truth or falsity of the gospel's account.

Likewise, the metaphoric chi theorist will say that chi exists and does all the things traditionally attributed to it, but it is ineffable, and as best we can do is say 'it is like energy'. Again, this gives some immediate relief from investigative discomfort, but the relief is temporary. Concepts in science can begin as metaphors; but if they last, if they do work and get incorporated into an established theory, or become the basis for a new one, then they have to be cashed out. The strength of a bond has to be specified, what it can and cannot bond needs to be specified, mechanisms for the bonding needs to be specified, and quantification, prediction, and experimentation all need to be undertaken. Without this, the chemical 'bond' concept is not scientific; it is just poetic talk. Ditto for natural selection, electron layers, and so on. To keep insisting that core concepts are metaphorical is just lazy science; their use cloaks ignorance and so prematurely halts investigation: 'Oh, so that is what it is like'.

For the purpose of this book, these fundamental philosophical arguments need just be noted, not solved. Students and teachers can themselves, with time and curricula allowing, follow up the arguments and literature.

---

[22] For some of the significant philosophical literature on metaphor in science, see at least contributions to Hintikka (1994) and Machamer (2000). For metaphorical thinking in a larger context, see Ortony (1979).

Conclusion

Chi (*qi*) is not supernatural; yet it is peculiar and unknown to science. It is not supernatural because it is supposed to be a part of nature and putatively has an enormous range of impacts and influences. Once this is acknowledged, chi claims are in the realm of legitimate scientific inquiry. As outlined above, the amount of scientific testing of chi theory is inversely related to the vastness of the extraordinary empirical claims made for it. But the testing that has been done confirms what is obvious: no chi-effect mechanism has been found or isolated; the bulk of scientific confirmations are simply repeating the fallacy of affirming the consequent; there are other explanations apart from chi for whatever effects are found; there is no tradition of sustained engagement with orthodox science, its community, or its research journals.

The educational issue is to see how, when, where, and why these scientific and philosophical investigations might be included in school or university science programmes. It is obvious from all of the above that feng shui is a field ripe for deception, fraud, and exploitation of the gullible, and any decent education should not turn the other way and give such fields a 'free pass' into society and culture. As mentioned earlier in the book, Matteo Ricci made just this observation about feng shui practice of his time: 'Fraud is so common and new methods of deceiving are of such daily occurrence that a simple and credulous people are easily led into error' (Ricci 1615/1953, p. 84). The international reach of the web, and ease of electronic cash transfer, provides boundless opportunities to lead even non-credulous people into error or worse (Huston 1995).

The world is awash with magicians, con artists, and fraudsters. In the USA televangelists Jimmy Swaggart and Jim and Tammy Bakker took multimillions of dollars from their gullible audiences by selling their prayers, healing power, lotions, and potions. The USA alone has thousands of such corrupt practitioners who are on 24/7 TV and cable cycles.[23] Other nations, cultures, and faiths have their own 'home-grown' fraudsters and 'god-men' that sell healing powers and potions to the desperately poor whose precious money could better be spent on food and medicine but also to the fabulously rich and well-educated who could do something better with their spare money.

Notoriously, India is awash with god-men who fleece villagers and sophisticates alike with empty promises and false hopes (Kovoor 1978). In 2017 the Buddhist world was rocked by revelations that Grand Master, and mega best-selling, multi-translated author, Sogyal Rinpoche was basically a sexual lecher who regarded devotees as groupies and treated them accordingly. In Brazil, the 76-year-old Joao Teixeira de Faria (John of God) has for 50 years been channeling spirits and energies in the isolated town of Abadiania in order to effect cures of every kind of illness

---

[23] A recent disturbing study of five centuries of 'crackpot delusion and make-believe' in the USA is Kurt Andersen's *Fantasyland* (Andersen 2017). Another study of magic and 'dark arts' belief in the USA is Wicker (2005). See also Schimmel's *The Tenacity of Unreasonable Beliefs* (Schimmel 2008).

and ailment (Pellegrino-Estich, 2001). International groups, including an annual group from Australia led by an 'energy therapist', go on packaged 'cure trips' to his 'healing sanctuary' and stay in his accommodations. As well as channeling spirits and energies, John of God channels money: Brazilian police arrested him in December 2018 after he transferred $9 million out of the country. When his residential estate was raided, false walls were found in which bags of jewels, gold, and silver were found. After his arrest, 300 women have come forward alleging they were sexually abused by him. His biographer, and business manager, also published *The Power to Heal: A Clear, Concise and Comprehensive Guide to Energy Healing*. Everyone has an interest in such fraud, in separating con artist chaff from evidence-based medical wheat.

The overarching claim of this book is that it is legitimate and beneficial to investigate these questions in science classes. Such investigation can be immediately beneficial for the students' understanding of the nature of science but also beneficial for the cultural health of their society. It is irresponsible for teachers and curriculum writers to give this family of nonsensical beliefs a 'free pass'. Appraisal does not mean condemnation; the former can be conducted without the latter, and there are good educational grounds for this option.

Chapter 13
Feng Shui as Pseudoscience

Feng shui is a multibillion-dollar industry, affecting millions of people. It has medical, health, architectural, building construction, town planning, interior design, and divination components. Given the extent of feng shui belief, and the personal, social, cultural, and economic impact that it has, everyone can benefit from judging its scientificity. Is it a science? Is it an immature or protoscience that might be on its way to science? Is it a pseudoscience? Making these judgements is in everyone's interest. When doing so in school programmes, especially when done in a coordinated, cross-disciplinary manner, more can be learnt about the nature of science and about the three-millennia-long cultural tradition that has historically sustained feng shui and that enables it to continue, and indeed flourish, in the contemporary world.

Alfred Hwangbo, a scholar who has documented the impact of feng shui on contemporary architecture, well cautioned that:

> The current sweeping phenomenon of feng shui seems to have come, along with the proliferation of new religions, in reaction to modern nihilism. It feeds the illusion of a modern panacea …. Looked at positively, it could herald a return of spiritual values, but only if understood at a much more profound level. Whether contemporary architecture can attain any benefit by extending the tenure of feng shui principles, or whether it is just another way of commercialising tradition, remains to be seen. (Hwangbo 1999, p. 198)

The beginning of an answer to these questions is to recognize that belief systems that make cognitive or 'truth-apt' claims about the world – that is, they generate propositions that can be true or false, probably true or probably false, or hypotheses that can be warranted or unwarranted – can usefully be categorized as:

- *Science* which in turn can be *natural* or *social* and either can be *mature* or *immature*; and these in turn can be *adopted*, *disputed*, or *rejected*.
- *Non-science* which in turn can be categorized as *claiming to be scientific* or *not claiming to be scientific*. In the first there will be *protoscience* and *pseudosci-*

ence; in the second there will be a range of *humanities* (art, theology, mathematics, philosophy, etc.) that make truth claims without claiming to be scientific (Table 13.1).

Such classification does depend upon demarcation criteria at each level, but these need not be sharp, and they need not be timeless and essentialist. Demarcation criteria can and have changed over time as scientific inquiry matures and takes new forms.[1] In 1938 Robert Merton provided the beginning of a modern classification when he characterized science as disinterested, universal, communal, and sceptical (Merton 1938/1973). The basic division is not all-or-nothing. Science contains non-empirical elements – mathematics, logic, ethics, and metaphysics, to name the obvious ones. Humanities contain scientific elements – historical research, biographical details, and sociological information, to name the obvious ones. Membership of a category is not cut and dried; it is a matter of family resemblance; there are clusters of criteria that mark out the categories, these can change over time, and the borders are to some extent porous.

In the following table, cold fusion research could reasonably be placed in the 'mature and rejected' category as it was originally a detailed theory proposed by competent and established scientists or now in the 'pseudoscience' category on account of persisting with a uniformly rejected research programme indicates a failure to be scientific (Huizenga 1992). Likewise, tobacco industry research showing that smoking is unrelated to lung cancer initially was scientific but over time became 'bad science' because its claims were disproved and now is pseudoscience because it has lost all credibility and contact with the scientific community.[2] To continue to pursue it indicates a lack of scientificity; it becomes a pseudoscientific not a scientific practice. It might be conducted in laboratories with scientific equipment, but it has ceased to be science because it has become a 'closed' system; its practitioners do not respect evidence, they do not have a scientific attitude.

All pseudoscience contains some scientific content – concepts, mathematics, instruments, and measurements – in order to give the practice credibility. It is of the essence of pseudoscience to appear to be scientific; its 'authority' depends on mimicking science. Science has journals, so pseudosciences commence their own or 'take over' established journals; science has peer review, so pseudoscience has the same; science has numbers and statistics, so pseudoscience has tables, figures, and correlations; science has experiments, so pseudoscientists conduct their own; and science has meetings and conferences, so pseudoscience does the same. The practical and philosophical task is to reliably separate the real from the mimic and the gimmick. The beginning of this task is first to distinguish science from non-science.

---

[1] There is an enormous amount of philosophical discussion on the 'demarcation problem'. See at least Bunge (2001, Chap. 8), Butts (1993), Hansson (2009, 2013), Mahner (2007, 2013), McIntyre (2019), Nickles (2013), Pigliucci (2013), and Shermer (2013).
[2] On the scientific research of the tobacco industry, see Brandt (2007) and Oreskes and Conway (2010).

Table 13.1 Classification of systemic cognitive claims

Systemic cognitive (truth-apt) claims						
Science			Non-science			
			Claims to be science		No claim to be science	
Physical science	*Social science*		*Protoscience*	*Pseudoscience*	*Humanities*	*Arts*
Physics, chemistry, biology	History, sociology, anthropology, economics					
Mature and are or were adopted	*Mature and disputed*	*Mature and rejected*				
Newtonianism Darwinism	Punctuated equilibria	Phlogiston theory; caloric theory; Velikovsky	String theory; cold fusion	Feng shui; Christian science; tobacco industry research	Literature	Music appreciation
					Poetry	Art criticism
					Theology	

Demarcation of Science from Non-science

Efforts to distinguish science from non-science, the original 'demarcation problem', have been pursued since at least David Hume's (1711–1776) time when in his *Inquiry* he advised that:

> When we entertain, therefore, any suspicion that a philosophical term is employed without any meaning or idea (as is but too frequent), we need but enquire, *from what impression is that supposed idea derived?* And if it be impossible to assign any, this will serve to confirm our suspicion. (Hume 1777/1902, p. 22, emphasis in original)

Hume was enunciating his empiricism and using the grounding in sensation as a way of separating 'sensible' ideas from the wide class of others. Ernst Mach took Hume's point seriously and, in conjunction with his philosophical phenomenalism, argued that a whole raft of central scientific concepts – mass, force, absolute space, absolute time, atom, and molecule – were not scientific as they went beyond their sensory anchors or the observation statements that grounded them (Mach 1910/1992). He famously said that he would 'leave the Church of Physics' if belief in atoms was required for its membership (Blackmore 1989).[3]

Karl Popper acknowledged the force of Mach's critique, but rather than accept the bulk of orthodox science as unscientific, he proposed in his 1934 *Logik der Forschung*[4] a new demarcation of science from non-science, namely, falsificationism or testability. Rejecting the Humean/Machian/Positivist experiential confirmatory criterion, he proposed instead:

> But I shall certainly admit a system as empirical or scientific only if it is capable of being tested by experience. These considerations suggest that not the *verifiability* but the *falsifiability* of a system is to be taken as a criterion of demarcation. (Popper 1934/1959, p. 40)

He addressed this foundational demarcation issue in a 1953 Cambridge lecture 'Science: Conjectures and Refutations' published in his 1963 anthology *Conjectures and Refutations: The Growth of Scientific Knowledge* (Popper 1963). He was adamant that his falsifiability criterion was not meant to separate meaningful from meaningless statements (Hume's project) but scientific from non-scientific statements or systems. There, dismissing the positivist link-to-experience (sensation) criterion as a demarcator of science, he says:

> But this criterion is too narrow (*and* too wide): it excludes from science practically everything that is, in fact, characteristic of it (while failing in effect to exclude astrology). No scientific theory can ever be deduced from observation statements, or be described as a truth-function of observation statements. (Popper 1963, p. 40)

and proposed instead:

> One can sum up all this by saying that *the criterion of the scientific status of a theory is its falsifiability, or refutability, or testability.* (Popper 1963, p. 37, emphasis in original)

---

[3] Mach's seemingly antediluvian position can be defended by saying he forsook committed belief in the then current 'plum pudding' picture of the atom that had been advanced by J.J. Thompson. This is an issue for Machian scholarship.

[4] First English translation in 1959, *The Logic of Scientific Discovery* (Popper 1934/1959).

and later:

> A system is to be considered as scientific only if it makes assertions which may clash with observations; and a system is, in fact, tested by attempts to produce such clashes, that is to say by attempts to refute it. (Popper 1963, p. 256)

Popper's original concern was to separate and defend good and revolutionary science, as manifest in Einstein's theory of general relativity that had spectacularly, and very publicly, been confirmed by Arthur Eddington's 1919 solar eclipse expedition, from popular belief systems of the time that were also being enthusiastically embraced: astrology, psychoanalytic theory, and historical materialism. For Popper, each of the latter was a pseudoscience, and his testability criterion was meant to separate them from the real thing.

But falsifiability did not quite work this way. On the one hand, many supposed pseudosciences made claims about the world that could be, and were, falsified – creationist science and astrology, for instance. So, these should be just 'bad' science, not 'pseudoscience'. On the other hand, many established sciences made claims that were falsified by empirical evidence, but this did not result in rejection of the theory. So, these should be pseudoscience.

Karl Popper was correct in identifying the *growth* of knowledge as a hallmark of the scientific tradition; a static tradition is not scientific. In a 1961 Presidential Address to the British Society for the Philosophy of Science, he stated:

> My aim in this lecture is to stress the significance of one particular aspect of science – its need to grow, or, if you like, its need to progress. ... I assert that continued growth is essential to the rational and empirical character of scientific knowledge; that if science ceases to grow it must lose that character. It is the way of its growth that makes science rational and empirical; the way, that is, in which scientists discriminate between available theories and choose the better one or (in the absence of a satisfactory theory) the way they give reasons for rejecting all the available theories, thereby suggesting some of the conditions with which a satisfactory theory should comply. (Popper 1963, p. 215)

Fifty years later, the German philosopher Paul Hoyningen-Huene concurred:

> One of the most astonishing facts about science, especially about modern natural science, is its remarkable growth, both in scope and in precision. Science is a dynamic enterprise through and through. This feature probably best distinguishes science from all other knowledge systems, past and present. (Hoyningen-Huene 2008, p. 176)

Consider the 2017 detection of gravity waves for which Barry Barish, Rainer Weiss, and Kip Thorne were awarded the Nobel Prize in physics. Since their initial postulation by Henri Poincaré in 1905, these waves had something of the appearance, or feel, of feng shui's chi waves: they seemed mysterious, there was no obvious indicator for them, and seemingly their only warrant was speculation. But this piece of metaphysics was different from feng shui metaphysics: it did not emerge from nowhere or from textual analysis. Poincaré thought gravity waves had to be a consequence of Lorentz's electron theory; the latter required the former. In 1916 Einstein cemented gravity waves' place on the scientific agenda by showing that they were a requirement of his own General Theory of Relativity, 'ripples in the fabric of space-time' as they have been called. But it was a full century of theoretical and experimental refinement, and finally millions of dollars spent in the LIGO project, before the 'gravity wave' agenda item was approved.

The contrast with feng shui speculation is dramatic. In all of the 3000+ years of chi talk and appeals to mysterious energies, there simply has been nothing comparable to the proposing and search for gravity waves. It was the continued and deep engagement with science that moved gravity waves from speculative metaphysics to tentative physics and then to being part of the accepted furniture of the world. There is no such movement in the feng shui tradition. Although there is a surfeit of spread, there is no intellectual depth. The same stories and mantras are endlessly repeated.

A great deal of late twentieth-century philosophy of science has been taken up with problems occasioned by using Popper's testability as a demarcation criterion for science and with efforts to find other more adequate criteria. Willard van Orman Quine, Thomas Kuhn, Imre Lakatos, Paul Feyerabend, Paul Thagard, and Larry Laudan all contributed to this debate.[5] Lakatos thought that his 'methodology of scientific research programmes' did provide a warranted demarcation in the way that Popper and Kuhn's had failed to do (Lakatos 1970). Further this was important because with his memories of Hungarian communism:

> … the problem of demarcation between science and pseudoscience is not a pseudo-problem of armchair philosophers: it has grave ethical and political implications. (Lakatos 1978, p. 7)

The mushrooming, internationalizing, billion-dollar feng shui industry, and its related alternative or holistic medicine industry, is an example of the ethical, political, and cultural consequences of failing to identify pseudoscience or saying that such identification is impossible. Being able to robustly identify feng shui as pseudoscience might put some brake on its spread and impact; it might redirect people's monies to effective treatments; and in some jurisdictions, it might enable conviction for false advertising or even fraud. And beyond this, familiarity with such identification procedures can engage citizens in a better understanding the nature of science.

Carl Hempel usefully offered a list of seven *desiderata* that identified good scientific theories and which can serve in characterizing good scientific practice:

- A theory should yield precise, preferably quantitative, predictions.
- It should be accurate in the sense that testable consequences derivable from it should be in good agreement with the results of experimental tests.
- It should be consistent with currently accepted theories in its own and neighbouring fields.
- It should have broad scope.
- It should predict phenomena that are novel in the sense of not having been known or taken into account when the theory was formulated.
- It should be simple.
- It should be fruitful (Hempel 1983, pp. 87–88; author formatting added).

---

[5] For an outline of the arguments and literature, see especially Ladyman (2002, Chap. 3) and Nickles (2013).

This account usefully employs a number of criteria to distinguish good theories from not-so-good or poor theories. Indeed, 'marks out of five' can be given to theories on the basis of how well they meet each criterion, with a maximum possible score of 35. Then discussion can occur about 'cut-off' marks for separating good from poor theories or from proto-theories. On this account, poor theories can be improved; they can raise their mark by attending to one or other deficiency.

Hempel's *desiderata* are meant to separate good science from not-so-good or poor science. But there comes a point where poor marks indicate something is other than a poor or a protoscience, but it is rather a pseudoscience. Minimally, a zero on the third *desideratum* – consistency with currently accepted theory – is a strong indicator that something belongs to a pseudoscience, rather than being just part of a poor science. This is, of course, a conservative criterion; it puts something that is entirely inconsistent with best established science in a domain, beyond the pale. There is an element of 'closed shop' here, but it can be justified. Over the span of about 400 years, the Galilean-Newtonian Paradigm (GNP) has developed and matured into modern science with its ontological, methodological, ethical, and sociological dimensions. If something is inconsistent with all of these core characteristics, then it may be something, but it is not science; to claim that it is amounts to being a pseudoscience.

In appraising feng shui, it is important to note that good theories, as Hempel characterizes them, are the expected outcome and indicator of good science; but science, as an organized, structured, historical-sociological entity, needs further characterization beyond what suffices for the identification of good theory. Extra ontological, methodological, and sociological criteria are required, the more so in order to separate science, as a historically situated, organized, knowledge-seeking activity, from pseudoscience. For a research group to be called a scientific group, or for it to be pursuing a scientific practice or inquiry, it needs to have the following characteristics[6]:

- It should reliably produce a 'quota' of good scientific theories.
- It needs to seek new knowledge and to do research: not be ossified, stand still, and repeat extant knowledge.
- It should be constituted as a research community pursuing cognitive goals and committed to finding out new things about the natural and/or social worlds, not just a community sharing beliefs, inquiring into texts, or formulating legislative laws.
- Its members need be trained or certified in such cognitive inquiry; science can be advanced by lay-people, but if no or few members of the community are suitably trained, then the community falls short of being a scientific community.
- It should appeal only to ontologically stable entities in its explanations and theorizing. Reference to 'here today, gone tomorrow' entities – or entities that come in and out of existence depending on who is thinking about them or for what culture they exist – diminish the scientific status of theories and communities that propose them.

---

[6] See Bunge (1991a, 2001, Chap. 8) who develops these points.

- It needs be committed to at least pragmatic methodological naturalism as the basis for evidence collection and theory appraisal; appeals to political, ideological, or religious authority are simply not allowed. Nor is deference to divine scripture or revelation permitted in justifying metaphysics or defending particular claims. Science simply does not allow such appeals to outside authority be it the College of Cardinals, the Central Committee, the Caliph, the Chief Rabbi or the First Presidency of the Mormon Church.

Feng shui fails both the scientific *theory* test and the scientific *organization* test; it lacks scientificity.

Rejecting the Demarcation Project

Larry Laudan, in a much commented-upon paper, hoped to bring this discussion to an end with his claim that the demarcation quest was hopelessly and in-principle contentious:

> … it is probably fair to say that there is no demarcation line between science and non-science, or between science and pseudo-science, which would win assent from a majority of philosophers. (Laudan 1983/1996, p. 211)

And further that the efforts were misdirected because they:

> managed to conflate two quite distinct questions: What makes a belief well founded (or heuristically fertile)? And what makes a belief scientific? (Laudan 1983/1996, p. 222)

He concluded his paper with the admonition:

> If we would stand up and be counted on the side of reason, we ought to drop terms like 'pseudo-science' and 'unscientific' from our vocabulary; they are just hollow phrases which do only emotive work for us. (Laudan 1983/1996, p. 222)

Laudan's paper is puzzling. He says that the term 'pseudoscience' is merely rhetorical and lacks specification. Yet two years earlier, he published a detailed contribution to the 'Science Wars' (Brown 2001) critical of the Edinburgh Strong Programme and its pretension to reduce philosophy of science to sociology of science and rationality to politics by other means. The title of his earlier paper was 'The Pseudo-Science of Science?' (Laudan 1981/1996). And Laudan there had recourse to a substantive view of pseudoscience that goes beyond just a banner headline. He rejects in its entirety David Bloor's, and the Strong Programme's, key work *Knowledge and Social Imagery* (Bloor 1976/1991), saying:

> … one must regard his [Bloor's] efforts at legitimation by assimilating himself to the scientist as rhetorical window dressing and nothing more than that. As for my calling his approach 'pseudo-scientific', the label comes to seem increasingly appropriate. A pseudo-scientist is, after all, one who claims himself to be a scientist but who is unable or unwilling to indicate what is scientific about his beliefs and his modus operandi. (Laudan 1981/1996, p. 207)

Why, two years later, 'pseudoscience' is relegated to mere rhetoric is not made clear.

Independently of Laudan, Roger Cooter a historian also argued that the label 'pseudoscience' has no epistemological value; it has only rhetorical value:

> ... it would be preferable to have the term 'pseudoscience' replaced in our vocabularies with something like 'unorthodox science' or 'non-establishment science'. (Cooter 1982, p. 138)

Another historian, in writing of the Velikovsky Dispute, makes the same claim:

> 'Pseudoscience' is an empty category, a term of abuse, and there is nothing that necessarily links those dubbed pseudoscientists besides their separate alienation from science at the hands of the establishment. (Gordin 2012, p. 206)

Cooter argued for his case on social constructivist grounds, maintaining that all knowledge claims are the result of social negotiation in which truthfulness or falsity does not play a determining role; the rise and fall of theories reflect differences in social and cultural power. Truth tracks power. Attaching labels is a matter of ideological contention, and the label's purpose is to either hide or serve social interests. So, the label 'pseudoscience' simply indicates 'sociopolitical deviance' (Cooter 1982, p. 137). Earlier he had written that whenever the label 'pseudoscience' is used, it is in the service of 'conserving social interests' (Cooter 1980, p. 237). This because:

> ... since all knowledge of external nature is made by men and socially constructed, the identification and criticism of any particular body of knowledge as 'pseudoscientific' must count as a defence of some other body of knowledge. (Cooter 1980, p. 259)

Cooter's papers are in the tradition of social constructivist history[7] that was energized by the philosophical claims of the Edinburgh Strong Programme. The founders of this programme were Barry Barnes (1977), David Bloor (1976/1991), Harry Collins (1985), Bruno Latour and Stephen Woolgar (1979/1986) who all explicitly appealed to Thomas Kuhn's account of science in order to get the programme off the ground. The strong programme predictably energized constructivism, relativism, and multiculturalism in education. It gave succour to those arguing for the public funding of Alternative Medicine research and the establishment of Non-Traditional Medicine departments in universities. After all, if scientific theories are a 'front' for social forces, then all such 'fronts' or representatives should be equally supported, with perhaps affirmative action for rejected theories and poorly supported programmes. The programme has been thoroughly criticized, including by Kuhn (1991/2000).[8]

Against Cooter, the later Laudan, and all other social constructivists who reject the use of 'pseudoscience' on account of its rhetorical function, it needs to be recognized that labels can have both rhetorical *and* epistemological functions; to acknowledge a rhetorical function is not to say that the term has no epistemological

---

[7] For scholarly and tightly argued refutation of this historicist programme, see Wootton's *The Invention of Science* (Wootton 2015). The book is reviewed in Matthews (2017).

[8] For critiques of the strong programme, see, among many, Brown (2001), Bunge (1991b, 1992), Nola (1991, 2000), and Slezak (1994a, b).

import; to say that some analysis supports a particular social group is not to say that the analysis is not correct or true or constitutes knowledge. To say of a football team that it is 'excellent' is to support the team, but at the same time, its use makes claims about the competence of the team. That part of the claim can be appraised in standard public ways. The theory of global warming might support the renewable energy lobby, but that does not mean it makes no truth claims or that its claims are false or compromised. Appraisal of those claims can be detached from appraisal of the political claims; the two appraisals are orthogonal.

Laudan's philosophical argument against demarcation and for the merely rhetorical function of 'pseudoscience' gained the assent of the majority of philosophers of science, not just the assent of the more general scholarly or educational constructivist community who could be expected to readily embrace it as Laudan was 'speaking their language'. Constructivists were very happy to hear prominent philosophers saying that 'everything is science; it is only politics, ideology, or culture that makes distinctions for their own purposes'. This view, of course, was well received by Creationists and Intelligent Design advocates who were outraged at Judge Overton's ruling in the US 1981 *McLean* vs. *Arkansas* trial that Creationism was not scientific and so had no place in US classrooms (Ruse 1988). Also happy with Laudan were proponents of multicultural science, specifically those wanting to recognize feng shui as science, not just as 'traditional' science, but as science and consequently warranting an endorsed place in the science programme, not just a showcase to illustrate non-science or pseudoscience.

Although many philosophers concurred with Laudan's arguments on the problems of demarcation, not all did so. Robert Pennock was one among many defending demarcation in the Humean-Popper tradition:

> Because Laudan's and Quinn's discussions of demarcation, which can only be described as histrionic and ill-considered, and those of their careless imitators continue to muddy the waters to the detriment of both science and philosophy of science. (Pennock 2011, p. 180)

Other philosophers felt the same way and engaged in careful, informed, and detailed refutation of Laudan's arguments.[9] His obituary for demarcation was premature. Mario Bunge provides both a broad and detailed account of the requirements for any cognitive field (i.e. any inquiry generating putatively true or false propositions or theories) to be scientific (Bunge 1967/1998, Chap. 1; 2001, Chap. 8). His account subsumes the central theses brought forward by different contributors to the Laudan debate. For Bunge, a mature science has ten features:

- A community (C) of appropriately trained inquirers with recognized public means of information exchange.
- A general outlook or philosophical background (G) that includes an ontology of discernible things, a realist epistemology, and an ethos supporting the free search for truth.

---

[9] See at least Bunge (1991a, 2001, Chap. 8), Butts (1993), Derksen (1993), Ladyman (2013), Mahner (2007, 2013), Pennock (2011), Pigliucci (2010, 2013), Shermer (2013), most of the 23 contributions to Pigliucci and Boudry (2013), and contributors to Boudry and Pigliucci (2017).

- Its domain of investigation (D) is real events and processes in the world, not texts and not ideas, though, of course, the latter are utilized.
- Its formal background (F) is a collection of current best logical and mathematical theories about (D).
- Its specific background (B) is a collection of up-to-date and reasonably well-confirmed data, hypotheses, and theories from other fields relevant to (F).
- Its problems or puzzles (P) consist of cognitive rather than practical matters concerning items and events in (D), being usually a quest for laws.
- Its fund of knowledge (K) is a collection of up-to-date and testable (though not final) theories, hypotheses, and data compatible with (B).
- Its aims or goals (A) are the discovery of laws or confirmed hypotheses about elements of (D).
- Its methods (M) consist exclusively of scrutable, checkable, and justifiable procedures; there need not be commitment to a single method.
- It has a significant overlap (O) with other scientific fields of inquiry such that there are overlaps in the respective G, D, F, B, P, K, A, and M sets. A mature science does not exist in cognitive isolation from other mature sciences; they learn from and feed off each other (Bunge 2001, pp. 170–171).

Pseudoscience as a Warranted Category

It helps the argument of this book to focus on just the first distinction, namely, science-pseudoscience, rather than the wider task of separating both from non-sciences such as art, history, mathematics, theology, music, and so on. What constitutes justifiable warrant for claims in these latter inquiries or disciplines is a matter for separate investigation as one finds in philosophical theology, philosophy of mathematics, aesthetics, and so on.

Different philosophical, sociological, and political indicators or markers of pseudoscience have been advanced. Pseudoscience can be identified by working through each of Bunge's foregoing ten identifiers of mature science and taking the degrees of their negation of each feature. Sven Hansson provided another such list whereby a corpus of belief and practice can be judged pseudoscientific in as much as:

- There is overdependence on authority figures.
- Unrepeatable experiments are too frequently adduced.
- Data selectivity or cherry-picking of evidence is too common.
- There is an unwillingness to seriously test claims and predictions.
- Confirmation bias is endemic and disconfirmation is neither sought nor recognized.
- Some explanations are changed without systematic consideration (Hansson 2009).

And when:

- They make claims about events and mechanisms in the natural world.
- The claims cannot be epistemically warranted, yet effort is made to show their scientificity.
- They too easily resort to auxiliary hypotheses to insulate claims from empirical refutation (Hansson 2009).

A further characteristic that can be added to Hansson's list is:

- The practice makes scientific claims but refuses to engage with the scientific community by publishing in established research journals and presenting at research conferences.

Pseudosciences violate the fundamental principle that 'no science is an island sufficient to itself'. All scientific endeavours and disciplines have contact with their neighbours; more than contact, they need to accommodate adjacent sciences. This is what drives the creation of cross-over or interdisciplinary sciences: biochemistry, electrochemistry, geophysics, palaeoanthropology, physical chemistry, and so on (Hoyningen-Huene 2013). Intellectual isolationism is a key marker of pseudoscience.

Consider, for instance, 'Black Hat' Tantric Esoteric Buddhist Feng Shui (BHB) which is enormously popular in the USA. Black Hat feng shui was brought from China by His Holiness Grandmaster Professor Thomas Lin Yun, who was feted on numerous university campuses and in many business boardrooms and has thousands and thousands of followers. According to the BHB website, 'His Holiness was not only revered as a religious leader, a scholar, but also the world's most prominent authority on Feng Shui'. The BHB website says its teachings:

> Represent a comprehensive integration of Buddhist teachings, yin-yang philosophy, I-Ching, Feng Shui, theory of Ch'I, holistic healing, Chinese folklore, transcendental cures, meditation, spiritual cultivation and development, Chinese poetry, etc.[10]

Clearly a full-hand of complex subjects. Further, understanding by outsiders of the teachings is difficult as BHB can only be communicated orally; listening is essential to the message.

> In the Black Sect Esoteric Buddhist tradition, the only legitimate path for instruction is the sacred discipline of oral transmission from teacher to pupil. Her Holiness Khadro Crystal Chu Rinpoche is the exclusive teacher and transmitter of the teachings of Black Sect Esoteric Buddhism, as appointed buy His Holiness the late Grandmaster professor Lin Yun.[11]

The fiat that 'only insiders can understand the teachings' is a common defensive ploy for charlatans and purveyors of pseudoscience. It protects the 'theory' and isolates the practice from relevant established sciences (Fig. 13.1).

Recall Jerry King, mentioned in Chap. 1, who studied Feng Shui and Four Pillars under various masters in Taiwan and Hong Kong. He specializes in Purple Star

---

[10] See http://www.yunlintemple.org/home

[11] See http://www.yunlintemple.org/home

Fig. 13.1 H.H. Lin Yun & H.H. Crystal C. Rinpoche

Astrology readings and travels extensively, consulting globally and obtaining research data and verifying theories of cosmic flow in the Four Pillars of Destiny.[12] The claim about 'obtaining research data and verifying theories' clearly invites the first question: But is it science? Having on all the above grounds, answered 'No'; the second question arises: Is it pseudoscience? And here the answer is 'Yes'.

Not surprisingly, cognitive isolationism makes feng shui sectarian. For instance, Kathryn Weber, a 'former Black Hat practitioner', writes in defence of her reversion to classical feng shui that:

> Here in the US, Black Hat is the preferred method because of how easy it is to apply – the front is always north, and the back left is always the wealth corner and the rear right is always the romance direction irrespective of the actual compass directions. Black Hat feng shui is often considered more spiritually-oriented versus classical feng shui. It's (sic) disadvantage is that it doesn't take into account unique compass directions or time and its influence.[13]

It is a mark of pseudoscience that these sectarian differences cannot be settled. There are controversies, even long-standing ones, in science, but there is a degree of agreement about how they can be solved, and reference to an authority figure as

---

[12] See https://www.fengshuitoday.com/feng-shui-of-the-hsbc-headquarters-building-in-hong-kong/
[13] See https://redlotusletter.com/classical-feng-shui-and-western-black-feng-shui-the-6-critical-differences-confessions-of-a-former-black-hat-practitioner/

being the ultimate arbiter is not among them.[14] Not so in feng shui. Grandmaster Lin Yun appointed his companion Her Holiness Khadro Crystal Chu Rinpoche to be the authoritative, and only, interpreter of Black Hat Sect teaching. This dependence on authority is characteristic of the worst of political and religious sects. It is another marker of pseudoscience. It is illustrative to juxtapose the supposedly scientific claims of Lin Yun and the qigong claims of Yan Xin detailed in Chap. 12, with Hansson's pseudoscience checklist above: all the identifying 'boxes' for pseudoscience are ticked.

The central feng shui ontological entity, chi – its sine qua non as Stephen Field rightly called it (Guo 2001) – does not appear in any reputable physics research journal or book. Chi is not mentioned anywhere in the hugely funded international search for new energy, renewable energy, or clean energy. The *Science in Contemporary China* handbook (Orleans 1980) has chapters on all the Chinese natural, social, and applied sciences, including biomedical sciences, but no chapter on either chi or feng shui. The only mention of either in the entire work is in discussion of acupuncture where the 'meridian' theory of freeing vital energy pathways, notionally 12 on either side of the body, by inserting needles into various of the 600+ meridian intersections (acupoints), is listed as the original explanation for the efficacy of acupuncture; but this is then passed over in favour of contemporary, routine physiological, or psychological accounts (placebo) of the highly debated effects of acupuncture, such as have been discussed in Chap. 5.

Yet 40 years later, the CCP is embracing chi-based TCM and qigong routine and opening feng shui architectural programmes. Thousands of 'Health Qigong Management Centers' have been established under Ministry of Health authority. The fine line for the CCP is to maintain its 'anti-superstition' campaign which provides cloverleaf justification for banning Falun Gong while maintaining that chi medicine and exercise are not superstition. This is a philosophical/political issue for the CCP and one that can readily be posed as a school exercise.

Ecology of Science and Pseudoscience

Science is not just the product of a thinking head, of a thoughtful and hard-working scientist. Science always occurs in a social-economic-technological context which has its own conceptual and philosophical characteristics. For Bunge, historical, sociological, and philosophical studies of science show that it requires social and intellectual environs characterized by the following political, ethical, and philosophical commitments. He usefully calls this the 'conceptual ecology of science' and represents it as a pentagram (Bunge 2012a, Chap. 2) (Fig. 13.2).

Humanism/Commercialism Scientists need to promote human welfare, not misery, business advancement, or political advantage. The latter purposes more easily

---

[14] See Engelhardt and Caplan (1987), Hellman (1998), and Machamer et al. (2000).

Fig. 13.2 Conceptual ecology for scientific progress (Bunge 2012a, p. 28)

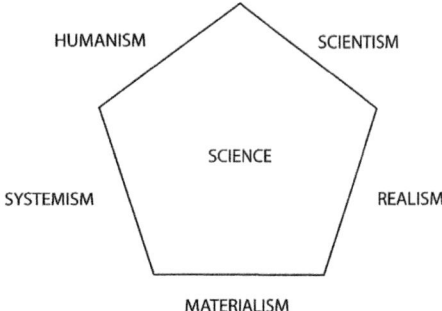

lead to corruption of science (witness Nazi Germany, Stalinist Russia, or current 'big business' tobacco, oil, and pharma science). But there are also less visible effects of commercialization on academic and industry research. These are effects that impact on directions of research, constriction of 'public knowledge' and access, the reward system in science and universities, communitarianism in science, and other considerations.[15] There can and should be applied science, but it ought to be for human welfare and improvement.

Systemism/Compartmentalism Competent well-informed scientists recognize that there are no isolated events, mechanisms, or problems in the world. Structures and events are parts of systematic causal wholes. John Donne famously wrote that 'no man is an island', so also no event, personal action, or social movement is a causal island; and no science is an island. Consequently, good science generates cross-disciplinary research fields: geophysics, astrophysics, biochemistry, astrochemistry, social psychology, molecular biology, psycholinguists, economic history, political economy, and so on. Because they do not emerge from science, hybrids such as astropsychology or creation science are just pseudolabels.

Materialism/Spiritualism Scientists seek for causes and explanations in the kinds of things and mechanisms that are within the accepted ontology of science. A materialist ontology is informed by science; hence gravitational and electrical fields are material though not physical. *Methodological naturalism* can satisfy this requirement, but evocation of *spiritualism, supernaturalism, occultism,* or *tradition-based* entities violates it. To the degree that a society believes that the gods, spirits, or the occult are responsible for earthquakes, then money for geophysical research will be limited; to the degree that societies are fatalistic, believing that 'everything happens for the better', then prevention of disaster and remediation will not be undertaken; and where illness is seen as spirit possession, then medical science does not develop.

Individuals can be spiritual without believing in or practising spiritualism; the latter involves belief in the intervention in worldly processes of spirits, supernatural entities, or the occult, and this impedes the growth of science. And as was docu-

---

[15] On this see contributions to Irzik (2013).

mented in Chaps. 3 and 7, modern science was created by Christian believers much of whose work was dependent on achievements of Islamic science; and religious scientists of all faiths contribute to the advance of modern science. These contributions were dependent on adoption, explicitly or implicitly, of methodological naturalism. As explained in Chap. 12, there is nothing per se about the chi construct that rules it out of science; it is just that as soon as it is examined, its scientificity disappears; it cannot be 'nailed down' and elusiveness is part of its nature.

Realism/Subjectivism Scientists recognize that there is an external world independent of human consciousness or experience; natural science attempts to provide knowledge of such a world; and these attempts are partially successful. Our concepts and theories are human creations, but apart from the social sciences the reality they conceptualize or explain is not a human creation. The external world judges the efforts of scientists to understand it; good theorizing is not just the prevailing of local or wider political power. Witness the ultimate collapse of Church-backed Ptolemaic astronomy, Nazi-backed German blood science, or Party-backed Soviet genetics.

Scientism/Irrationalism Scientists believe that science is rational; indeed it provides a model for social rationality; further, Enlightenment-influenced scientists believe that scientific methods are applicable outside the laboratory and are the only way in which knowledge of the world and society is attained. Without this commitment, social and cultural problems are addressed in wholly ineffective ways: praying for the end of Middle East conflict can be a comforting cultural engagement, but it can shed no light on the conflict, its history, or its remediation. Prayer might motivate such investigation, but equally it can, and often does, bypass a naturalistic and scientific investigation.

For any society, to the degree that the first member of the above couples is maximized, then science can flourish. To the degree that the second member is elevated, then the society allows and promotes the growth of pseudosciences (see Fig. 13.3).

In societies and cultures where spiritualism, non-systematism, commercialism, irrationalism, and subjectivism (phenomenalism or instrumentalism) prevail, then science cannot thrive, but pseudoscience surely can and does.

Contemporary USA provides a case study for this claim. Spiritualism is pervasive. God and gods are evoked everywhere, including on dollar bills, and for every purpose, including the killing of declared enemies, the prevention of natural disasters and the amelioration of their effects. Megachurches, attended by tens of thousands of 'happy clapping' congregants, are common; televangelism, with in-studio and at-home miracles, functions non-stop, 24/7, on TV and cable networks. Bookshop aisles and websites are filled with paranormal, alternative, and esoteric literatures.[16]

---

[16] Kurt Andersen's *Fantasyland* (2017) provides extensive, if depressing, documentation of the 500-year history of what counts as spiritualism in the USA. Parts of the Roman Catholic and Protestant traditions do their best to separate themselves from this spiritualism which they see as commercialized, corrupt, and theologically ill-informed.

Fig. 13.3 Conceptual ecology for pseudoscience (Bunge 2012a, p. 33)

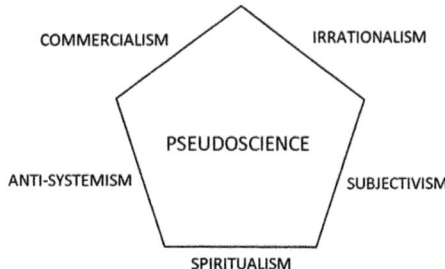

Anti-systematism is routine. Life is compartmentalized. A general or liberal education is progressively harder to get; specialization is the academic norm; there are career, funding, and disciplinary barriers to cross-disciplinary research. The much leapt-upon NOMA bandwagon launched by Stephen Jay Gould formalized the separation of science from other disciplines, specifically theology (Gould 1997). They cannot judge each other. But if theology makes claims about the world, then the claims can and needs be judged by science (Boudry 2017).

Commercialisation and moneymaking are preoccupations of dominant US groups; if this was not their preoccupation, they would not be dominant. Commercialisation is captured in everyone's image of Wall Street, where excess, self-interest, and pursuit of the bottom line are just normal business activities. It is equally captured in the Walmarting of hundreds of towns where whole downtown business and residential communities have been destroyed by the Walton family's pursuit of extra millions of dollars being spent in their own edge-of-town megastores.[17] Powerful mining, agriculture, transport, tobacco, and oil interests have always put commercial interest above community and environmental interest. President Trump rode US commercialisation all the way to the White House.

The malignance of commercialisation goes beyond the elevation of profit over social interest; this is at least objective, public, and debatable; worse, it is giving epistemological warrant to commercial interest. This is what was so depressing about the 'scientists for hire' and 'research for sale' realities in the tobacco and oil industries that were so well and depressingly documented by Oreskes and Conway (2010). Truth is bent or just invented for commercial interest. This mirrors the same degrading of truth for political, party, ideological, and religious interests. If truth claims about the world are not settled by reference to the world, then obviously the claims can be settled by other considerations. American Pragmatism has something to answer for on this score.

Irrationalism is now a respected and examined subject in US universities; goodly portions of whole faculties and colleges are given over to its promotion.[18]

---

[17] Apart from numerous books, the Robert Greenwald documentary *The High Cost of Low Price* (2005) well captures the Walmarted experience of the USA.

[18] David Stove provides a nice, informed, and witty introduction to how irrationalism took root in contemporary philosophy of science (Stove 1982).

Universalism is everywhere rejected in favour of gender, race, religious, political, sexual, economic, and cultural localism. Supposedly, all truth is local, all rationality is local, and all ethics are local. Antirationalism or postmodernism is the Philosophy Department, Cultural Studies Department, and College of Education norm. The tension, if not contradiction, between localism and rationalism is seldom explored; their consistency is assumed. But how local can rationality be before it becomes irrationality? And what features can be identified to define the local? Husserl, Kuhn, Feyerabend, Heidegger, Derrida, Latour, Lyotard, Irigaray, Barnes, Collins, Pinch, Harding, von Glasersfeld, Giroux, and others are among the most read and most cited authors in sociology, education, philosophy, and humanities programmes. Senior figures in science education routinely advance outrageous, silly, and discredited positions and are cheered and awarded for doing so.[19] What unites them all is the rejection of truth and the discounting of rationality. A steady diet of these authors does have an effect on education and most other things dependent on clear thinking and some appreciation of intellectual traditions. The enormity of the Sokal Hoax is testament to the diet's effect in academia (Sokal 2009). That the president of the USA can say that 'truth does not matter' and that his agents can say 'there are alternative realities' speak to its down-stream effect in society and politics.[20]

Subjectivism and empiricism are also deeply entrenched in US culture and academies. Epistemologically these are the claim that the test of truth is how things appear to the individual; individual experience is the epistemological bottom line. Hence the intellectual ground is prepared for when President Trump says that no matter what his experts advise, he goes with his 'gut feelings'. Citizens nod their head on hearing this. In one book by a leading science educator, the personal pronoun occurs 96 times on one page, and it is not an autobiography. This level of narcissism and self-absorption flows easily from subjectivist and empiricist doctrine. The doctrine is profoundly antiscientific. The whole history of instrumentation in science is the history of making intersubjective appraisals of temperature, heat, speed, duration, pressure, pulse rate, rain fall, voltage, wind speed, weight, and so on (Crump 2001). Objective, impersonal, nonsubjective measurement is a precondition for science. At every step, progress in science has meant the overcoming of everyday experience. When a dash of Kant is added to empiricism, then reality becomes the unknowable 'thing in itself'. With just the slightest extra intellectual nudge, even this disappears, and we are left with ontological idealism: there is no reality, just our experience.

Subjectivism was turbo-charged in the 1920s by the common, but mistaken, philosophical interpretation of quantum theory, otherwise known as the Copenhagen Interpretation. Niels Bohr and Werner Heisenberg were among the first to bring the

---

[19] On the inroads, if not capture, of universities by irrationalism, see Bunge (1994, 1996); and contributions to Gross, Levitt, and Lewis (1996), Koertge (1998), and Kurtz and Madigan (1994). Latour now regrets the irrationalist wave his earlier writing unleashed on the academy (Latour 2004).

[20] A good and informed account of the attack on truth in both the academy and society is *Respecting Truth* by philosopher and social scientist Lee McIntyre (2015).

observer into measurement processes at the quantum level and thus to make physics subjective. Although rejected by Einstein, the Copenhagen Interpretation became for decades the norm in physics; it was advanced by von Neumann, Wigner, and countless senior figures and textbook writers. In the Anglo-world, it was popularized by two knighted physicists Sir James Jeans and Sir Arthur Eddington. Jeans, in a widely read and influential book, wrote:

> As the subject developed, it became clear that the phenomena of nature were determined by us and our experiences rather than by a mechanical universe outside us and independent of us. (Jeans 1948, p. 294)

One can imagine the enthusiasm with which such claims were and still are greeted. They gave a scientific green light for every imaginable brand of idealism, mysticism, obscurantism, and gender, race, class, and cultural localism. The philosopher David Mermin opined that quantum physics has taught us that 'the Moon is not there when nobody looks' (Mermin 1981, p. 405). Such is the grip of nonsense that many read this and just nodded their head. Sadly, Copenhagen subjectivism, though in retreat among physicists and philosophers, is still being repeated at the highest levels in science education. So:

> science as public knowledge is not so much a "discovery" as a carefully checked "construction" ... and that scientists construct theoretical entities (magnetic fields, genes, electron orbitals ...) which in turn take on a "reality". (Driver 1988, p. 137)

And:

> ...For constructivists, observations, objects, events, data, laws, and theory do not exist independently of observers. The lawful and certain nature of natural phenomena are properties of us, those who describe, not of nature, that is described. (Staver 1998, p. 503)

Such claims are made from the podium at international science education research conferences where they receive, if not standing ovations, then little critical comment.

But beginning with Einstein, there has always been substantial scientific and philosophical opposition to this scientific epistemological and ontological subjectivism. Susan Stebbing was among the first philosophers to voice criticism (Stebbing 1937/1958). At the core of Mario Bunge's philosophical project has been the rejection of subjectivism, phenomenalism, instrumentalism, and positivism in physics and counterwise the defence of realism. As he writes in his autobiography:

> I believe that my main contribution to physics has been my book *Foundations of Physics* (1967), which had a strong philosophical motivation. This was my attempt to prove, not just state, that quantum and relativistic theories are realistic (observer-free) and that their subjectivist (observer-centered) interpretations are illegitimate philosophical grafts. (Bunge 2016, p. 406)

He points out that none of the founders of quantum mechanics practised the subjectivism they preached:

> In fact, when calculating energy levels, transition probabilities, scattering cross-sections, and the like, all quantum physicists assume tacitly that no reference to the measurement device, much less to the observers' mind, occurs in their calculations. (Bunge 2006, p. 68)

Many philosophers and physicists share Bunge's critical estimation of both epistemological and ontological subjectivism in physics (Hobson 2019; Romero 2019).

The foregoing elaborations of the ecological pentagram for science and pseudoscience can be made for all societies and nations. And more detailed mechanisms can be described for the effects of each of the five cultural factors in the pentagram. But the above should suffice to show that the movement up or down of each of the five factors influences the growth of science or supports the proliferation of pseudoscience.

Conclusion

To say that feng shui practitioners are engaged in and informed by 'unorthodox' or 'alternative' science is too generous. Too easily feng shui exponents resort to this 'mysterium' defence as is well illustrated by the following authors:

> Life is defined by *Qi* even though it is impossible to grasp, measure, quantify, see, or isolate. Immaterial yet essential, the material world is formed by it. (Beinfield and Korngold 1991, p. 30)

And,

> To subject alternative therapies to sterile, impersonal double-blind conditions strips them of intrinsic qualities that are part of their power. (Sampson 1996, p. 195)

The mysterium defence is ruled out of science. It might function as a short-term place holder, but it cannot be entrenched beyond that. Failure to find and measure chi in 3000 years means it is an unscientific concept, yet it is the very heart of the whole feng shui enterprise.

Leaving aside philosophical considerations, on just sociological or externalist grounds alone, chi-based feng shui is not a science. It does not meet the criteria of 'playing the game'; practitioners might be playing *a* game, but it is not the *science* game. The situation is akin to a local rugby team saying they are playing 'alternative baseball' or 'unorthodox baseball'. For the latter names to be meaningful, there has to be an overlap in key elements with baseball. And the latter games cannot be called 'poor rugby' as that suggests that by incremental improvements baseball could become rugby. It cannot. Rugby is an alternative to baseball; it is not unorthodox baseball or even neo-baseball.

Similarly, science and feng shui are not on a continuum; the difference is not like that between full-strength and lite beer. When the measurements, columns of numbers, experiments, journals, peer reviews, conferences, and the rest of feng shui's 'scientific' clothing are examined, they all fall short of the standards of science. A critical appreciation of, and respect for evidence is missing; practitioners individually and as a community do not have a scientific habit of mind. Charitably, this makes feng shui bad science. But historical, sociological, economic, and psychological perspectives on feng shui show the whole belief system has been and is exploited for fraudulent purposes. As Ricci in the sixteenth century, Eitel in the nineteenth century, and Ch'en Duxiu at the beginning of the twentieth century so clearly stated, the whole system, by promising so much for so little, is tailor made for delusion and fraud. It is pseudoscience not science. Where, as in Asia, the beliefs are commonplace, science students can usefully and with great benefit come to appreciate the inherent problems with the practice. Indeed, such examination should be seen as a professional obligation of science teachers and curriculum writers. Where feng shui is not commonplace, its examination is still educationally beneficial; it can be a case study that sheds light on important psychological, scientific, philosophical, and cultural dimensions of human life.

Part V
Conclusion

Chapter 14
Concluding Remarks

Feng shui theorists and consultants are practising something that superficially appears like science and is infused with scientific terminology – witness the title of a recent book *Scientific Feng Shui for the Built Environment* (Mak and So 2015) or the numerous feng shui claims that have been documented in this work, of which the following is representative:

> This flow that regulates our lives is an invisible energy known as ch'I (or *qi*). To partake in this energy, we can arrange our inner nature and our outer environment to allow it to flow like water or drift like the wind, and provide us with benefits rather than harm. We cannot control the wind, but we can however arrange our lives so this 'energy' benefits us. (Rolnick 2004, p. 9)

But despite the incessant waving of the science banner, beating of the science drum, and liberal use of scientific terms, despite the significant number of science-trained practitioners, and despite employment of instruments such as Meridian Energy Analysis Devices (MEAD) connected to computer monitors, feng shui is not a scientific practice. Further, it is not just poor science; it is pseudoscientific. 'Poor' science suggests it can get better and that if a few things (measurements, readings, data collection) are done more accurately, then feng shui can progress along to 'fair' or 'good' science. It cannot do this because it is fundamentally not science at all; it is outside the scientific pale.

The key elements of science – content, methodology, experiment, mathematization, theoretical and conceptual growth and refinement, a scientific habit of mind, and social organization – are present only as simulacrums. There is no tradition of controlled and reproducible experiment; there is no recognition of the defect of ad hoc rescuing of failed hypotheses; there is no effort to disentangle variables and study their contributions; there is a dramatic inconsistency with the core of established scientific knowledge, most especially the conservation of energy postulate; there is no effort to explain this inconsistency by engagement with the scientific community; there are no contributions to established, peer-reviewed, scientific research journals; there is altogether unwarranted dependence upon individual or sectarian interpretation of basic feng shui principles; despite 3000–4000 years of adherence and cultivation, there is no cognitive growth; there is a resolute refusal to subject feng shui to serious empirical test, instead there are 'get out of jail' cards built into the very core of the belief system; and finally there is a radical disjunct between the law-governed, deterministic worldview of science and the chaotic, idiosyncratic 'fortune-telling', 'auspicious times', 'good or bad luck' worldview of feng shui promoters and qigongists. Affirming all or even some of the latter requires that the fundamental laws of causation for macroobjects are jettisoned; and if that happens, then science is also abandoned. Whatever legitimate debate there is about causality at the micro- or subatomic level (Bunge 1982; Weinert 2005, Chap. 5), there is none at the macro-level where feng shui interventions and random chi-caused events supposedly take place.

Unfortunately, the feng shui and qigong communities do not have a monopoly on the juxtaposition of scientific competence and antiscientific commitments and beliefs. For example, Edgar Dean Mitchell the NASA astronaut who was the sixth person to walk on the moon after piloting the Apollo 14 craft and who has science and engineering doctorate degrees from MIT holds a constellation of 'extra scientific' beliefs. Mitchell has claimed that on his way back from the moon, he had a Savikalpa Samadhi experience during which his soul absorbed the fire of Spirit-Wisdom that 'roasts' or destroys the seeds of body-bound inclinations. After this experience he conducted in-flight ESP experiments with his friends back home. These experiments were published in the *Journal of Parapsychology*. Mitchell believes a remote healer, Adam Dreamhealer, cured his kidney cancer over the telephone. He also believes in UFOs, interplanetary visitations, and that he has had personal encounters with these extraterrestrials.

There are hundreds of thousands, if not millions, of Mitchells for whom first-rate science education seems to have had little if any flow-on effect for the rest of their beliefs or conduct of life. Such rigidly compartmentalized thinking is a particular problem for those believing that science education should beneficially impact student's personal life and should contribute more generally to the improvement of society and culture. This was the expectation of the Enlightenment philosophers and educators; it was John Dewey's hope; and it is the expectation of the American Association for the Advancement of Science who maintained that:

> the scientifically literate person is one who is aware that science, mathematics, and technology are interdependent human enterprises with strengths and limitations; understands key concepts and principles of science; is familiar with the natural world and recognises both its diversity and unity; *and uses scientific knowledge and scientific ways of thinking for individual and social purposes*. (AAAS 1989, p. 4, italics added)

Without using the expression, this AAAS Enlightenment-informed expectation is the hope of all people going into science teaching as a career; it is the reason that governments put resources and money into science education. The assumption underwriting all components of a school curriculum is that they individually and collectively make personal and social life better; there is a flow-on effect from good education. Richard Peters was correct in saying that education is like reform; it has improvement built into its very meaning (Peters 1966).

The unique contribution of the science programme to this more general problem-solving, society-improving, and personal flourishing educational goal is, as described in Chap. 2, the cultivation and refinement of what the AAAS call scientific *habits of mind* and what others have called a scientific attitude (McIntyre 2019). These are meant to 'flow on' from the classroom and laboratory bench to the home, workplace, and community. For the AAAS, the wider 'planetary' problems are not just material – they are social, cultural, and ideological – but application of a 'scientific habit of mind' is necessary for solving these wider problems. They are not solved by feng shui consultations or qigong exercises. This is a restatement of the fundamental Enlightenment principle of scientism.

The beginning of scientism can be seen in the once-revolutionary claim of Newton, Condorcet, and the early Enlightenment philosophers that the methods and outlook of the new science should be applied outside the laboratory; they should be harnessed in understanding and solving other pressing social and cultural problems including ones associated with superstitions and the exercise of unjustified ecclesial and feudal powers (see Chap. 2).

Many reject this characterization of science and of science education fearful that it leads to scientism (Haack 2016; Sorell 1991), but only caricatures of scientism need be feared. For instance, Massimo Pigliucci is simply wrong when he says scientism is the belief that 'science is able to answer whatever meaningful questions we may wish to pose' (Pigliucci 2010, p. 235). This is pseudoscientism. Scientism has had a bad press in social science, in 'critical' and postmodernist philosophy, and especially in constructivist science education. In these circles, 'scientism' is regarded as a synonym for reductionism, dogmatism, closedmindedism, superficialism, colonialism, scientific imperialism, intellectual overreachism, and most other -isms with which no sensible and well-informed defender of science would wish to be associated.

Components of this constellation may accurately reflect some who wear the 'scientism' badge, but these are outliers; they represent a caricature of the position. Just as pseudoscience is a reality, so also is pseudoscientism. This is often just vulgar or ill-informed scientism, the kind of pretense that led Robert McNamara to assert that the Vietnam War would be won by science. Science is itself fallible, tentative, and self-correcting, so also is any extension of science and of scientific method. Mario Bunge (1986, 2010, Chap. 13; 2014, Chap. 2) and James Ladyman (2011, 2018) are just two such informed defenders of science who are also proponents of scientism. As with all such proponents, they reject reductionism, positivism, and dogmatism, and, although advancing scientific approaches in social science and psychology, they reject scientific imperialism. Physicists should not be deferred to in determining government tariff or cultural policy, but should be heard on climate change policy, just as medical researchers should have a voice, but not an exclusive one, in determining government health policy.[1] Ladyman speaks for most upholders of scientism when he writes:

> In sum, humane scientism takes science to be authoritative in respect of objective knowledge, including about human beings and society. It recognizes no limits to science in principle, but is also antithetical to scientific hubris and hype. However, humane scientism

---

[1] Economics, demography, and ethics are just some of the other things entering into health policy.

holds the best of the arts and humanities in high esteem and recognises the role that culture and custom, and religion and tradition, play in a good human life. (Ladyman 2018)

Scientism is the view that *only* the methods of natural science are capable of providing knowledge of the natural, social, and personal worlds; there are no other routes to such knowledge. Listening to gurus, holding Ouija boards, invoking mediums, remembering dreams, interpreting sacred texts, or consulting astrologers simply gives no knowledge of nature (earthquakes), social circumstances (collapse of economies), public events (the outbreak of war), personal episodes (sudden illness or death), or even psychic episodes (delusions, emotional states, and so on). Such sources might provoke hypotheses or ideas to be tested, but they do not provide knowledge. Thus stated, scientism is not nearly as 'beyond the pale' as it is usually taken to be (Ross et al. 2007; Boudry and Pigliucci 2017; de Ridder et al. 2019).

In as much as the modernization of thought about the natural and social world depends upon its reconciliation with science, then feng shui ideology and qigong theory are barriers to the modernization of thought. Considering the Rolnick quotation at the opening of this chapter, it is clear that everyone benefits from appropriately arranging their lives, environment and social circumstances. But this is a difficult and complex enough task just considering naturalistic, economic, and graspable factors; to add completely unmeasurable, ungraspable, imaginary factors such as chi flow and accumulation into the equation of life is a thoroughly unhelpful distraction. Moreover, it is not merely unhelpful, it can be positively dangerous and damaging; it sends people down a false path and allows harm to flourish. Thousands of children needlessly die each year because their parents shun established medicine in favour of any of the smorgasbord of alternative or complementary medicines.

Further, it is manifest from the most cursory of web searches that feng shui practice has been overtaken by charlatans and fraudsters, by practitioners of the 'dark arts' as Matteo Ricci said in the sixteenth century, Ernst Eitel in the nineteenth century, Ch'en Duxiu in the early twentieth century, and so many others down to the present day have said. Charlatans and fraudsters are not absent from science, but they can be identified and called out.[2] It is not at all clear how this can be done for feng shui practice – especially when the heavy artillery of 'multilogicality', 'many worlds', 'parallel truths', 'incommensurability', 'paradigm change', or 'cultural autonomy' – provided by some philosophical traditions can be called upon to rescue any position once its error and weakness is exposed. It then becomes a philosophical task to neuter the artillery, to show that it does not do what its champions claim it to do.

All institutions, belief systems, and ideologies benefit from historical study, from understanding themselves in an historical sequence and context. To the discomfort and distress of fundamentalists, all the major religions have gained from developing such historical perspectives. Both the development and corruption of religious and political institutions and beliefs over time are common historical realities. So too science has changed over time. Famously Thomas Kuhn said that it was his

---

[2] On fraud in contemporary science, see at least Bell (1992), Gardner (1981, pp. 123–130), Gratzer (2000), Oreskes and Conway (2010), Park (2000), and Silverberg (1965).

unplanned teaching of a history of science course at Harvard University that opened his eyes to the manner in which scientific theories changed or were supplanted in the history of science. He credits the development of his whole novel theory of scientific change to this exposure to the historical dimension of science (Kuhn 1970, p. v). Science students, and all other students, have much to gain by becoming familiar with the history and philosophy of their own discipline.

Students can benefit from applying the same historical-philosophical analysis to feng shui. Many episodes, transitions, and debates have been discussed in this book. The cases of Matteo Ricci and Ernst Eitel, the 'New Youth' movement in early twentieth-century China, and the claims made for Traditional Chinese Medicine and acupuncture can all be examined in science classes. And even better in cross-disciplinary teaching where science, technology, history, social studies, philosophy, and religion faculty can coordinate their programmes. Across the spectrum of features of science – experimentation, authority, prediction, precision, mathematization, idealization, coherence, and testimony (Matthews 2011) – feng shui can be juxtaposed with science and similarities and differences drawn out.

Richard Feynman in a 1974 Commencement Address at Caltech University titled 'Cargo Cult Science: Some Remarks on Science, Pseudoscience, and Learning How to Not Fool Yourself' well illustrated this book's argument about feng shui being a pseudoscience:

> We really ought to look into theories that don't work, and science that isn't science. I think the educational and psychological studies I mentioned are examples of what I would like to call Cargo Cult Science. In the South Seas there is a Cargo Cult of people. During the war they saw airplanes with lots of good materials, and they want the same thing to happen now. So they've arranged to make things like runways, to put fires along the sides of the runways, to make a wooden hut for a man to sit in, with two wooden pieces on his head for headphones and bars of bamboo sticking out like antennas—he's the controller—and they wait for the airplanes to land. They're doing everything right. The form is perfect. It looks exactly the way it looked before. But it doesn't work. No airplanes land. So I call these things Cargo Cult Science, because they follow all the apparent precepts and forms of scientific investigation, but they're missing something essential, because the planes don't land. (Feynman 1974, p. 11)

For Feynman, the plane does not land for the Tanna Island villagers because they do not have the special and demanding integrity required when acting as a scientist; namely 'bending over backwards to show how you're maybe wrong' (Feynman 1974, p. 13). Lee McIntyre identifies the same quality as the core requirement of a scientific attitude; the extra-methodological feature that separates science from pseudoscience (McIntyre 2019). Feynman, Bunge, Popper, Dewey, Huxley, Mach, and countless others continue the Enlightenment commitment to the primacy of evidence in establishing, maintaining, and defending both theoretical and empirical scientific beliefs. This is something that needs be passed on in science classrooms.

Feng shui belief in society and in classrooms presents not so much a problem for teachers as an opportunity. Its considered and informed examination is a way for students to learn about the nature of science and other important social processes – the impact of marketing, the cultural determiners of gullibility, and so on. It will be apparent that feng shui violates all constitutive and procedural components of science. Its ontology is evasive, ill-determined, and unbound; its epistemology is empiricist and subjectivist. Such learning can be science education's contribution to the cultural health of society which is inversely related to the degree that gullibility, credulity, superstition, and unwarranted beliefs prevail in the society. Chinese people need only think of the Cultural Revolution to have this truth driven home, while US citizens need only reflect on the election of Donald Trump to have the same lesson.

The quality of such learning will depend on the quality, sensitivity, and informedness of the teaching. At all points of classroom contact with feng shui, and its chi-based worldview, the issues should be problematized, questions asked, claims examined, and alternatives investigated. Little is gained by a didactic, catechism-like approach to the issues. This is the deadening and useless approach to religion so frequently taken both by evangelists for religion in all religious traditions and by opponents of religion in Soviet and communist-states. Over time, and by engagement with problematic aspects of feng shui, the strengths and advantages of a scientific outlook should become apparent to students, along with appreciation of the methods and achievements in non-scientific intellectual and cultural domains.

References

Aarnio, K., & Lindeman, M. (2005). Paranormal beliefs, education, and thinking styles. *Personality and Individual Differences, 39*, 1227–1236.

Ackermann, R. J. (1989). The new experimentalism. *British Journal for the Philosophy of Science, 40*, 185–190.

Aczel, A. D. (2001). *The riddle of the compass: The invention that changed the world*. Orlando: Harcourt.

Agassi, J. (1964). The nature of scientific problems and their roots in metaphysics. In M. Bunge (Ed.), *The critical approach*. Glencoe: Free Press. Reprinted in J. Agassi, *Science in flux*, Reidel, Boston, 1975, pp. 208–239.

Agassi, J. (2013). *The very idea of modern science: Francis Bacon and Robert Boyle*. Dordrecht: Springer.

Agazzi, E. (2014). *Scientific objectivity and its contexts*. Dordrecht: Springer.

Agazzi, E. (Ed.). (2017). *Varieties of scientific realism: Objectivity and truth in science*. Dordrecht: Springer.

Aikenhead, G. S. (1996). Cultural assimilation in science classrooms: Border crossings and other solutions. *Studies in Science Education, 7*, 1–52.

Alcock, J. E. (2018). *Belief: What it means to believe and why our convictions are so compelling*. Amherst: Prometheus Books.

Alexander, C. (1987). *A timeless way of building*. New York: Oxford University Press.

Alexander, S. C. (2015). *Victorian literature and the physics of the imponderable*. New York: Routledge.

Alston, W. P. (1985). Divine foreknowledge and alternative conceptions of human freedom. *International Journal for Philosophy of Religion, 18*(1), 19–32.

Alston, W. P. (1996). *A realist conception of truth*. Ithaca: Cornell University Press.

Althusser, L. (1969). *For Marx*. Harmondsworth: Penguin.

American Association for the Advancement of Science (AAAS). (1989). *Project 2061: Science for all Americans*. Washington, DC: AAAS. Also published by Oxford University Press, 1990.

Amorth, G. (2010). *The memoirs of an exorcist*. Rome: Ediciones Urano.

Amsterdamski, S. (1975). *Between experience and metaphysics: Philosophical problems in the evolution of science*. Dordrecht: Reidel Publishing Company.

Andersen, K. (2017). *Fantasyland: How America went Haywire, a 500-year history*. London: Ebury Press.

Angle, S. C., & Tiwald, J. (2017). *Neo-Confucianism: A philosophical introduction*. Cambridge: Polity Press.

Arons, A. B. (1983). Achieving wider scientific literacy. *Daedalus, 112*(2), 91–122.

Arons, A. B. (1999). Development of energy concepts in introductory physics courses. *American Journal of Physics, 67*(12), 1063–1067.

Bachelard, G. (1934/1984). *The new scientific spirit*. Boston: Beacon Books.

Bachman, D. (1991). *Bureaucracy, economy, and leadership in China: The institutional origins of the great leap forward*. Cambridge: Cambridge University Press.

Bächtold, M., & Guedj, M. (2014). Teaching energy informed by the history and epistemology of the concept with implications for teacher education. In M. R. Matthews (Ed.), *International handbook of research in history, philosophy and science teaching* (pp. 211–244). Dordrecht: Springer.

Bacon, F. (1620/1939). Novum organum. In E. A. Burt (Ed.), *The English philosophers from Bacon to Mill* (pp. 24–123). New York: Random House.

Bailin, S., & Siegel, H. (2003). Critical thinking. In N. Blake, P. Smeyers, R. Smith, & P. Standish (Eds.), *The Blackwell guide to the philosophy of education* (pp. 181–193). Oxford: Blackwell Publishers.

Baillie, J. (1951). *The belief in progress*. New York: Charles Scribner's Sons.

Barnes, B. (1977). *Interests and the growth of knowledge*. London: Routledge & Kegan Paul.

Barnes, B., & Bloor, D. (1982). Relativism, rationalism and the sociology of knowledge. In M. Hollis & S. Lukes (Eds.), *Rationality and relativism* (pp. 21–47). Cambridge: MIT Press.

Baum, G. (Ed.). (1967). *The future of belief debate*. New York: Herder & Herder.

Bays, D. H. (1978). *China enters the twentieth century: Chang Chih-Tung and the issues of a new age, 1895–1909*. Ann Arbor: University of Michigan Press.

Beinfield, H., & Korngold, E. (1991). *Between heaven and earth: A guide to Chinese medicine*. New York: Random House.

Bell, R. (1992). *Impure science: Fraud, compromise and political influence in scientific research*. New York: Wiley.

Bennett, S. J. (1978). Patterns of the sky and earth: A Chinese science of applied cosmology. *Chinese Science, 3*, 1–26.

Bergson, H. (1911). *Creative evolution* (A. Mitchell, Trans.). New York: Random House.

Berkman, M. B., Pancheco, J. S., & Plutzer, E. (2008, May). Evolution and creationism in America's classrooms: A national portrait. *PLOS: Biology*. https://doi.org/10.1371/journal.pbio.0060124.

Bernal, J. D. (1939). *The social function of science*. London: Routledge & Kegan Paul.

Bernal, J. D. (1939/1949). Science teaching in general education. *Science Education*. Reproduced in *School Science Review* (1946) *27*, 150–158. And in his *The freedom of necessity* (pp. 135–146). London: Routledge & Kegan Paul, 1949.

Bernard, H. (1935). *Matteo Ricci's scientific contribution to China* (French E. C. Werner, Trans.). Hyperion Press: Westport.

Berthrong, J. H. (1998). *Transformation of the Confucian way*. Boulder: Westview Press.

Berthrong, J. H. (2003). Confucian views of nature. In H. Selin (Ed.), *Nature across cultures: Views of nature and the environment in non-Western cultures* (pp. 373–392). Dordrecht: Springer.

Bevilacqua, F. (Ed.). (2014a). Energy conservation: History, philosophy and education. *Science & Education, 23*(6).

Bevilacqua, F. (2014b). Energy: Learning from the past. *Science & Education, 23*(6), 1231–1243.

Beyerstein, B. L., & Sampson, W. (1996a). Traditional medicine and pseudoscience in China: A report of the second CSICOP delegation (part 1). *Skeptical Inquirer, 20*(4). https://www.csicop.org/si/show/china_conference_1

Beyerstein, B. L., & Sampson, W. (1996b). Traditional medicine and pseudoscience in China: A report of the second CSICOP delegation (part 2). *Skeptical Inquirer, 20*(5). https://www.csicop.org/si/show/china_conference_2

Bhakthavatsalam, S. (2019). The value of false theories in science education. *Science & Education, 28*(1–2), 5–23.

Bhargava, P. M., & Chakrabarti, C. (1995). The scientific temper and the scientific method in science in India through history, with special reference to biology. In D. P. Chattopadhyaya &

R. Kumar (Eds.), *Mathematics, astronomy and biology in Indian tradition: Some conceptual preliminaries* (pp. 28–55). New Delhi: Indian Council of Philosophical Research.

Bhargava, P. M., & Chakrabarti, C. (Eds.). (2010). *Devils and science: A collection of articles on scientific temper*. New Delhi: National Book Trust.

Binet, A., & Fèrè, C. (1890). *Animal magnetism*. New York: D. Appleton.

Birdsall, G. (1995). *Feng shui: The ancient art of placement*. Sydney: Waterwood Management Proprietary Ltd..

Black, M. (1962). *Models and metaphors*. Ithaca: Cornell University Press.

Blackmore, J. T. (1989). Ernst Mach leaves "The Church of Physics". *British Journal for Philosophy of Science, 40*, 519–540.

Bloembergen, N. (1980). Physics. In L. A. Orleans (Ed.), *Science in contemporary China* (pp. 85–109). Stanford: Stanford University Press.

Bloor, D. (1976/1991). *Knowledge and social imagery*. London: Routledge & Kegan Paul (Second edition, 1991).

Blue, G. (1998). Joseph Needham, heterodox Marxism and the social background to Chinese science. *Science & Society, 62*(2), 195–217.

Blüh, O., & Elder, J. (1955). *Principles and applications of physics*. Edinburgh: Oliver & Boyd.

Bodde, D. (1938). *China's first unifier: A study of the Ch'in dynasty as seen in the life of Li Ssu*. Leiden: E.J. Brill.

Bodde, D. (1991). *Chinese thought, society, and science: The intellectual and social background of science and technology in pre-modern China*. Honolulu: University of Hawaii Press.

Boerschmann, E. (1912). *Chinese architecture and its relation to Chinese culture* (pp. 539–577). Smithsonian Institute Report. Washington, DC: Smithsonian Institute, 1911 (published 1912).

Boerschmann, E. (1924). *Picturesque China, architecture and landscape: A journey through twelve provinces*. London: Unwin.

Bolton, H. C. (Ed.). (1892). *Scientific correspondence of Joseph Priestley*. New York: Collins printing House.

Bossi, M., & Poggi, S. (Eds.). (1994). *Romanticism in science: Science in Europe, 1790–1840*. Boston: Kluwer.

Boudry, M. (2017). Plus ultra: Why science does not have limits. In M. Boudry & M. Pigliucci (Eds.), *Science unlimited? The challenges of scientism* (pp. 31–52). Chicago: University of Chicago Press.

Boudry, M., & Pigliucci, M. (Eds.). (2017). *Science unlimited? The challenges of scientism*. Chicago: University of Chicago Press.

Boudry, M., Blancke, S., & Braeckman, J. (2012). Grist to the mill of anti-evolutionism: The failed strategy of ruling the supernatural out of science by philosophical fiat. *Science & Education, 21*, 1151–1165.

Boulos, P. J. (2006). Newton's path to universal gravitation: The role of the pendulum. *Science & Education, 15*(6), 577–595.

Brandon, R. (1983). *The spiritualists: The passion for the occult in the nineteenth and twentieth centuries*. New York: Alfred A. Knopf.

Brandt, A. (2007). *The cigarette century: The rise, fall, and deadly persistence of the product that defined America*. New York: Basic Books.

Brière, O. J. (1956). *Fifty years of Chinese philosophy 1898–1950* (L. G. Thompson, Trans.). London: G. Allen & Unwin.

Brockey, L. M. (2008). *Journey to the East: The Jesuit mission to China, 1579–1724*. Cambridge, MA: Harvard University Press.

Brooke, J. H. (1991). *Science and religion: Some historical perspectives*. Cambridge: Cambridge University Press.

Brooke, J. H. (1995). *Thinking about matter: Studies in the history of chemical philosophy*. Aldershot: Variorum Press.

Brooke, J. H. (2001). Religious belief and the content of the sciences. *Osiris, 16*, 3–28.

Brown, J. R. (1994). *Smoke and mirrors: How science reflects reality*. New York: Routledge.

Brown, J. R. (2001). *Who rules in science: An opinionated guide to the science wars*. Cambridge, MA: Harvard University Press.
Brown, S. (2005). *The feng shui bible: The definitive guide to practising feng shui*. London: Octupus Publishing Group.
Bruun, O. (2003). *Fengshui in China: Geomantic divination between state orthodoxy and popular religion*. Honolulu: University of Hawaii Press.
Bruun, O. (2008). *An introduction to feng shui*. Cambridge: Cambridge University Press.
Bruun, O., & Kalland, A. (Eds.). (1995). *Asian perceptions of nature – A critical approach*. London: Curzon Press.
Buchdahl, G. (1969). *Metaphysics and the philosophy of science*. Oxford: Basil Blackwell.
Bunge, M. (1959). *Metascientific queries*. Springfield: Charles C. Thomas Publisher.
Bunge, M. (1967/1998). *Scientific research 1, the search for system*. New Brunswick: Transaction Publishers.
Bunge, M. (1973). *Method, model and matter*. Dordrecht: Reidel.
Bunge, M. (1977). *Treatise on basic philosophy. Vol. 3, The furniture of the world*. Dordrecht: Reidel.
Bunge, M. (1982). The revival of causality. In G. Floistad (Ed.), *Contemporary philosophy* (Vol. 2, pp. 133–155). The Hague: Martinus Nijhoff. Reproduced in Martin Mahner (ed.) *Selected essays of Mario Bunge*, 2001, Prometheus Books, Amherst, pp. 57–74.
Bunge, M. (1986). In defence of realism and scientism. *Annals of Theoretical Psychology, 4*, 23–26.
Bunge, M. (1988/2001). The nature of applied science and technology. In V. Cauchy (Ed.), *Philosophie et Culture* (pp. 599–605). Laval: Editions Montmorency. Reproduced in Martin Mahner (Ed.) *Scientific realism: Selected essays of Mario Bunge* (pp. 345–351). Amherst: Prometheus Books, 2001.
Bunge, M. (1991a). What is science? Does it matter to distinguish it from pseudoscience? *New Ideas in Psychology, 9*, 245–283.
Bunge, M. (1991b). A critical examination of the new sociology of science: Part 1. *Philosophy of the Social Sciences, 21*(4), 524–560.
Bunge, M. (1992). A critical examination of the new sociology of science: Part 2. *Philosophy of the Social Sciences, 22*(1), 46–76.
Bunge, M. (1994). Counter-enlightenment in contemporary social studies. In P. Kurtz & T. J. Madigan (Eds.), *Challenges to the enlightenment: In defense of reason and science* (pp. 25–42). Buffalo: Prometheus Books.
Bunge, M. (1996). In praise of intolerance to charlatanism in academia. In P. R. Gross, N. Levitt, & M. W. Lewis (Eds.), *The flight from science and reason* (pp. 96–115). New York: New York Academy of Sciences.
Bunge, M. (1998). *Philosophy of science: From problem to theory* (Vol. 1). New Brunswick: Transaction Publishers.
Bunge, M. (2000). Energy: Between physics and metaphysics. *Science & Education, 9*(5), 457–461.
Bunge, M. (2001). *Philosophy in crisis: The need for reconstruction*. Amherst: Prometheus Books.
Bunge, M. (2003). *Emergence and convergence: Qualitative novelty and the unity of knowledge*. Toronto: University of Toronto Press.
Bunge, M. (2006). *Chasing reality: Strife over realism*. Toronto: University of Toronto Press.
Bunge, M. (2009). Advantages and limits of naturalism. In J. R. Shook & P. Kurtz (Eds.), *The future of naturalism* (pp. 43–63). Amherst: Humanity Books.
Bunge, M. (2010). From philosophy to physics and back. In S. Nuccetelli, O. Schutte, & P. Bueno (Eds.), *A companion to Latin American philosophy* (pp. 525–539). Malden: Wiley-Blackwell.
Bunge, M. (2011). *Scientific philosophy: Lectures to the Chinese Communist Party, Marxism school*. Unpublished.
Bunge, M. (2012a). *Evaluating philosophies* (Boston studies in the philosophy of science) (Vol. 295). Dordrecht: Springer.

Bunge, M. (2012b). The correspondence theory of truth. *Semiotica, 188*, 65–76.
Bunge, M. (2014). In defense of scientism. *Free Inquiry, 35*(1), 24–28.
Bunge, M. (2016). *Between two worlds: Memoirs of a philosopher-scientist*. Dordrecht: Springer.
Burtt, E. A. (1932). *The metaphysical foundations of modern physical science* (2nd ed.). London: Routledge & Kegan Paul. (First edition, 1924).
Bury, J. B. (1920). *The idea of progress: An inquiry into its origin and growth*. New York: Macmillan. Reprinted 1982, Greenwood Press, Westport.
Bussmann, B., & Kötter, M. (2018). Between scientism and relativism: Epistemic competence as an important aim in science and philosophy of education. *RISTAL: Research in Subject-Matter Teaching and Learning, 1*(1), 82–101.
Butterfield, J., & Pagonis, C. (Eds.). (1999). *From physics to philosophy*. Cambridge: Cambridge University Press.
Butts, R. E. (1993). Sciences and pseudosciences: An attempt at a new form of demarcation. In J. Earman, A. I. Janis, G. J. Massey, & N. Rescher (Eds.), *Philosophical problems of the internal and external worlds: Essays on the philosophy of Adolf Grünbaum* (pp. 163–185). Pittsburgh: University of Pittsburgh Press.
Callicott, J. B., & Ames, R. T. (1989). *Nature in Asian traditions of thought: Essays in environmental philosophy*. Albany: State University of New York Press.
Campion, N. (2016). *Astrology and popular religion in the modern west: Prophecy, cosmology and the new age movement*. New York: Routledge.
Capra, F. (1982). *The turning point*. New York: Bantam Books.
Capra, F. (1984). *The Tao of physics* (2nd revised edition of 1975 original). New York: Bantam Books.
Carson, R. (1997). Science and the ideals of liberal education. *Science & Education, 6*(3), 225–238.
Cartwright, N. (1983). *How the laws of physics lie*. Oxford: Clarendon Press.
Chalmers, A. F. (2009). *The scientist's atom and the philosopher's stone: How science succeeded and philosophy failed to gain knowledge of atoms*. Dordrecht: Springer.
Chan, W.-T. (1957). Neo-Confucianism and Chinese scientific thought. *Philosophy East and West, 6*(4), 309–332.
Chan, W.-T. (1963). *A source book in Chinese philosophy*. Princeton: Princeton University Press.
Chan, W.-T. (1967). *Chinese philosophy, 1949–1963: An annotated bibliography of Mainland China publications*. Honolulu: East-West Center Press.
Chang, C. (1957). *The development of neo-Confucian thought*. New York: Bookman Associates.
Chang, S. T. (1976). *The complete book of acupuncture*. Berkeley: Celestial Arts Press.
Chen, K. (2002). Review of Lu Zuyin 'Scientific Qigong exploration: The wonders and mysteries of Qi'. *Journal of Scientific Exploration, 16*(3), 483–489.
Chen, Y. J. (2014). Legitimation discourse and the theory of the five elements in imperial China. *Journal of Song-Yuan Studies, 44*, 325–364.
Chen, X., & Wu, J. (2009). Sustainable landscape architecture: Implications of the Chinese philosophy of 'unity of man with nature' and beyond. *Landscape Ecology, 24*(8), 1015–1026.
Chen, R. F., Eisenkraft, A., Fortus, D., Krajcik, J., Neumann, K., Nordine, J., & Scheff, A. (Eds.). (2014). *Teaching and learning of energy in K-12 education*. Dordrecht: Springer.
Chow, T.-T. (1960). *The May Fourth Movement: Intellectual revolution in modern China*. Cambridge, MA: Harvard University Press.
Chow, K.-W. (1993). Ritual, cosmology and ontology: Chang Tsai's moral philosophy and neo-Confucian ethics. *Philosophy East & West, 43*(2), 201–228.
Clark, A. (1985). Psychological causation and the concept of psychosomatic disease. In D. Stalker & C. Glymour (Eds.), *Examining holistic medicine* (pp. 67–106). Buffalo: Prometheus Books.
Coelho, R. L. (2009). On the concept of energy: How understanding its history can improve physics teaching. *Science & Education, 18*(8), 961–983.
Cohen, I. B. (1956). *Franklin and Newton*. Philadelphia: American Philosophical Society.
Cohen, H. F. (1994). *The scientific revolution: A historiographical inquiry*. Chicago: University of Chicago Press.

Cohen, I. B. (1995). *Science and the founding fathers: Science in the political thought of Jefferson, Franklin, Adams, and Madison*. New York: W. W. Norton.

Collins, H. M. (1985). *Changing order: Replication and induction in scientific practice*. London: Sage.

Colquhoun, D. (2011). *Acupuncturists show that acupuncture doesn't work, but conclude the opposite: Journal fails*. Available at: http://www.dcscience.net/?p=4439

Colquhoun, D., & Novella, S. (2013). Acupuncture is theatrical placebo. *Anesthesia & Analgesia, 116*(6), 1360–1363.

Commager, H. S. (1977). *The empire of reason: How Europe imagined and America realized the enlightenment*. New York: Doubleday.

Conant, J. B. (1945). *General education in a free society: Report of the Harvard Committee*. Cambridge: Harvard University Press.

Condorcet, N. (1976). In K. M. Baker (Ed.), *Selected writings*. Indianapolis: Bobbs-Merrill.

Cook, D. J., & Rosemont, H. (Eds.). (1994). *Gottfried Wilhelm Freiherr von Leibniz: Writings on China*. La Salle: Open Court.

Coopersmith, J. (2010). *Energy, the subtle concept: The discovery of Feynman's blocks from Leibniz to Einstein*. Oxford: Oxford University Press.

Cooter, R. (1980). Deploying "pseudoscience": Then and now. In M. P. Hanen, M. J. Osler, & R. G. Weyant (Eds.), *Science, pseudoscience and society* (pp. 237–272). Calgary: Wilfrid Laurier University Press.

Cooter, R. (1982). The conservatism of "pseudoscience". In P. Grim (Ed.), *Philosophy of science and the occult* (pp. 130–143). Albany: State University of New York Press.

Crombie, A. C. (1952). *Augustine to Galileo*. London: Heinemann.

Crombie, A. C. (1953). *Robert Grosseteste and the origins of experimental science 100–1700*. Oxford: Clarendon Press.

Crombie, A. C. (1955/1990). Grosseteste's position in the history of science. In D. A. Callus (Ed.), *Robert Grosseteste, scholar and bishop* (pp. 98–120). Oxford: Clarendon Press. Reproduced in A.C. Crombie *Science, optics and music in medieval and early modern thought* (pp. 115–137). London: The Hambledon Press.

Crombie, A. C. (1959/1990). The significance of medieval discussions of scientific method for the scientific revolution. In M. Claggett (Ed.), *Critical problems in the history of science* (pp. 66–101). Madison: University of Wisconsin Press. Reproduced in A.C. Crombie *Science, optics and music in medieval and early modern thought* (pp. 139–160). London: The Hambledon Press.

Crombie, A. C. (1994). *Styles of scientific thinking in the European tradition* (3 Vols.). London: Duckworth.

Cronin, V. (1955). *The wise man from the west: Matteo Ricci and his mission to China*. New York: Harper Collins.

Cross, T. (2005). What is a disposition? *Synthese, 144*, 321–341.

Crump, T. (2001). *A brief history of science: As seen through the development of scientific instruments*. London: Robinson.

Cunningham, A., & Jardine, N. (Eds.). (1990). *Romanticism and the sciences*. Cambridge: Cambridge University Press.

Cyranoski, D. (2017, November 29). China to roll back regulations for traditional medicine despite safety concerns. *Nature*.

d'Elia, P. M. (1960). *Galileo in China. Relations through the Roman College between Galileo and the Jesuit Scientist-Missionaries (1610–1640)* (R. Suter, Trans.). Cambridge: Harvard University Press.

d'Espagnat, B. (2006). *On physics and philosophy*. Princeton: Princeton University Press.

Dai, N. (1996). The development of modern physics in China: The 50th anniversary of the founding of the Chinese Physical Society. In F. Dainian & R. S. Cohen (Eds.), *Chinese studies in the history and philosophy of science* (pp. 207–218). Dordrecht: Kluwer Academic Publishers.

Dainian, F., & Cohen, R. S. (Eds.). (1996). *Chinese studies in the history and philosophy of science and technology*. Dordrecht: Kluwer Academic Publishers.

Dallal, A. (2010). *Islam, science, and the challenge of history*. New Haven: Yale University Press.

Darnton, R. (1968). *Mesmerism and the end of the enlightenment in France*. Cambridge, MA: Harvard University Press.

Davies, M., & Barnett, R. (Eds.). (2015). *The Palgrave handbook of critical thinking in higher education*. New York: Palgrave Macmillan.

Dawes, R. (2001). *Everyday irrationality: How pseudo-scientists, lunatics, and the rest of us systematically fail to think rationally*. Boulder: Westview Press.

de Groot, J. J. M. (1892–1910). *The religious system of China* (6 Vols.). Leiden: E.J. Brill.

de Ridder, J., Peels, R., & van Woudenberg, R. (Eds.). (2019). *Scientism: Prospects and problems*. Oxford: Oxford University Press.

de Tocqueville, A. (1835/1953). *Democracy in America, two vols*. New York: Alfred A. Knoff.

DeBoer, G. E. (2000). Scientific literacy: Another look at its historical and contemporary meanings, and its relationship to science education reform. *Journal of Research in Science Teaching, 37*(6), 582–601.

Department of Energy (DoE) USA. (2012). *Energy literacy: Essential principles an fundamental concepts for energy education*. Washington, DC: US Department of Energy.

Derksen, A. A. (1993). The seven sins of pseudoscience. *Journal for General Philosophy of Science, 24,* 17–42.

Devitt, M. (1991). *Realism & truth* (2nd ed.). Oxford: Basil Blackwell.

Dewey, J. (1938). Unity of science as a social problem. In O. Neurath, R. Carnap, & C. W. Morris (Eds.), *International encyclopedia of unified science* (Vol. 1, pp. 29–38). London: University of Chicago Press.

Dijksterhuis, E. J. (1961/1986). *The mechanization of the world picture*. Princeton: Princeton University Press.

Dikotter, F. (2010). *Mao's great famine: The history of China's most devastating catastrophe 1958–1962*. New York: Bloomsbury.

Dilworth, C. (1996/2006). *The metaphysics of science. An account of modern science in terms of principles, laws and theories*. Dordrecht: Kluwer Academic Publishers (Second edition 2006).

Donaldson, I. M. L. (2005). Mesmer's 1780 proposal for a controlled trial to test his method of treatment using "animal magnetism". *Journal of the Royal Society of Medicine, 98*(12), 572–575.

Dong, G. (1996). The Book of changes and mathematics. In F. Dainian & R. S. Cohen (Eds.), *Chinese studies in the history and philosophy of science and technology* (pp. 125–135). Dordrecht: Kluwer Academic Publishers.

Doorly, J. W. (1946). *The pure science of Christian science*. London: The Foundational Book Company.

Driver, R. (1988). A constructivist approach to curriculum development. In P. Fensham (Ed.), *Development and dilemmas in science education* (pp. 133–149). New York: Falmer Press.

Dubs, H. H. (1928). *The works of Hsün Tzu*. London: Probsthain Publishers.

Duit, R. (1986). Search of an energy concept. In R. Driver & R. Millar (Eds.), *Energy matters* (pp. 67–102). Leeds: University of Leeds.

Duit, R. (1987). Should energy be introduced as something quasi-material? *International Journal of Science Education, 9,* 139–145.

Duit, R. (1991). The role of analogy and metaphor in learning science. *Science Education, 75,* 649–672.

Dukes, E. J. (1885). Feng shui: The biggest of all bugbears. In *Everyday life in China* (pp. 145–159). London: Religious Tract Society.

Dunne, G. H. (1962). *Generation of giants: The story of the Jesuits in China in the last decades of the Ming dynasty*. London: Burns and Oates.

Dupree, A. H. (1986). Christianity and the scientific community in the age of Darwin. In D. G. Lindberg & R. L. Numbers (Eds.), *God and nature: Historical essays on the encounter between Christianity and science* (pp. 351–368). Berkeley: University of California Press.

Eddy, M. B. (1875/1990). *Science and health with key to the scriptures*. Boston: Christian Science Publishing.

Eder, E., Turic, K., Milasowszky, N., van Adzin, K., & Hergovich, A. (2010). The relationships between paranormal belief, creationism, intelligent design and evolution at secondary schools in Vienna (Austria). *Science & Education, 20*(5–6), 517–534.

Edis, T. (2007). *An illusion of harmony: Science and religion in Islam*. Amherst: Prometheus Books.

Educational Policies Commission (EPC). (1966). *Education and the spirit of science*. Washington, DC: National Education Association.

Einstein, A. (1922). Ether and theory of relativity. In *Sidelights on relativity*. London: Metheun.

Eitel, E. J. (1873). *Buddhism: Its historical, theoretical and popular aspects*. London: Trübner.

Eitel, E. J. (1873/1987). *Feng shui: The rudiments of natural science in China*. Hong Kong: Lane Crawford (Graham Brash, Singapore).

Eitel, E. J. (1895/1983). *Europe in China: A history of Hong Kong* (H. J. Lethbridge, Ed.). Hong Kong: Oxford University Press.

Elliot, T. (2013, April 20). Witch-hunt. *Sydney Morning Herald*, Good Weekend, pp. 16–21.

Ellis, B. D. (1976). The existence of forces. *Studies in History and Philosophy of Science, 7*(2), 171–185.

Elman, B. (2000). *A cultural history of civil examinations in late imperial China*. Berkeley: University of California Press.

Emmons, C. F. (1992). Hong Kong's *Feng Shui:* Popular magic in a modern urban setting. *The Journal of Popular Culture, 26*(1), 39–50.

Engelfriet, P. M. (1998). *Euclid in China: The genesis of the first Chinese translation of Euclid's 'elements' books 1–VI, and its reception up to 1723*. Leiden: Brill.

Engelhardt, H. T., & Caplan, A. L. (Eds.). (1987). *Scientific controversies: Case studies in the resolution and closure of disputes in science and technology*. Cambridge, MA: Cambridge University Press.

Engels, F. (1875–82/1934). *Dialectics of nature*. Moscow: Progress Publishers.

Ernst, E., Lee, M. S., & Choi, T.-Y. (2011). Acupuncture: Does it alleviate pain and are there serious risks? A review of reviews. *Pain, 152*(4), 755–764.

Eve, R. A., & Dunn, D. (1990). Psychic powers, astrology, and creationism in the classroom? *American Biology Teacher, 52*, 10–21.

Fang, L. (1982/1992). From "water is the origin of all things" to "space-time is the form of the existence of matter". In L. Fang (Ed.), *Bringing down the Great Wall: Writings on science, culture, and democracy in China* (pp. 15–19). New York: Norton & Company.

Fang, L. (1987/1992). Interview with Tiziano Terzani. In L. Fang (Ed.), *Bringing down the Great Wall: Writings on science, culture, and democracy in China* (pp. 207–217). New York: Norton & Company.

Fang, L. (1989/1992). A note on the interface between science and religion. In L. Fang (Ed.), *Bringing down the Great Wall: Writings on science, culture, and democracy in China* (pp. 30–37). New York: Norton & Company.

Fang, L. (1990/1992). Free to speak: Second interview with Tiziano Terzani. In L. Fang (Ed.), *Bringing down the Great Wall: Writings on science, culture, and democracy in China* (pp. 276–298). New York: Norton & Company.

Fang, L. (1992). *Bringing down the Great Wall: Writings on science, culture and democracy in China* (J. H. Williams, Ed. & Trans.). New York: Alfred A. Knopf.

Fang, L. (2016). *The most wanted man in China: My journey from scientist to enemy of the state* (P. Link, Trans.). New York: Henry Holt & Co.

Fara, P. (2005). *Fatal attraction: Magnetic mysteries of the enlightenment*. New York: MJF Books.

Fara, P. (2009). *Science: A four thousand year history*. Oxford: Oxford University Press.

Fasce, A., & Picó, A. (2019). Science as a vaccine: The relation between scientific literacy and unwarranted beliefs. *Science & Education, 28*(1–2), 109–125.

Feekes, G. B. (1986). *The hierarchy of energy systems: From atom to society*. Oxford: Pergamon Press.

Fenton, P. H. (1996). *Shaolin Nei Jin Qi Gong: Ancient healing in the modern world*. York Beach: Samuel Weiser.

Ferguson, C. (2012). *Determined spirits: Eugenics, heredity and racial regeneration in Anglo-American spiritualist writings 1848–1930*. Edinburgh: Edinburgh University Press.

Fernflores, F. (2012). The equivalence of mass and energy. In E. N. Zalta (Ed.), *The Stanford encyclopedia of philosophy* (Spring 2012 ed.). https://plato.stanford.edu/archives/spr2012/entries/equivME/

Feuchtwang, S. (2003). *An anthropological analysis of Chinese geomancy*. Bangkok: White Lotus.

Feyerabend, P. K. (1975). *Against method*. London: New Left Books.

Feynman, R. P. (1963). *The Feynman lectures on physics, Vol. I: Mainly mechanics, radiation, and heat*. Pasadena: California Institute of Technology.

Feynman, R. P. (1974). Cargo cult science: Some remarks on science, pseudoscience, and learning how to not fool yourself. *Engineering and Science, 37*, 10–13.

Figala, K. (2002). Newton's alchemy. In I. B. Cohen & G. E. Smith (Eds.), *The Cambridge companion to Newton* (pp. 370–386). Cambridge: Cambridge University Press.

Finocchiaro, M. A. (1989). *The Galileo affair: A documentary history*. Berkeley: University of California Press.

Finocchiaro, M. A. (2010). *Defending Copernicus and Galileo: Critical reasoning in the two affairs*. Dordrecht: Springer.

Fishman, Y. I. (2009). Can science test supernatural worldviews? *Science & Education, 18*(6–7), 813–837.

Fishman, Y. I., & Boudry, M. (2013). Does science presuppose naturalism (or, indeed, anything at all)? *Science & Education, 22*(5), 921–949.

Franke, W. (1963). *The reform and abolition of the traditional Chinese examination system*. Cambridge, MA: Harvard University Press.

Franklin, A. (1986). *The neglect of experiment*. Cambridge: Cambridge University Press.

Franklin, B., Le Roy, M., Bailly, S. J.-S., Guillotin, J.-I., de Bory, D.' A., & Lavoisier, A. (1784/2002). Report of the commissioners charged by the King with the examination of animal magnetism. *International Journal of Clinical and Experimental Hypnosis, 50*(4), 332–363.

Franklin, B., Le Roy, M., Bailly, S. J.-S., Guillotin, J.-I., de Bory, d'A., & Lavoisier, A. (1784/2014). Report of the commissioners charged by the king with the examination of animal magnetism. In I. M. L. Donaldson (Trans. & Ed.), *The reports of the Royal Commission on Mesmer's system of animal magnetism and other contemporary documents* (pp. 39–76). Edinburgh: James Lind Library and Royal College of Physicians of Edinburgh.

Freedman, M. (1974). On the sociological study of Chinese religion. In A. P. Wolf (Ed.), *Religion and ritual in Chinese society*. Stanford: Stanford University Press.

Friedlander, M. W. (1995). *At the fringes of science*. Boulder: Westview Press.

Fung, Y.-L. (1922). Why China has no science – An interpretation of the history and consequences of Chinese philosophy. *International Journal of Ethics, 32*(3), 237–263.

Fung, Y.-L. (1947). *The spirit of Chinese philosophy* (E. R. Hughes, Trans.). London: Kegan Paul.

Fung, Y.-L. (1949). *Short history of Chinese philosophy* D. Bodde, (Trans.). Princeton: Princeton University Press.

Fung, Y.-L. (1952–1953). *History of Chinese philosophy* (2 Vols., D. Bodde, Trans.). Princeton: Princeton University Press.

Fung, Y.-L. (2008). *Selected philosophical writings of Feng Yu-Lan*. Beijing: Foreign Languages Press.

Furth, C. (1983). Intellectual change: From the Reform movement to the May Fourth Movement, 1895–1920. In J. K. Fairbank (Ed.), *Republican China 1912–1949* (pp. 322–405). Cambridge: Cambridge University Press.

Galileo, G. (1638/1954). *Dialogues concerning two new sciences* (H. Crew & A. de Salvio, Trans.). New York: Dover Publications (orig. 1914).
Galison, P. (1987). *How experiments end*. Chicago: University of Chicago Press.
Gallagher, W. (1993). *The power of place: How our surroundings shape our thoughts, emotions & actions*. New York: Simon & Schuster.
Gallup, G. H., Jr., & Newport, F. (1991). Belief in paranormal phenomena among adult American. *Skeptical Inquirer, 15*, 137–147.
Gardner, M. (1981). *Science: Good, bad and bogus: A skeptical look at extraordinary claims*. Buffalo: Prometheus Books.
Gauch, H. G., Jr. (2003). *Scientific method in practice*. Cambridge: Cambridge University Press.
Gaukroger, S. (2001). *Francis Bacon and the transformation of early-modern philosophy*. Cambridge: Cambridge University Press.
Gauld, C. F. (1982). The scientific attitude and science education: A critical reappraisal. *Science Education, 66*(1), 109–121.
Gauld, C. F. (2005). Habits of mind, scholarship and decision making in science and religion. *Science & Education, 14*(4–5), 291–308.
Giere, R. N. (1999). *Science without laws*. Chicago: University of Chicago Press.
Gill, H. V. (1943). *Fact and fiction in modern science*. Dublin: M.H. Gill & Son.
Gillies, M. A. (1996). *Henri Bergson and British modernism*. Montreal: McGill Queen's University Press.
Gjertsen, D. (1989). *Science and philosophy: Past and present*. Harmondsworth: Penguin.
Glymour, C., & Stalker, D. (1982). Winning through pseudoscience. In P. Grim (Ed.), *Philosophy of science and the occult* (pp. 75–86). Albany: State University of New York Press.
Goldring, H., & Osborne, J. (1994). Students' difficulties with energy and related concepts. *Physics Education, 29*, 26–31.
Gong, Y. (1996a). *Dialectics of nature in China*. Beijing: Beijing University Press.
Gong, Y. (1996b). Historical development of the Chinese Communist Party's scientific policy (prior to the founding of the People's Republic). In F. Dainian & R. S. Cohen (Eds.), *Chinese studies in the history and philosophy of science* (pp. 13–25). Dordrecht: Kluwer Academic Publishers.
Goode, E. (2011). *The paranormal: Who believes, why they believe, and why it matters*. Amherst: Prometheus Books.
Gooding, D., Pinch, T., & Schaffer, S. (Eds.). (1989). *The uses of experiment*. Cambridge: Cambridge University Press.
Goodman, L. (1971). *Linda Goodman's star-signs*. New York: Bantam Books.
Gordin, M. D. (2012). *The pseudoscience wars: Immanuel Velikovsky and the birth of the modern fringe*. Chicago: University of Chicago Press.
Gould, S. J. (1997). Nonoverlapping magisteria. *Natural History, 106*, 16–22. Reprinted in R. Pennock (ed.), *Intelligent design creationism and its critics: Philosophical, theological, and scientific perspectives*. Cambridge, MA, MIT Press, 2001, 737–749.
Graham, A. C. (1973). China, Europe, and the origins of modern science: Needham's the grand titration. In S. Nakayama & N. Sivin (Eds.), *Chinese science: Explorations of an ancient tradition* (pp. 45–69). Cambridge, MA: MIT Press.
Graham, L. R. (1973). *Science and philosophy in the Soviet Union*. New York: Alfred A. Knopf.
Graham, A. C. (1989). *Disputers of the Tao: Philosophical argument in ancient China*. LaSalle: Open Court.
Gratzer, W. B. (2000). *The undergrowth of science: Delusion, self-deception and human frailty*. Oxford: Oxford University Press.
Greenblatt, S. (2011). *The swerve: How the world became modern*. New York: W.W. Norton.
Grim, P. (Ed.). (1990). *Philosophy of science and the occult*. Albany: State University of New York Press.
Gross, H. (2004). *Rome in the age of enlightenment: The post-tridentine syndrome and the ancien régime*. Cambridge: Cambridge University Press.

Gross, P. R., Levitt, N., & Lewis, M. W. (Eds.). (1996). *The flight from science and reason*. New York: New York Academy of Sciences, (distributed by Johns Hopkins University Press, Baltimore).
Grosser, M. (1962). *The discovery of Neptune*. New York: Dover.
Grove, W. R. (1866/1970). 'Force and energy in an industrial society'. Presidential Address to 1866 BAAS Nottingham. In G. Basalla, W. Coleman, & R. H. Kargon (Eds.), *Victorian science: A self-portrait from the presidential addresses to the British Association for the Advancement of Science* (pp. 89–97). Garden City: Doubleday.
Grove, J. W. (1985). Rationality at risk: Science against pseudoscience. *Minerva, 23*(2), 216–240.
Grove, J. W. (1989). *In defence of science. Science, technology and politics in modern society*. Toronto: University of Toronto Press.
Guo, P. (2001). *The Zangshu, or book of burial* (S. Field, Trans.). web source. (original ≈ 300BC).
Guo, P. (2004). *The Zangshu, or book of burial* (J. Zhang, Trans.). Lewiston: Edwin Mellen Press (original ≈ 300BC).
Guo, Y. (2014). The philosophy of science and technology in China: Political and ideological influences. *Science & Education, 23*(9), 1835–1844.
Haack, S. (2016). *Scientism and its discontents*. London: Rounded Globe Publishers.
Hacking, I. (1988). Philosophers of experiment. In A. Fine & J. Leplin (Eds.), *PSA, 2*, 147–156.
Hakfoort, C. (1992). Science deified: Wilhelm Ostwald's energeticist world-view and the history of scientism. *Annals of Science, 49*(2), 525–544.
Hall, A. R. (1970). On the historical singularity of the scientific revolution of the seventeenth century. In *The diversity of history: Essays in honour of Sir Herbert Butterfield*. Ithaca: Cornell University Press.
Hall, A. R. (1983). *The scientific revolution: 1500–1800* (3rd ed.). Boston: Beacon Press. (First edition 1954, second 1962.).
Hall, H. (2006). Teaching pigs to sing: An experiment in bringing critical thinking to the masses. *Skeptical Inquirer, 30*(3), 36–39.
Han, K.-T. (2001). Traditional Chinese site selection-feng shui: An evolutionary/ecological perspective. *Journal of Cultural Geography, 19*(1), 75–96.
Hankins, T. L. (1985). *Science and the enlightenment*. Cambridge: Cambridge University Press.
Hansson, S. O. (2009). Cutting the Gordian Knot of demarcation. *International Studies in the Philosophy of Science, 23*, 237–243.
Hansson, S. O. (2013). Defining pseudoscience and science. In M. Pigliccci & M. Boudry (Eds.), *Philosophy of pseudoscience: Reconsidering the demarcation problem* (pp. 61–77). Chicago: University of Chicago Press.
Harding, S. G. (Ed.). (1976). *Can theories be refuted? Essays on the Duhem-Quine thesis*. Dordrecht: Reidel.
Harman, P. M. (1982). *Energy, force and matter: The conceptual development of nineteenth-century physics*. Cambridge: Cambridge University Press.
Harper, W. L. (2011). *Isaac Newton's scientific method: Turning data into evidence about gravity and cosmology*. Oxford: Oxford University Press.
Harper, D. (2017). Science in ancient China. In I. R. Morus (Ed.), *The Oxford illustrated history of science* (pp. 45–71). Oxford: Oxford University Press.
Harré, R. (1964). *Matter and method*. London: Macmillan.
Harré, R. (1981). *Great scientific experiments: Twenty experiments that changed our view of the world*. Oxford: Oxford University Press.
Harrison, M. (2000). From medical astrology to medical astronomy: Sol-lunar and planetary theories of disease in British medicine, c. 1700–1850. *The British Journal for the History of Science, 33*(1), 25–48.
Haygarth, J. (1800). *Of the imagination as a cause and as a cure of disorders of the body*, R. Cruttwell, Bath. http://www.jameslindlibrary.org/haygarth-j-1800/
Hellman, H. (1998). *Great feuds in science*. New York: Wiley.

Hempel, C. G. (1983). Valuation and objectivity in science. In R. S. Cohen & L. Laudan (Eds.), *Physics, philosophy and psychoanalysis: Essays in honor of Adolf Grünbaum* (pp. 111–127). Dordrecht: Reidel.
Henderson, J. R. (2010). Cosmology and concepts of nature in ancient China. In H. U. Vogel & G. Dux (Eds.), *Concepts of nature: A Chinese-European cross-cultural perspective* (pp. 198–218). Leiden: Brill.
Hersey, F. (2017, September 14). Traditional medicine courses rolled out in Chinese schools as 12-year-olds learn acupuncture. *The Telegraph.*
Hershey, D. H. (1991). Digging deeper into Helmont's famous willow tree experiment. *The American Biology Teacher, 53*(8), 458–460.
Hesse, M. B. (1966). *Models and analogies in science.* South Bend: University of Notre Dame Press.
Hessen, B. M. (1931). The social and economic roots of Newton's *Principia*. In *Science at the crossroads*. London: Kniga. Reprinted in G. Basalla (ed.) *The rise of modern science: External or internal factors?* D.C. Heath & Co., New York, 1968, pp. 31–38.
Hill, K. (2013, September 25). The feng shui way: A catastrophe for city planning. *Skeptical Inquirer.*
Hines, T. (2003). *Pseudoscience and the paranormal* (2nd ed.). Amherst: Prometheus Books.
Hintikka, J. (Ed.). (1994). *Aspects of metaphor*. Dordrecht: Kluwer Academic Publishers.
Ho, P.-T. (1967). *The ladder of success: Aspects of mobility in China 1368–1911*. New York: Columbia University Press.
Ho, P.-Y. (1969). The astronomical bureau of Ming China. *Journal of Asian History, 3-4*, 139–153.
Hobson, A. (2019). A realist analysis of six controversial quantum issues. In M. R. Matthews (Ed.), *Mario Bunge: A centenary festschrift*. Dordrecht: Springer.
Hodson, D. (2008). *Towards scientific literacy: A teachers' guide to the history, philosophy and sociology of science*. Rotterdam: Sense Publishers.
Hodson, D. (2014). Nature of science in the science curriculum: Origin, development and shifting emphases. In M. R. Matthews (Ed.), *International handbook of research in history, philosophy and science teaching* (pp. 911–970). Dordrecht: Springer.
Holmes, F. L. (2009). *The age of wonder: How the romantic generation discovered the beauty and terror of science*. London: Pantheon.
Holton, G. (Ed.). (1967). *Science & culture*. Boston: Beacon Press.
Holton, G. (1986). Metaphors in science and education. In *The advancement of science and its burdens* (pp. 229–252). Cambridge: Cambridge University Press.
Hooykaas, R. (1999). *Fact, faith and fiction in the development of science*. Dordrecht: Kluwer Academic Publishers.
Horton, R. (1971). African traditional thought and Western science. In M. F. D. Young (Ed.), *Knowledge and control* (pp. 208–266). London: Collier-Macmillan.
Hoyningen-Huene, P. (2008). Systematicity: The nature of science. *Philosophia, 36*, 167–180.
Hoyningen-Huene, P. (2013). *Systematicity: The nature of science*. New York: Oxford University Press.
Hsia, P.-C. R. (2012). *A Jesuit in the forbidden city: Matteo Ricci 1552–1610*. Oxford: Oxford University Press.
Hsu, S.-T. (2003). *Yin & Yang of love: Feng shui for relationships*. St. Paul: Llewellyn.
Hua, S. (1995). *Science and humanism: Two cultures in post-Mao China (1978–1989)*. Albany: State University of New York Press.
Huang, S.-C. (1968). Chang Tsai's concept of *Ch'i*. *Philosophy East and West, 18*(4), 247–260.
Huang, S.-C. (1999). *Essentials of new-Confucianism: Eight major philosophers of the Song and Ming periods*. Westport: Greenwood Press.
Huang, A. (2010). *The complete I Ching*. Rochester: Inner Traditions.
Huff, T. E. (1993). *The rise of early modern science: Islam, China, and the West*. Cambridge: Cambridge University Press.
Huizenga, J. (1992). *Cold fusion: The scientific fiasco of the century*. Rochester: University of Rochester Press.
Hulskramer, G. (2004). *I Ching in plain English*. London: Souvenir Press.

Hume, D. (1739/1888). *A treatise of human nature: Being an attempt to introduce the experimental method of reasoning into moral subjects*. Oxford: Clarendon Press.

Hume, D. (1777/1902). In L. A. Selby-Bigge (Ed.), *Enquiries concerning the human understanding and concerning the principles of morals*. Oxford: Clarendon Press.

Hume, D. (1826). *Philosophical works of David Hume* (Vol. 111). London: Black, Tait & Tait.

Hunt, I. E., & Suchting, W. A. (1969). Force and "Natural Motion". *Philosophy of Science, 36*, 233–251.

Huston, P. (1995). China, chi and chicanary: Examining traditional Chinese medicine and chi theory. *Skeptical Inquirer, 19*(5), 38–42.

Hwangbo, A. B. (1999). A new millennium and feng shui. *The Journal of Architecture, 4*(2), 191–198.

Hwangbo, A. B. (2002). An alternative tradition in architecture: Conceptions in feng shui and its continuous tradition. *Journal of Architectural and Planning Research, 19*(2), 110–130.

Hyman, R. (1985). A critical historical overview of parapsychology. In P. Kurtz (Ed.), *A skeptics handbook of parapsychology* (pp. 3–96). Amherst: Prometheus Books.

Iltis, C. (1971). Leibniz and the *vis viva* controversy. *Isis, 62*(1), 21–35.

Irzik, G. (Ed.). (2013). Commercialisation and commodification of science: Educational responses. *Science & Education, 22*(10), 2375–2384.

Israel, J. I. (2001). *Radical enlightenment: Philosophy and the making of modernity 1650–1750*. Oxford: Oxford University Press.

Israel, J. I. (2011). *Democratic enlightenment: Philosophy, revolution, and human rights, 1750–1790*. Oxford: Oxford University Press.

Jacobs, J. A. (2012). *Reason, religion, and natural law: From Plato to Spinoza*. Oxford: Oxford University Press.

Jaki, S. L. (1969a). Introductory essay. In P. Duhem (Ed.), *To save the phenomena: An essay on the idea of physical theory from Plato to Galileo* (pp. ix–xxvi). Chicago: University of Chicago Press.

Jaki, S. L. (1969b). Goethe and the physicists. *American Journal of Physics, 37*, 195–203.

Jaki, S. L. (1974). *Science & creation*. Edinburgh: Scottish Academic Press.

Jaki, S. L. (1978). *The road of science and the ways to god*. Chicago: University of Chicago Press.

Jammer, M. (1957). *Concepts of force. A study in the foundations of dynamics*. Cambridge: Harvard University Press.

Jastrow, J. (1935/1962). *Error and eccentricity in human belief*. New York: Dover Books.

Jeans, J. (1948). *The growth of physical science*. Cambridge: Cambridge University Press.

Jefferson, T. (1787/1982). *Notes on the state of Virginia*. New York: W.W. Norton.

Jeffreys, D. (2008). *Asprin: The remarkable story of a wonder drug*. Philadelphia: Chemical Heritage Foundation.

Jegede, O. J. (1997). School science and the development of scientific culture: A review of contemporary science education in Africa. *International Journal of Science Education, 19*(1), 1–20.

Jenkins, E. W. (2019). *Science for all: The struggle to establish school science in England*. London: University College of London Institute of Education Press.

Jin, G., Fan, H., & Liu, Q. (1996). The structure of science and technology in history: On the factors delaying the development of science and technology in china in comparison with the west since the 17th century (part one). In F. Dainian & R. C. Cohen (Eds.), *Chinese studies in the history and philosophy of science and technology* (pp. 137–164). Dordrecht: Kluwer Academic Publishers.

Johansson, L.-G. (2016). *Philosophy of science for scientists*. Dordrecht: Springer.

Johnson, M., & Pigliucci, M. (2004). Is knowledge of science associated with higher skepticism of pseudoscientific claims? *The American Biology Teacher, 66*(8), 536–548.

Joravsky, D. (1970). *The Lysenko affair*. Chicago: University of Chicago Press.

Kahneman, D. (2013). *Thinking Fast and Slow, Farrar, Straus and Giroux*. New York.

Kahneman, D., Slovic, P., & Tversky, A. (Eds.). (1982). *Judgment under uncertainty: Heuristics and biases*. New York: Cambridge University Press.

Kaminer, W. (1999). *Sleeping with extra-terrestrials: The rise of irrationalism and the perils of piety*. New York: Pantheon Books.
Kant, I. (1787/1933). *Critique of pure reason* (2nd ed., N. K. Smith, Trans.). London: Macmillan. (First edition, 1781).
Kaptchuk, T. J. (2000). *Chinese medicine: The web that has no weaver*. London: Rider.
Karchmer, E. (2002). Magic, science, and qigong in contemporary China. In S. D. Blum & L. M. Jensen (Eds.), *China off center: Mapping the margins of the middle kingdom* (pp. 311–322). Honolulu: University of Hawaii Press.
Karl, R. E., & Zarrow, P. G. (Eds.). (2002). *Rethinking the 1898 reform period: Political and cultural change in late Qing China*. Cambridge, MA: Harvard University Press.
Kasoff, I. E. (1984). *The thought of Chang Tsai (1020–1077)*. Cambridge: Cambridge University Press.
Kelly, I. W. (1982). Astrology, cosmobiology and humanistic astrology. In P. Grim (Ed.), *Philosophy of science and the occult* (pp. 47–68). Albany: State University of New York Press.
Kelly, D. A. (1985). Controversies over the guiding role of philosophy over science. *Australian Journal of Chinese Affairs, 14*, 21–35.
Kelly, D. A. (1990). Chinese intellectuals in the 1989 democracy movement. In G. Hicks (Ed.), *The broken mirror: China after Tiananmen* (pp. 24–51). Chicago: St. James Press.
Kenny, A. (1985). *A path from Rome: An autobiography*. London: Sidgwick & Jackson.
Kim, Y. S. (1982). Natural knowledge in a traditional culture: Problems in the study of the history of Chinese science. *Minerva, 20*(1–2), 83–104.
Kim, P. W. (2014). The wheel model of convergence STEAM education based on traditional scientific contents. *Eurasia Journal of Mathematics, Science & Technology Education, 12*(9), 2353–2371.
Kim, J.-Y. (2015). *Zhang Zai's philosophy of Qi*. Lanham: Lexington Books.
King, G., Pan, J., & Roberts, M. (2013, May). How censorship in China allows government criticism but silences collective expression. *American Political Science Review*, 1–18. https://doi.org/10.1017/S0003055413000014.
Knight, N. (2005). *Marxist philosophy in China: From Qu Qiubai to Mao Zedong, 1923–1945*. Dordrecht: Springer.
Knoop, A. (2001). *Feng shui for architects*. American Institute of Architects, Newsletter. http://www.lezley.info/feng-shui.htm
Koertge, N. (Ed.). (1998). *A house built on sand: Exposing postmodern myths about science*. New York: Oxford University Press.
Kolstø, S. D. (2001). Scientific literacy for citizenship: Tools for dealing with the science dimension of controversial socioscientific issues. *Science Education, 85*(3), 291–310.
Kovoor, A. (1978). *Begone godmen! Encounters with spiritual frauds*. Mumbai: Jaico.
Koyré, A. (1943/1968). Galileo and Plato. *Journal of the History of Ideas, 4*, 400–428. Reprinted in his *Metaphysics and Measurement*, 1968, pp. 16–43.
Koyré, A. (1968). *Metaphysics and measurement*. Cambridge: Harvard University Press.
Kristof, N. D., & Wudunn, S. (1994). *China awakes: The struggle for the soul of a rising power*. New York: Random House.
Kuhn, T. S. (1959/1977). Energy conservation as an example of simultaneous discovery. In M. Claggett (Ed.), *Critical problems in the history of science: Proceedings of the Institute for the History of Science* (pp. 321–356). Madison: University of Wisconsin Press. Reprinted in his *The essential tension: Selected studies in scientific tradition and change* (pp. 66–104). Chicago: University of Chicago Press.
Kuhn, T. S. (1970). *The structure of scientific revolutions* (2nd ed.). Chicago: Chicago University Press. (First edition, 1962).
Kuhn, T. S. (1991/2000). 'The trouble with historical philosophy of science', The Robert and Maurine Rothschild lecture, Department of History of Science, Harvard University. In J. Conant & J. Haugeland (Eds.), *The road since structure: Thomas S. Kuhn* (pp. 105–120). Chicago: University of Chicago Press.

Kurtz, P. (1985). Spiritualists, mediums, and psychics: Some evidence of fraud. In P. Kurtz (Ed.), *A skeptics handbook of parapsychology* (pp. 177–223). Amherst: Prometheus Books.

Kurtz, P., & Madigan, T. J. (1994). *Challenges to the enlightenment: In defense of reason and science*. Buffalo: Prometheus Books.

Kwok, D. W. Y. (1965). *Scientism in Chinese thought: 1900–1950*. New Haven: Yale University Press.

Ladyman, J. (2002). *Understanding philosophy of science*. London: Routledge.

Ladyman, J. (2011). The scientistic stance: The empirical and materialist stances reconciled. *Synthese, 178*(1), 87–98.

Ladyman, J. (2013). Toward a demarcation of science from pseudoscience. In M. Pigliucci & M. Boudry (Eds.), *Philosophy of pseudoscience: Reconsidering the demarcation problem* (pp. 45–59). Chicago: University of Chicago Press.

Ladyman, J. (2018). Scientism with a humane face. In J. de Ridder, R. Peels, & R. van Woudenberg (Eds.), *Scientism: Prospects and problems*. New York: Oxford University Press.

Lakatos, I. (1970). Falsification and the methodology of scientific research programmes. In I. Lakatos & A. Musgrave (Eds.), *Criticism and the growth of knowledge* (pp. 91–196). Cambridge: Cambridge University Press.

Lakatos, I. (1978). Introduction: Science and pseudoscience. In J. Worrall & G. Currie (Eds.), *The methodology of scientific research programmes: Volume I* (pp. 1–7). Cambridge: Cambridge University Press.

Lamont, P. (2013). *Extraordinary beliefs: A historical approach to a psychological problem*. Cambridge: Cambridge University Press.

Lange, M. (2001). The most famous equation. *Journal of Philosophy, 98*, 219–238.

Latour, B. (2004). Why has critique run out of steam? From matters of fact to matters of concern. *Critical Inquiry, 30*, 225–248.

Latour, B., & Woolgar, S. (1979/1986). *Laboratory life: The social construction of scientific facts* (2nd ed.). London: Sage.

Laudan, L. (1981). A confutation of convergent realism. *Philosophy of Science, 48*, 19–49.

Laudan, L. (1981/1996). The pseudo-science of science? *Philosophy of Social Science, 11*, 173–198. In L. Laudan *Beyond positivism and relativism*, Westview Press, Boulder, 1996, pp. 183–209.

Laudan, L. (1983/1996). The demise of the demarcation problem. In L. Laudan (Ed.), *Beyond positivism and relativism: Theory, method and evidence* (pp. 210–222). Boulder: Westview Press.

Laugksch, E. (2000). Scientific literacy: A conceptual overview. *Science Education, 84*(1), 71–94.

Laven, M. (2011). *Mission to China: Matteo Ricci and the Jesuit encounter with the east*. London: Faber & Faber.

Lederman, N. G., Bartos, S. A., & Lederman, J. (2014). The development, use, and interpretation of nature of science assessments. In M. R. Matthews (Ed.), *International handbook of research in history, philosophy and science teaching* (pp. 971–997). Dordrecht: Springer.

Lee, O. (1999). Equity implications based on the conceptions of science achievement in major reform documents. *Review of Educational Research, 69*, 83–115.

Lee, M. S., Pittler, M. H., & Ernst, E. (2008). Effects of Reiki in clinical practice: A systematic review of randomised clinical trials. *International Journal of Clinical Practice, 62*(6), 947–954.

Legge, J. (1887/2006). *The life and teachings of Confucius* (Vol. 1). New York: Cosimo Classics.

Leibniz, G. W. (1668/1956). On transubstantiation. In L. E. Loemker (Ed.), *Gottfried Wilhelm Leibniz: Philosophical papers and letters* (Vol. 1, pp. 178–185). Chicago: University of Chicago Press.

Leibniz, G. W. (1686/1956). A brief demonstration of a notable error of Descartes and others concerning a natural law. In L. E. Loemker (Ed. & Trans.), *Gottfried Wilhelm Leibniz: Philosophical papers and letters* (Vol. 1, pp. 455–463). Chicago: University of Chicago Press.

Lemish, L. (2008, August 19). Why is Falun Gong banned. *The Economist*.

Lenin, V. I. (1908/1970). *Materialism and empirio-criticism: Critical comments on a reactionary philosophy* (2nd ed.). Moscow: Progress Publishers.

Leung, K. (2010). Beliefs in Chinese culture. In *Oxford handbook of Chinese psychology* (pp. 221–240). Oxford: Oxford University Press.
Lewis, W. H., & Elvin-Lewis, M. P. F. (1977). *Medical botany: Plants affecting health*. New York: Wiley-Interscience.
Li, Z. (1996). *The private life of chairman Mao: The memoirs of Mao's personal physician*. New York: Random House.
Li, H. (1999). *Zhuan Falun*. New York: The Universe Publishing Company.
Lieber, A. (1978). *The lunar effect: Biological tides and human emotions*. Garden City: Doubleday.
Liftig, I. F. (2009, September). Nurturing scientific habits of mind. *Science Scope*.
Lim, J. T. Y. (2003). *Feng shui for business and office*. Toronto: Warwick Publishing.
Lin, J. Y. (1995). The Needham puzzle: Why the industrial revolution did not originate in China. *Economic Development and Cultural Change, 43*(2), 269–292.
Lin, Z. X., Yu, L., Guo, Y. Z., Zhang, H. L., Shen, Z. Y., & Zhang, T. L. (2000). *Qigong: Chinese medicine or pseudoscience?* Amherst: Prometheus Books.
Lindberg, D. C., & Numbers, R. L. (Eds.). (1986). *God and nature: Historical essays on the encounter between Christianity and science*. Berkeley: University of California Press.
Lindeman, M., & Aarnio, K. (2007). Superstitious, magical, and paranormal beliefs: An integrative model. *Journal of Research in Personality, 41*(4), 731–744.
Lindsay, J. (1974). *Blast-power and ballistics: Concepts of force and energy in the ancient world*. Frederick: Muller.
Lip, E. (2008). *Feng shui for success in business*. Singapore: Marshall Cavendish.
Lipman, M. (1991). *Thinking in education*. Cambridge: Cambridge University Press.
Littlejohn, R. (2014). First contact: The earliest Western views of Daoism in Matteo Ricci's journals. In D. Jones & M. Marion (Eds.), *Cultural counterpoint in Asian studies* (pp. 111–125). Albany: State University of New York Press.
Liu, J. L. (2010). 'Wang Fuzhi's philosophy of principle (*Li*) inherent in *Qi*. In J. Makeham (Ed.), *Dao companion to neo-Confucian philosophy* (pp. 355–380). Dordrecht: Springer.
Liu, J. L. (2015). In defense of Chinese Qi-naturalism. In C. Li, F. Perkins, K. Alan, & L. Chan (Eds.), *Chinese metaphysics and its problems* (pp. 33–53). Cambridge: Cambridge University Press.
Liu, J. L. (2018). The basic constituent of things: Zhang Zai's monist theory of *Qi*. In J. L. Liu (Ed.), *Neo-Confucianism: Metaphysics, mind, and morality* (pp. 61–83). New York: Wiley.
Liu, J. L., & Berger, D. (Eds.). (2014). *Nothingness in Asian philosophy*. New York: Routledge.
Lodge, O. (1909). *The ether of space*. London: Harper & Bros.
Lodge, O. (1916). *Raymond or life and death*. London: Metheun.
Lodge, O. (1930). *The reality of a spiritual world*. London: Ernst Benn.
Losh, S. C., & Nzekwe, B. (2011). Creatures in the classroom: Preservice teacher beliefs about fantastic beasts, magic, extrterrstrials, evolution and creationism. *Science & Education, 20*(5–6), 473–489.
Low, M. F. (1998). Beyond Joseph Needham: Science, technology, and medicine in East and Southeast Asia. *Osiris, 13*, 1–8.
Lu, G.-D. (1982). The first half-life of Joseph Needham. In G. Li, M. Zhang, & T. Cao (Eds.), *Explorations in the history of science and technology in China* (pp. 1–38). Shanghai: Shanghai Chinese Classics Publishing House.
Lu, J. (1996). Studies of the south-pointing chariot. In F. Dainian & R. S. Cohen (Eds.), *Chinese studies in the history and philosophy of science* (pp. 267–278). Dordrecht: Kluwer Academic Publishers.
Lu, Z. (1997). *Scientific Qigong exploration: The wonders and mysteries of Qi*. Malvern: Amber Leaf Press.
Lucas, J. R., & Hodgson, P. E. (1990). *Spacetime and electromagnetism: An essay on the philosophy of the special theory of relativity*. Oxford: Claredon Press.
MacCorquodale, K., & Meehl, P. (1948). On a distinction between hypothetical constructs and intervening variables. *Psychological Review, 55*, 95–107.

MacFarquhar, R., & Schoenhals, M. (2006). *Mao's last revolution*. Cambridge, MA: Harvard University Press.
Mach, E. (1893/1974). *The science of mechanics* (6th ed., T. McCormack, Trans.). LaSalle: Open Court Publishing Company.
Mach, E. (1910/1992). Sensory elements and scientific concepts. In J. Blackmore (Ed.), *Ernst Mach: A deeper look* (pp. 118–126). Dordrecht: Kluwer Academic Publishers.
Machamer, P. (2000). The nature of metaphor and scientific descriptions. In F. Hallyn (Ed.), *Metaphor and analogy in the sciences* (pp. 35–52). Dordrecht: Kluwer Academic Puüishers.
Machamer, P., Pera, M., & Baltas, A. (Eds.). (2000). *Scientific controversies: Philosophical and historical perspectives*. New York: Oxford University Press.
Magiels, G. (2010). *From sunlight to insight: Jan IngenHousz, the discovery of photosynthesis and science in the light of ecology*. Brussels: Brussels University Press.
Mahner, M. (2007). Demarcating science from pseudoscience. In T. Kuipers (Ed.), *Handbook of the philosophy of science: General philosophy of science-focal issue* (pp. 515–575). Amsterdam: Elsevier.
Mahner, M. (2012). The role of metaphysical naturalism in science. *Science & Education, 21*(10), 1437–1459.
Mahner, M. (2013). Science and pseudoscience: How to demarcate after the (alleged) demise of the demarcation problem. In M. Pigliucci & M. Boudry (Eds.), *Philosophy of pseudoscience: Reconsidering the demarcation problem* (pp. 29–59). Chicago: University of Chicago Press.
Mak, M. Y., & Ng, S. T. (2005). The art and science of feng shui—A study on architects' perception. *Building and Environment, 40*(3), 427–434.
Mak, M. Y., & So, A. T. (2015). *Scientific feng shui for the built environment: Theories and applications*. Hong Kong: City University of Hong Kong Press.
Makeham, J. (Ed.). (2010). *Dao companion to neo-Confucian philosophy*. Dordrecht: Springer.
Mamiani, M. (2002). Newton on prophecy and the apocalypse. In I. B. Cohen & G. E. Smith (Eds.), *The Cambridge companion to Newton* (pp. 387–408). Cambridge: Cambridge University Press.
Marafa, L. M. (2003). Integrating natural and cultural heritage: The advantage of feng shui landscape resources. *International Journal of. Heritage Studies, 9*(Part 4), 307–324.
March, A. L. (1968). An appreciation of Chinese geomancy. *Journal of Asian Studies, 27*(2), 253–267.
Martin, M. (1971). The use of pseudo-science in science education. *Science Education, 55*, 53–56.
Martin, M. (1994). Pseudoscience, the paranormal, and science education. *Science & Education, 3*(4), 357–372.
Mascall, E. L. (1956). *Christian theology and natural science: Some questions in their relations*. London: Longmans, Green & Co.
Matthews, M. R. (1988). A role for history and philosophy in science teaching. *Educational Philosophy and Theory, 20*(2), 67–81.
Matthews, M. R. (Ed.). (1989). *The scientific background to modern philosophy*. Indianapolis: Hackett Publishing Company.
Matthews, M. R. (1994). *Science teaching: The role of history and philosophy of science*. New York: Routledge.
Matthews, M. R. (1995). *Challenging New Zealand science education*. Palmerston North: Dunmore Press.
Matthews, M. R. (Ed.). (1998). *Constructivism and science education: A philosophical examination*. Dordrecht: Kluwer Academic Publishers.
Matthews, M. R. (2000). Constructivism in science and mathematics education. In D. C. Phillips (Ed.), *National society for the study of education, 99th yearbook* (pp. 161–192). Chicago: University of Chicago Press.
Matthews, M. R. (Ed.). (2009a). *Science, worldviews and education*. Dordrecht: Springer.
Matthews, M. R. (2009b). Teaching the philosophical and worldview components of science. *Science & Education, 18*(6–7), 697–728.

Matthews, M. R. (2009c). Science and worldviews in the classroom: Joseph Priestley and photosynthesis. *Science & Education, 18*(6–7), 929–960.

Matthews, M. R. (2009d). The Philosophy of Education Society of Australasia (PESA) and My intellectual growing-up. *Educational Philosophy and Theory, 41*(7), 777–781.

Matthews, M. R. (2011). From nature of science (NOS) to features of science (FOS). In M. S. Khine (Ed.), *Nature of science research: Concepts and methodologies* (pp. 1–26). Dordrecht: Springer.

Matthews, M. R. (2012). Philosophical and pedagogical problems with constructivism in science education. *Tréma, 38*, 41–56.

Matthews, M. R. (Ed.). (2014a). *International handbook of research in history, philosophy and science teaching* (3 Vols.). Dordrecht: Springer.

Matthews, M. R. (2014b). Pendulum motion: A case study in how history and philosophy can contribute to science education. In M. R. Matthews (Ed.), *International handbook of research in history, philosophy and science teaching* (pp. 19–56). Dordrecht: Springer.

Matthews, M. R. (2015a). *Science teaching: The contribution of history and philosophy of science: 20th anniversary revised and enlarged edition*. New York: Routledge.

Matthews, M. R. (2015b). Reflections on 25-years of journal editorship. *Science & Education, 24*(5–6), 749–805.

Matthews, M. R. (2017). In praise of philosophically-engaged history of science. *Science & Education, 26*(1–2), 175–184.

Matthews, M. R. (2018). Feng shui: Educational responsibilities and opportunities. In M. R. Matthews (Ed.), *History, philosophy and science teaching: New perspectives* (pp. 3–41). Dordrecht: Springer.

Matthews, M. R. (Ed.). (2019a). *Mario Bunge: A centenary Festschrift*. Dordrecht: Springer.

Matthews, M. R. (2019b). The contribution of philosophy to science teacher education. In A. D. Colgan & B. Maxwell (Eds.), *Philosophical thinking in teacher education*. New York: Routledge.

Matthews, M. R., Gauld, C. F., & Stinner, A. (Eds.). (2005). *The pendulum: Scientific, historical, philosophical and educational perspectives*. Dordrecht: Springer.

Maxwell, J. C. (1873). *Treatise on electricity and magnetism*. Oxford: Clarendon Press.

Mayo, D. G. (1996). *Error and the growth of experimental knowledge*. Chicago: University of Chicago Press.

Mbali, M. (2004). AIDS discourses and the South African state: Government denialism and post-apartheid AIDS policy-making. *Transformation: Critical Perspectives on Southern Africa, 54*(1), 104–122. https://doi.org/10.1353/trn.2004.0023.

McCabe, J. (1914). *The religion of Sir Oliver Lodge*. London: Watts & Co.

McCabe, J. (1935). *The social record of Christianity*. London: Watts & Co.

McCarthy, C. L. (2018). Cultural studies of science education: An appraisal. In M. R. Matthews (Ed.), *History, philosophy and science teaching: New perspectives* (pp. 99–136). Dordrecht: Springer.

McIntyre, L. (2015). *Respecting truth: Willful ignorance in the internet age*. New York: Routledge.

McIntyre, L. C. (2019). *The scientific attitude: Defending science from denial, fraud, and pseudoscience*. Cambridge, MA: MIT Press.

McMullin, E. (Ed.). (1963). *The concept of matter in modern philosophy*. Notre Dame: University of Notre Dame Press.

McMullin, E. (1987). Scientific controversy and its termination. In H. T. Engelhardt & A. L. Caplan (Eds.), *Scientific controversies: Case studies in the resolution and closure of disputes in science and technology* (pp. 49–92). Cambridge, MA: Cambridge University Press.

Mermin, D. N. (1981). Quantum mysteries for anyone. *Journal of Philosophy, 78*, 397–408.

Merton, R. K. (1938/1973). Science and the social order. In *The sociology of science* (pp. 254–266). Chicago: University of Chicago Press.

Mesmer, F. A. (1766/1980). Physical-medical treatise on the influence of the planets. In G. Bloch (Ed. & Trans.), *Mesmerism: A translation of the original scientific and medical writings of F.A. Mesmer* (pp. 3–29). Los Altos: Kaufmann.

Mesmer, F. A. (1775/1980). Letter to A.M. Unzer, Doctor of Medicine, on the medicinal usage of the magnet. In G. Bloch (Ed. & Trans.), *Mesmerism: A translation of the original scientific and medical writings of F.A. Mesmer* (pp. 25–29). Los Altos: Kaufmann.

Mesmer, F. A. (1779/1980). Dissertation on the discovery of animal magnetism. In G. Bloch (Ed. & Trans.), *Mesmerism: A translation of the original scientific and medical writings of F.A. Mesmer* (pp. 43–70). Los Altos: Kaufmann.

Mesmer, F. A. (1781/1980). Extract from *Précis Histrorique des Faitss Relatifs au Magnétisme Animal*. In G. Bloch (Ed. & Trans.), *Mesmerism: A translation of the original scientific and medical writings of F.A. Mesmer* (pp. 135–37). Los Altos: Kaufmann.

Miller, H. L. (1996). *Science and dissent in post-Mao China: The politics of knowledge*. Seattle: University of Washington Press.

Miller, J. D. (2004). Public understanding of, and attitudes toward, scientific research: What we know and what we need to know. *Public Understanding of Science, 13*, 273–294.

Miner, E. D. (1998). *Uranus: The planet, rings and satellites*. New York: Wiley.

Minzer, C. (2018). *End of an era: How China's authoritarian revival is undermining its rise*. Oxford: Oxford University Press.

Miyazaki, I. (1976). *China's examination hell: The civil service examinations of imperial China*. New Haven: Yale University Press.

Mnookin, S. (2011). *The panic virus: The true story behind the vaccine– Autism controversy*. New York: Simon and Schuster.

Moody, E. A. (1951). Galileo and Avempace: The dynamics of the leaning tower experiment. *Journal of the History of Ideas, 12*, 163–193, 375–422. Reprinted in his *Studies in medieval philosophy, science and logic* (pp. 203–286). Berkeley: University of California Press, 1975.

Moody, E. A. (1966). Galileo and his precursors. In C. L. Golino (Ed.), *Galileo reappraised*. Berkeley: University of California Press. Reprinted in his *Studies in medieval philosophy, science and logic*, University of California Press, Berkeley, 1975, pp. 393–408.

Moody, E. A. (1975). *Studies in medieval philosophy, science and logic*. Berkeley: University of California Press

Mooney, C., & Kirshenbaum, S. (2009). *Unscientific America: How American scientific illiteracy threatens our future*. Basic Books.

Moore, M. (2011, March 8). Feng shui scandals cause 20 per cent drop in business. *The Telegraph*.

Mugaloglu, E. Z. (2014). Constructivism and pseudoscience in the science classroom. *Science & Education, 23*(4), 829–842.

Musella, D. P. (2005). Gallup poll shows that Americans' belief in the paranormal persists. *The Skeptical Inquirer, 29*(5), 5.

Nakayama, S. (1973). Joseph Needham: Organic philosopher. In S. Nakayama & N. Sivin (Eds.), *Chinese science: Explorations of an ancient tradition* (pp. 23–44). Cambridge, MA: MIT Press.

Nakayama, S., & Sivin, N. (Eds.). (1973). *Chinese science: Explorations of an ancient tradition*. Cambridge, MA: MIT Press.

Nanda, M. (1998). The epistemic charity of the social constructivist critics of science and why the third world should refuse the offer. In N. Koertge (Ed.), *A house built on sand: Exposing postmodernist myths about science* (pp. 286–311). New York: Oxford University Press.

Nanda, M. (2003). *Prophets facing backward. Postmodern critiques of science and Hindu nationalism in India*. New Brunswick: Rutgers University Press.

Nanda, M. (2005). *The wrongs of the religious right: Reflections on science, secularism and Hindutva*. New Delhi: Three Essays Collective.

Nanda, M. (2016). *Science in saffron: Skeptical essays on the history of science*. Gurgaon: Three Essays Collective.

Nash, L. K. (1948). Plants and the atmosphere. In J. B. Conant (Ed.), *Harvard case histories in experimental science* (2 Vols, pp. 325–436). Cambridge, MA: Harvard University Press.

National Academy of Science (NAS). (1998). *Teaching about evolution and the nature of science*. Washington, DC: National Academy Press.
National Research Council (NRC). (2013). *Next generation science standards*. Washington, DC: National Academies Press.
National Science Foundation (NSF). (2002). *Science and engineering: Indicators 2002*. Science and technology: Public attitudes and public understanding. Science Fiction and Pseudoscience. www.nsf.gov/sbe/srs/
Needham, J. (1925). *Science, religion and reality*. London: The Sheldon Press.
Needham, J. (1936/1968). *Order and life: The Terry lectures*. Cambridge, MA: MIT Press.
Needham, J. (1963). Poverties and triumphs of the Chinese scientific tradition. In A. C. Crombie (Ed.), *Scientific change: Historical studies in the intellectual, social and technical conditions for scientific discovery and technical invention, from antiquity to the present* (pp. 117–153). New York: Basic Books.
Needham, J. (1964). Science and China's influence on the world. In R. Dawson (Ed.), *The legacy of China* (pp. 234–308). Oxford: Clarendon Press.
Needham, J. (1969). *The grand titration: Science and society in east and west*. London: Allen & Unwin.
Needham, J. & others. (1954–2004). *Science and civilisation in China* (Vols. 1–7). Cambridge: Cambridge University Press.
Needham, J., & Ling, W. (1956). *Science and civilisation in China, Vol. 2, History of scientific thought*. Cambridge: Cambridge University Press.
Needham, J., & Ling, W. (1959). *Science and civilisation in China, Vol. 3, Mathematics and the sciences of the heavens and the earth*. Cambridge: Cambridge University Press.
Needham, J., & Ling, W. (1962). *Science and civilisation in China, Vol. 4, Physics*. Cambridge: Cambridge University Press.
Nehru, J. L. (1946/1981). *The discovery of India*. New Delhi: Oxford University Press.
Newton, I. (1726/1999). *Mathematical principles of mathematical philosophy* (3rd ed., I. Bernard Cohen & A. Whitman, Trans.). Berkeley: University of California Press.
Newton, I. (1730/1979). *Opticks or a treatise of the reflections, refractions, inflections & colours of light*. New York: Dover Publications.
Nickles, T. (2013). The problem of demarcation: History and future. In M. Pigliucci & M. Boudry (Eds.), *Philosophy of pseudoscience: Reconsidering the demarcation problem* (pp. 101–120). Chicago: University of Chicago Press.
Noakes, R. (2004). Spiritualism, science and the supernatural in mid-Victorian Britain. In N. Brown, C. Burdett, & P. Thurschwell (Eds.), *The Victorian supernatural* (pp. 23–43). Cambridge: Cambridge University Press.
Nola, R. (1991). Ordinary human inference as refutation of the strong programme. *Social Studies of Science, 21*, 107–129.
Nola, R. (2000). Saving Kuhn from the sociologists of science. *Science & Education, 9*(1–2), 77–90.
Nola, R., & Sankey, H. (2000). A selective survey of theories of scientific method. In R. Nola & H. Sankey (Eds.), *After Popper, Kuhn and Feyerabend* (pp. 1–65). Dordrecht: Kluwer Academic Publishers.
Norris, C. M. (2001). *Acupuncture: Treatment of musculoskeletal disorders*. Oxford: Butterworth-Heinemann.
Offit, P. (2013). *Do you believe in magic? The sense and nonsense of alternative medicine*. New York: HarperCollins.
Oppenheim, J. (1985). *The other world: Spiritualism and psychical research in England, 1850–1914*. Cambridge: Cambridge University Press.
Orenstein, A. (2002). Religion and paranormal belief. *Journal for the Scientific Study of Religion, 41*, 301–311.
Oreskes, N., & Conway, E. M. (2010). *Merchants of doubt: How a handful of scientists obscured the truth on issues from tobacco smoke to global warming*. New York: Bloomsbury Press.

Orleans, L. A. (Ed.). (1980). *Science in contemporary China*. Stanford: Stanford University Press.
Ortony, A. (Ed.). (1979). *Metaphor and thought*. Cambridge: Cambridge University Press.
Ostwald, W. (1901). *Natural philosophy* (T. Selzer, Trans.). New York: Henry Holt and Co.
Otto, S. (2016). *The war on science: Who is waging it, why it matters, what we can do about it*. Minneapolis: Milkweed.
Outram, D. (2005). *The enlightenment* (2nd ed.). New York: Cambridge University Press.
Ownby, D. (2008). *Falun Gong and the future of China*. Oxford: Oxford University Press.
Padgen, A. (2007). The immobility of China: Orientalism and Occidentalism in the enlightenment. In L. Wolff & M. Cipolloni (Eds.), *The anthropology of the enlightenment* (pp. 57–78). Stanford: Stanford University Press.
Paine, T. (1797/2004). *The age of reason, being an investigation of true and fabulous theology*. New York: Dover Publications.
Palmer, M. (2001). *The Jesus Sutras: Rediscovering the lost scrolls of Taoist Christianity*. New York: Ballantine Books.
Palmer, D. A. (2007). *Qigong fever: Body, science, and utopia in China*. New York: Columbia University Press.
Papadouris, N., & Constantinou, C. (2011). A philosophically informed teaching proposal on the topic of energy for students aged 11–14. *Science & Education, 20*, 961–979.
Park, R. L. (2000). *Voodoo science: The road from foolishness to fraud*. Oxford: Oxford University Press.
Park, R. L. (2008). *Superstition: Belief in the age of science*. Princeton: Princeton University Press.
Parkes, G. (2003). Winds, waters, and earth energies: *Fengshui* and awareness of place. In H. Selin (Ed.), *Nature across cultures: Views of nature and the environment in non-Western cultures* (pp. 185–209). Dordrecht: Springer.
Passmore, J. A. (1970). *The perfectibility of man*. London: Duckworth.
Paton, M. J. (2007). Feng shui: A continuation of the art of swindlers? *Journal of Chinese Philosophy, 34*(3), 427–445.
Paton, M. J. (2013). *Five classics of feng shui: Chinese spiritual geography in historical and environmental perspective*. Leiden: Brill.
Peake, C. H. (1934). Some aspects of the introduction of modern science into China. *Isis, 22*, 173–219.
Pellegrino-Estich, R. (2001). *The miracle man: The life story of Joao de Deus*. Cairns: Triad Publishers.
Pennick, N. (1979). *The ancient science of geomancy: Man in harmony with the earth*. London: Thames and Hudson.
Pennock, R. T. (1999). *Tower of Babel: The evidence against the new creationism*. Cambridge, MA: MIT Press.
Pennock, R. T. (2002). Should creationism be taught in the public schools? *Science & Education, 11*(2), 111–133.
Pennock, R. T. (2011). Can't philosophers tell the difference between science and religion? Demarcation revisited. *Synthese, 178*(2), 177–206.
Penny, B. (2012). *The religion of Falun Gong*. Chicago: University of Chicago Press.
Perry-Hobson, J. S. (1994). Feng shui: Its impacts on the Asian hospitality industry. *International Journal of Contemporary Hospitality Management, 6*(6), 21–26.
Peters, R. S. (1966). *Ethics and education*. London: George Allen & Unwin.
Pigliucci, M. (2007). The evolution-creation wars: Why teaching more science just is not enough. *McGill Journal of Education, 42*(2), 285–306.
Pigliucci, M. (2010). *Nonsense on stilts: How to tell science from bunk* (2nd ed.). Chicago: University of Chicago Press.
Pigliucci, M. (2013). The demarcation problem: A (belated) response to Laudan. In M. Pigliucci & M. Boudry (Eds.), *Philosophy of pseudoscience: Reconsidering the demarcation problem* (pp. 9–28). Chicago: University of Chicago Press.

Pigliucci, M., & Boudry, M. (Eds.). (2013). *Philosophy of pseudoscience: Reconsidering the demarcation problem.* Chicago: University of Chicago Press.

Platt, S. R. (2018). *Imperial twilight: The opium war and the end of China's last golden age.* New York: Alfred A. Knopf.

Polanyi, M. (1958). *Personal Knowledge.* London: Routledge and Kegan Paul.

Polo, M. (1958). *The travels of Marco Polo* (R. Latham, Trans.). Harmondsworth: Penguin Books.

Pomeroy, D. (1992). Science across cultures: Building bridges between traditional Western and Alaskan native cultures. In S. Hills (Ed.), *History and philosophy of science in science education* (Vol. 2, pp. 257–268). Kingston: Queen's University.

Popper, K. R. (1934/1959). *The logic of scientific discovery.* London: Hutchinson.

Popper, K. R. (1963). *Conjectures and refutations: The growth of scientific knowledge.* London: Routledge & Kegan Paul.

Porkert, M. (1974). *The theoretical foundations of Chinese medicine.* Cambridge: MIT Press.

Porkert, M. (1982). The difficult task of blending Chinese and Western science: The case of the modern interpretations of traditional Chinese medicine. In G. Li, M. Zhang, & T. Cao (Eds.), *Explorations in the history of science and technology in China* (pp. 553–572). Shanghai: Shanghai Chinese Classics Publishing House.

Porter, R., & Teich, M. (Eds.). (1981). *The enlightenment in national context.* Cambridge: Cambridge University Press.

Preece, P. F. W., & Baxter, J. H. (2000). Scepticism and gullibility: The superstitious and pseudo-scientific beliefs of secondary school students. *International Journal of Science Education, 22*(11), 1147–1156.

Priestley, J. (1772). Observations on different kinds of air. *Philosophical Transactions, 60,* 147–264.

Priestley, J. (1775–1777). *Experiments and observations on different kinds of air* (3 Vols., 2nd ed.). London: J. Johnson. Sections of the work have been published by the Alembic Club with the title *The discovery of oxygen* (Edinburgh, 1961).

Psillos, S. (1999). *Scientific realism: How science tracks truth.* London: Routledge.

Puro, J. (2002). Feng shui. In M. Shermer (Ed.), *The skeptic encyclopedia of pseudoscience* (pp. 102–112). Santa Barbara: ABC-CLIO.

Qian, W.-Y. (1985). *The great inertia: Scientific stagnation in traditional China.* London: Croom Helm.

Radder, H. (Ed.). (2003). *The philosophy of scientific experimentation.* Pittsburgh: University of Pittsburgh Press.

Radner, D., & Radner, M. (1982). *Science and unreason.* Belmont: Wadsworth Publishing Company.

Radner, D., & Radner, M. (1989). Holistic methodology and pseudoscience. In D. Stalker & C. Glymour (Eds.), *Examining holistic medicine* (pp. 149–159). Buffalo: Prometheus Books.

Randi, J. (1987). *The faith healers.* Buffalo: Prometheus Books.

Randi, J. (1992). *Conjuring: Being a definitive history of the venerable arts of sorcery, prestidigitation, wizardry, deception, & chicanery, and of the mountebanks & scoundrels who have perpetrated these subterfuges on a bewildered public, in short, MAGIC!* New York: St Martin's Press.

Randi, J. (1995). *An encyclopedia of claims, frauds, and hoaxes of the occult and supernatural.* New York: St. Martin's Press.

Ray, C. (1987). *The evolution of relativity.* Bristol: Adam Hilger.

Reifler, S. (1974). *I Ching: A new interpretation for modern times.* New York: Bantam Books.

Ricci, M. (1615/1953). On the Christian mission among the Chinese. In L. L. Gallagher (Ed.), *China in the sixteenth century: The journals of Matthew Ricci* (pp. 1583–1610). New York: Random House.

Ricci, M. (2001). *Lettere (1580–1609)* (F. D'Arelli, Ed.). Macerata: Quodlibet.

Ricci, M. (2016). *The true meaning of the lord of heaven* (D. Lanacashire sj & P. Hu Kuo-chen sj, Trans.). Boston College: Institute of Jesuit Sources.

References

Rice, T. W. (2003). Believe it or not: Religious and other paranormal beliefs in the United States. *Journal for the Scientific Study of Religion, 42*, 95–106.
Roberts, D. A. (2007). Scientific literacy/science literacy. In S. K. Abell & N. G. Lederman (Eds.), *Handbook of research in science education* (pp. 729–779). Mahwah: Erlbaum.
Rolnick, H. (2004). *Feng shui: The Chinese system of elements*. Hong Kong: FormAsia Books.
Romero, G. (2019). Physics and philosophy of physics in the work of Mario Bunge. In M. R. Matthews (Ed.), *Mario Bunge: A Centenary Festschrift*. Dordrecht: Springer.
Ronan, C. A. (1978). *The shorter science and civilization in China* (Vol. 1, abridgement of Vols. 1, 2 of Joseph Needham's original). Cambridge: Cambridge University Press.
Rosicka, C. (2016). *From concept to classroom: Translating STEM education research into practice*. Melbourne: Australian Council for Educational Research.
Ross, D., Ladyman, J., & Spurrett, D. (2007). In defense of scientism. In D. Ross & J. Ladyman (Eds.), *Everything must go: Metaphysics naturalized* (pp. 1–65). Oxford: Oxford University Press.
Rossbach, S. (1984). *Feng shui*. London: Rider.
Rossbach, S. (1987). *Interior design with feng shui*. New York: E.P. Dutton.
Rotton, J., & Kelly, I. (1985). Much Ado about the full moon: A meta-analysis of lunar-lunacy research. *Psychological Bulletin, 97*, 286–306.
Rowbotham, A. H. (1942). *Missionary and mandarin: The Jesuits at the court of China*. Berkeley: University of California Press.
Rowland, H. (1883, August 24). A Plea for pure science. *Science, 2*(29), 242–250.
Rule, P. A. (1972). *K'ung-tzu or Confucius? The Jesuit interpretation of Confucianism*. PhD thesis, Australian National University.
Rule, P. A. (1986). *K'ung-tzu or Confucius? The Jesuit interpretation of Confucianism*. Sydney: Allen & Unwin.
Ruse, M. (Ed.). (1988). *But is it science? The philosophical question in the creation/evolution controversy*. Albany: Prometheus Books.
Russell, B. (1922). *The problem of China*. London: George Allen & Unwin.
Rutt, R. (1996). *The book of changes (Zhouyi): A Bronze Age document*. Richmond: Curzon Press.
Sacks, B. (2014, May 16). Reiki goes mainstream: Spiritual touch practice now commonplace in hospitals. *The Washington Post*.
Sagan, C. (1996). *The demon-haunted world: Science as a candle in the dark*. London: Headline Book Publishing.
Sambursky, S. (1975). *Physical thought from the Presocratics to the quantum physicists*. New York: Pica Press.
Sampson, W. (1996). Antiscience trends in the rise of the "alternative medicine" movement. In P. R. Gross, N. Levitt, & M. W. Lewis (Eds.), *The flight from science and reason* (pp. 188–197). Baltimore: The Johns Hopkins University Press.
Schimmel, S. (2008). *The tenacity of unreasonable beliefs: Fundamentalism and the fear of truth*. Oxford: Oxford University Press.
Schofield, R. E. (Ed.). (1966). *A scientific autobiography of Joseph Priestley (1733–1804): Selected scientific correspondence*. Cambridge: MIT Press.
Schofield, R. E. (1970). *Mechanism and materialism: British natural philosophy in an age of reason*. Princeton: Princeton University Press.
Schofield, R. E. (1997). *The enlightenment of Joseph Priestley: A study of his life and work from 1733 to 1773*. University Park: Penn State Press.
Schofield, R. E. (2004). *The enlightened Joseph Priestley: A study of his life and work from 1773 to 1804*. University Park: Penn State Press.
Schwab, J. J. (1949/1978). The nature of scientific knowledge as related to liberal education. *Journal of General Education, 3*, 245–266. Reproduced in I. Westbury & N. J. Wilkof (Eds.). *Joseph J. Schwab: Science, curriculum, and liberal education* (pp. 68–104). Chicago: University of Chicago Press, 1978.

Schwarcz, V. (1986). *The Chinese enlightenment: Intellectuals and the legacy of the May Fourth Movement of 1919*. Berkeley: University of California Press.
Schwartz, B. I. (1951). Ch'en Tu-hsiu and the acceptance of the modern west. *Journal of the History of Ideas, 12*, 61–72.
Schwartz, B. I. (Ed.). (1971). *Reflections on the May Fourth Movement: A symposium*. Cambridge, MA: Harvard University Press.
Schweitzer, A. (1910/1954). *The quest of the historical Jesus: A critical study of its Progress from Reimarus to Wrede* (3rd ed.). London: Adam and Charles Black.
Selin, H. (Ed.). (2003). *Nature across cultures: Views of nature and the environment in non-Western cultures*. Dordrecht: Springer.
Semple, D., & Smyth, R. (2013). *Oxford handbook of psychiatry* (3rd ed.). Oxford: Oxford University Press.
Shamos, M. (1995). *The myth of scientific literacy*. New Brunswick: Rutgers University Press.
Shank, J. B. (2008). *The Newton wars and the beginning of the French enlightenment*. Chicago: University of Chicago Press.
Shapin, S. (1982). History of science and its sociological reconstructions. *History of Science, 22*, 157–211.
Shapiro, J. (2001). *Mao's war against nature: Politics and environment in revolutionary China*. Cambridge: Cambridge University Press.
Shaughnessy, E. L. (1997). *I Ching, the classic book of changes*. New York: Ballantine Books.
Sherman, T. F. (2018). *Energy, entropy, and the flow of nature*. Oxford: Oxford University Press.
Shermer, M. (1997). *Why people believe weird things: Pseudoscience, superstition, and other confusions of our time*. New York: W.H. Freemand.
Shermer, M. (2001). *The borderlands of science: Where sense meets nonsense*. Oxford: Oxford University Press.
Shermer, M. (2013). Science and pseudoscience: The difference in practice and the difference it makes. In M. Pigliucci & M. Boudry (Eds.), *Philosophy of pseudoscience: Reconsidering the demarcation problem* (pp. 203–223). Chicago: University of Chicago Press.
Shimony, A. (1965). Quantum physics and the philosophy of whitehead. In R. S. Cohen & M. W. Wartofsky (Eds.), *Boston studies in the philosophy of science* (pp. 307–330). New York: Humanities Press.
Shimony, A. (1989). Search for a worldview which can accommodate our knowledge of microphysics. In J. T. Cushing & E. McMullin (Eds.), *Philosophical consequences of quantum physics* (pp. 25–37). Notre Dame: University of Notre Dame Press.
Shimony, A. (1993a). *Search for a naturalistic world view Vol. I Scientific method and epistemology*. Cambridge: Cambridge University Press.
Shimony, A. (1993b). *Search for a naturalistic world view Vol. II Natural sciences and metaphysics*. Cambridge: Cambridge University Press.
Shinji, A. (1996). Astronomical studies by Zhao Youqin. *Taiwanese Journal for Philosophy and History of Science, 5*(1), 59–102.
Shirk, S. L. (2007). *China: The fragile superpower*. Oxford: Oxford University Press. HAVE.
Siegfried, R. (2002). *From elements to atoms: A history of chemical composition*. Philadelphia: American Philosophical Society.
Silverberg, R. (1965). *Scientists and scoundrels: A book of hoaxes*. New York: Ty Crowell.
Sivin, N. (1984). Why the scientific revolution did not take place in China – Or didn't it? In E. Mendelsohn (Ed.), *Transformation and tradition in the sciences* (pp. 531–554). Cambridge: Cambridge University Press.
Skinner, S. (1980). *Terrestrial astrology: Divination by geomancy*. London: Routledge & Kegan Paul.
Skinner, S. (1982). *The living earth manual of feng-shui: Chinese geomancy*. London: Routledge & Kegan Paul.
Skinner, S. (2008). *Guide to the feng shui compass: A compendium of classical feng shui*. Singapore: Golden Hoard Press.

Skinner, S. (2011). *Geomancy in theory and practice.* Kuala Lumpur: Golden Hoard Press. (Republication of Skinner 1980, *Terrestrial astrology: Divination by geomancy.*).

Skinner, S. (2015). *Advanced flying star feng shui.* Singapore: Golden Hoard Press.

Skrabanek, P. (1985). Acupuncture: Past, present and future. In D. Stalker & C. Glymour (Eds.), *Examining holistic medicine* (pp. 181–196). Buffalo: Prometheus Books.

Slack, G. (2007). *The battle over the meaning of everything: Evolution, intelligent design, and a school board in Dover, PA.* San Francisco: Jossey-Bass.

Slezak, P. (1994a). Sociology of science and science education: Part I. *Science & Education, 3*(3), 265–294.

Slezak, P. (1994b). Sociology of science and science education. Part II: Laboratory life under the microscope. *Science & Education, 3*(4), 329–356.

Slezak, P. (2012). Review of Michael Ruse Science and spirituality: Making room for faith in the age of science. *Science & Education, 21*, 403–413.

Smith, R. J. (1991). *Fortune-tellers and philosophers. Divination in Chinese society.* Boulder: Westview Press.

Smith, G. H. (1992). *Kura Kaupapa Maori schooling: Implications for the teaching of science in New Zealand.* Unpublished paper, Education Department, University of Auckland.

Smith, R. J. (1998, Winter). The place of Yi Jing (classic of changes) in world culture. *Journal of Chinese Philosophy*, pp. 391–422.

Smith, J. C. (2010). *Pseudoscience and extraordinary claims of the paranormal: A critical Thinker's toolkit.* Chichester: Wiley-Blackwell.

Sokal, A. (2009). *Beyond the hoax: Science, philosophy and culture.* Oxford: Oxford University Press.

Song, J. (1996). The historical value of the *Nine chapters on the mathematical art* in society and the economy. In F. Dainian & R. S. Cohen (Eds.), *Chinese studies in the history and philosophy of science and technology* (pp. 261–266). Dordrecht: Kluwer Academic Publishers.

Sorell, T. (1991). *Scientism: Philosophy and the infatuation with science.* London: Routledge.

Sorrell, R., & Sorrell, A. M. (1994). *The I Ching made easy.* San Francisco: Harper.

Spear, W. (1995). *Feng shui made easy.* London: HarperCollins.

Specter, M. (2009). *Denialism: How irrational thinking prevents scientific Progress, harms the planet and threatens our lives.* London: Duckworth.

Spence, J. D. (1969/1980). *To change China: Western advisors in China 1620–1960* (2nd ed.). New York: Penguin.

Spence, J. D. (1982). *The gate of heavenly peace: The Chinese and their revolution, 1895–1980.* London: Penguin.

Spence, J. D. (1984). *The memory palace of Matteo Ricci.* New York: Penguin.

Sprod, T. (2014). Philosophical inquiry and critical thinking in primary and secondary science education. In M. R. Matthews (Ed.), *International handbook of research in history, philosophy and science teaching* (pp. 1531–1564). Dordrecht: Springer.

Squires, N. (2017, October 3). Italians turn to fortune-tellers and occult as economy slumps. *The Telegraph*. https://www.brisbanetimes.com.au/world/italians-turn-to-fortunetellers-and-occult-as-economy-slumps-20171003-gyt347.html.

Stalker, D., & Glymour, C. (1989). *Examining holistic medicine.* Amherst: Prometheus Books.

Standage, T. (2000). *The Neptune file.* New York: Walker.

Starck, M. (1982). *Astrology: Key to holistic health.* Birmingham: Seek-It Publications.

Stark, E. (2012). Enhancing and assessing critical thinking in a psychological research methods course. *Teaching of Psychology, 39*, 107–112.

State Council Information Office (SCIO). (2016). *Traditional Chinese medicine in China.* Beijing: SCIO.

Staver, J. (1998). Constructivism: Sound theory for explicating the practice of science and science teaching. *Journal of Research in Science Teaching, 35*(5), 501–520.

Stebbing, L. S. (1937/1958). *Philosophy and the physicists.* New York: Dover Publications.

Stenger, V. J. (1990). *Physics and psychics: The search for a world beyond the senses*. Amherst: Prometheus Books.
Stenger, V. J. (2007). *God: The failed hypothesis: How science shows that god does not exist*. Amherst: Prometheus Books.
Stoljar, D. (2010). *Physicalism*. New York: Routledge.
Stove, D. C. (1982). *Popper and after: Four modern irrationalists*. Oxford: Pergamon Press.
Suchting, W. A. (1986). *Marx and philosophy: Three studies*. London: Macmillan.
Suchting, W. A. (1994). Notes on the cultural significance of the sciences. *Science & Education, 3*(1), 1–56.
Sutherland, S. (1992). *Irrationality: The enemy within*. London: Penguin.
Suttmeier, R. P. (1980). *Science, technology, and China's drive for modernization*. Stanford: Hoover Institute Press.
Tang, Y. (2015a). Some reflections on new Confucianism in Chinese Mainland culture of the 1990s. In Y. Tang (Ed.), *Confucianism, Buddhism, Daoism, Christianity and Chinese culture* (pp. 67–78). Dordrecht: Springer.
Tang, Y. (2015b). The attempt of Matteo Ricci to link Chinese and Western cultures. In Y. Tang (Ed.), *Confucianism, Buddhism, Daoism, Christianity and Chinese culture* (pp. 179–189). Dordrecht: Springer.
Tang, Y. (2015c). The enlightenment and its difficult journey in China. In Y. Tang (Ed.), *Confucianism, Buddhism, Daoism, Christianity and Chinese culture* (pp. 279–284). Dordrecht: Springer.
Tang, Y. (2015d). On the clash and coexistence of human civilizations. In Y. Tang (Ed.), *Confucianism, Buddhism, Daoism, Christianity and Chinese culture* (pp. 291–307). Dordrecht: Springer.
Tang, Y. (2015e). *Confucianism, Buddhism, Daoism, Christianity and Chinese culture*. Dordrecht: Springer.
Taylor, R. (2002). *Feng shui for the modern city: A practical guide*. Hod Hasharon: Astrolog Publishing House.
Thompson, B. (Count of Rumford). (1798). An inquiry concerning the source of the heat which is excited by friction. *Philosophical Transactions of Royal Society of London, 88*, 80–102.
Thomson, W. (Lord Kelvin). (1891–1894). *Popular lectures and addresses* (3 Vols.). London: Macmillan.
Thurs, D. P., & Numbers, R. L. (2013). Science, pseudoscience, and science falsely so-called. In M. Pigliucci & M. Boudry (Eds.), *Philosophy of pseudoscience: Reconsidering the demarcation problem* (pp. 121–144). Oxford: Oxford University Press.
Tobacyk, J. (2004). A revised paranormal belief scale. *International Journal of Transpersonal Studies, 23*, 94–98.
Tobin, K. (Ed.). (1993). *The practice of constructivism in science and mathematics education*. Washington, DC: AAAS Press.
Tobin, K. (2000). Constructivism in science education: Moving on. In D. C. Phillips (Ed.), *Constructivism in education* (pp. 227–253). Chicago: National Society for the Study of Education.
Tobin, K. (2015). Connecting science education to a world in crisis. *Asia-Pacific Science Education, 1*, 2. https://doi.org/10.1186/s41029-015-0003-z.
Too, L. (1994). *Practical applications of feng shui*. Adelaide: Oriental Publications.
Too, L. (1998). *Essential feng shui: A step-by step guide to enhancing your relationships, health and prosperity*. London: Random House.
Trefil, J. S. (1996). Scientific literacy. In P. R. Gross, N. Levitt, & M. W. Lewis (Eds.), *The flight from science and reason* (pp. 543–550). New York: New York Academy of Sciences.
Tresmontant, C. (1965). *Christian metaphysics*. New York: Sheed and Ward.
Tristram, H. (Ed.). (1952). *The idea of a liberal education: A selection from the works of Newman*. London: Harrap.
Trusted, J. (1991). *Physics and metaphysics: Theories of space and time*. London: Routledge.

Tsai, M. Y., Chen, C. Y., & Lin, C. C. (2017). Theoretical basis, application, reliability, and sample size estimates of a meridian energy analysis device for traditional Chinese medicine research. *Clinics Sao Paulo, 72*(4), 254–257. https://www.ncbi.nlm.nih.gov/pubmed/28492726.

Turgut, H. (2011). The context of demarcation in nature of science teaching: The case of astrology. *Science & Education, 20*(5–6), 491–515.

Tyler, V. E. (1985). Hazards of herbal medicine. In D. Stalker & C. Glymour (Eds.), *Examining holistic medicine* (pp. 323–339). Buffalo: Prometheus Books.

Udías, A. (1994). Jesuit astronomers in Beijing, 1601–1805. *Quarterly Journal of the Royal Astronomical Society, 35*, 463–478.

Udías, A. (2015). *Jesuit contribution to science: A history*. Dordrecht: Springer.

Uglow, J. (2002). *The lunar men: Five friends whose curiosity changed the world*. London: Faber & Faber.

Urbach, P. (1987). *Francis Bacon's philosophy of science*. La Salle: Open Court.

Urevbu, A. O. (1987). School science in South Africa: An assessment of the pedagogic impact of third world investment. *International Journal of Science Education, 9*(1), 3–12.

van Fraassen, B. C. (2002). *The empirical stance*. New Haven: Yale University Press.

Vickers, A. J., Cronin, A. M., Maschino, A. C., Lewith, G., MacPherson, H., Victro, N., Foster, N. E., Sherman, K. J., Witt, C. M., & Linde, K. (2012). Acupuncture for chronic pain: Individual patient data meta-analysis. *Archives of Internal Medicine, 172*(19), 1444–1453.

Vitzthum, R. C. (1995). *Materialism: An affirmative history and definition*. Amherst: Prometheus.

Vivekananda, S.. (2016). *Complete works of Swami Vivekananda* (8 Vols.). Delhi: Advaita Ashrama. www.advaitaashrama.org

Volkov, A. (1996a). Science and Daoism: An introduction. *Taiwanese Journal for Philosophy and History of Science, 5*(1), 1–58.

Volkov, A. (1996b). The mathematical work of Zhao Youqin: Remote surveying and the computation of π. *Taiwanese Journal for Philosophy and History of Science, 5*(1), 129–189.

Volkov, A. (1997). Zhao Youqin and his calculation of π. *Historia Mathematica, 24*, 301–331.

von Helmholtz, H. (1862/1995). On the conservation of force. In *Science and culture: Popular and philosophical essays* (pp. 96–126) (Edited with Introduction by David Cahan; original essays 1853–1892). Chicago: Chicago University Press.

Walker, W. R., Hoekstra, S, J,, & Vogl, R. J. (2002). Science education is no guarantee of skepticism. *Skeptic, 9*, 24–28.

Wallace, W. A. (1981). *Prelude to Galileo: Essays on medieval and sixteenth-century sources of Galileo's thought*. Dordrecht: Reidel Publishing Company.

Wan, Z. H., Wong, S. L., & Zhan, Y. (2013). When nature of science meets Marxism: Aspects of nature of science taught by Chinese science teacher educators to prospective science teachers. *Science Education, 22*(5), 1115–1140.

Wan, D., Zhang, H., & Wei, B. (2018). Impact of Chinese culture on pre-service science teachers' views of the nature of science. *Science & Education, 27*(3–4), 321–355.

Wang, Chong (Chhung). (ca.80/1963). Discourses weighed in the balance. In W.-T. Chan (Ed.), *A source book in Chinese philosophy*. Princeton: Princeton University Press.

Wang, Y. C. (1966). *Chinese intellectuals and the west 1872–1949*. Chapel Hill: University of North Carolina Press.

Wang, Y. (1996). Li Shanlan: Forerunner of modern science in China. In F. Dianian & R. S. Cohen (Eds.), *Chinese studies in the history and philosophy of science and technology* (pp. 345–368). Dordrecht: Kluwer Academic Publishers.

Wang, J. C.-S. (2007). *John Dewey in China: To teach and to learn*. Albany: State University of New York Press.

Wang, H.-S. (2014). The predicament of scientific culture in ancient China. In M. Burguete & L. Lam (Eds.), *All about science: Philosophy, history, sociology & communication* (pp. 116–136). Singapore: World Scientific.

Wang, Y. (2016). Why did *elements* have no influence on Chinese thinking in the Yuan dynasty? *Studies in Dialectics of Nature, 32*(9), 67–73.

Wang, R. R., & Ding, W. (2010). Zhang Zai's theory of vital energy. In J. Makeham (Ed.), *Dao companion to neo-Confucian philosophy* (pp. 39–57). Dordrecht: Springer.
Wartofsky, M. (1968). Metaphysics as a heuristic for science. In R. S. Cohen & M. W. Wartofsky (Eds.), *Boston Studies in the Philosophy of Science 3*, 123–172. Republished in his *Models* (pp. 40–89). Dordrecht: Reidel, 1979.
Webb, J. (1971). *The flight from reason*. London: Macdonald.
Wegman, M. E. (1980). Biomedical research: Clinical and public health aspects. In L. A. Orleans (Ed.), *Science in contemporary China* (pp. 269–294). Stanford: Stanford University Press.
Wei, J. (1997). Fifth modernization. In *The courage to stand alone: Letters from prison and other writings* (pp. 208–210). New York: Penguin.
Weinberg, S. (2015). *To explain the world: The discovery of modern science*. London: Penguin Books.
Weinert, F. (1995). The Duhem-Quine thesis revisited. *International Studies in Philosophy of Science, 9*(2), 147–156.
Weinert, F. (2005). *The scientist as philosopher: Philosophical consequences of great scientific discoveries*. Berlin: Springer.
Weisheipl, J. A. (1968). The revival of Thomism as a Christian philosophy. In R. M. McInerny (Ed.), *New themes in Christian philosophy* (pp. 164–185). South Bend: University of Notre Dame Press.
Weisheipl, J. A. (1985). *Nature and motion in the Middle Ages* (W. E. Carroll, Ed.). Washington, DC: Catholic University of America Press.
Welch, H. (1957). *Taoism: The parting of the way*. Boston: Beacon Press.
Wensel, L. O. (1980). *Acupuncture for Americans*. Reston: Reston Publishing.
Westfall, R. S. (1971). *The construction of modern science: Mechanisms and mechanics*. Cambridge: Cambridge University Press.
Westfall, R. S. (2000). The scientific revolution reasserted. In M. J. Osler (Ed.), *Rethinking the scientific revolution* (pp. 42–55). Cambridge: Cambridge University Press.
Weyant, R. G. (1980). Protoscience, pseudoscience, metaphors and animal magnetism. In M. P. Hanen, M. J. Osler, & R. G. Weyant (Eds.), *Science, pseudoscience and society* (pp. 77–114). Calgary: Wilfrid Laurier University Press.
Whitehead, A. N. (1925). *Science and the modern world*. New York: Macmillan.
Whitehead, A. N. (1929). *Process and reality: An essay in cosmology*. New York: Macmillan.
Wicker, C. (2005). *Not in Kansas anymore: A curious tale of how magic is transforming America*. New York: HarperCollins.
Wilhelm, R. (1950). *I Ching or book of changes* (C. F. Baynes, Trans.). Princeton: Princeton University Press.
Wilhelm, H. (1960). *Change: Eight lectures on the I Ching*. New York: Pantheon Books.
Wilhelm, H. (1977). *Heaven, earth, and man in the book of changes*. Seattle: University of Washington Press.
William, G., & Reeves, T. (2003). Another go at energy. *Physics Education, 18*, 150–155.
Wilson, J. A. (2018). Reducing pseudoscientific and paranormal beliefs in university students through a course in science and critical thinking. *Science & Education*. https://doi.org/10.1007/s11191-018-9956-0.
Winchester, S. (2008). *The man who loved China: The fantastic story of the eccentric scientist who unlocked the mysteries of the middle kingdom*. New York: HarperCollins.
Winchester, S. (2018). *Exactly: How precision engineers created the modern world*. London: William Collins.
Wise, M. N. (Ed.). (1995). *The values of precision*. Princeton: Princeton University Press.
Wong, G. (1963). China's opposition to Western science during the late Ming and early Ch'ing. *Isis, 54*(1), 29–49.
Wong, E. (1997). *Taoism*. Boston: Shambhala Publications.
Wong, T. M. K. (2000). The limits of ambiguity in the German identity in nineteenth century Hong Kong, with special reference to Ernest John Eitel (1938–1908). In R. K. S. Mak & D. S.

L. Paau (Eds.), *Sino-German relations since 1800: Multidisciplinary explorations* (pp. 73–91). Frankfurt: Peter Lang.
Wootton, D. (2015). *The invention of science: A new history of the scientific revolution*. London: Penguin Random House.
Wozniak, J. A., Wu, S., & Wang, H. (2001). *Yan Xin Qigong and contemporary sciences*. Champaign: International Yan Xin Qigong Association.
Wright, J. (2010). *Mission to China: Matteo Ricci and the Jesuit encounter with the east*. London: Faber.
Xu, L. (1981/1996). Essay on the role of science and democracy in society. In F. Dainian & R. S. Cohen (Eds.), *Chinese studies in the history and philosophy of science and technology* (pp. 5–11). Dordrecht: Kluwer Academic Publishers.
Yan, X. (2015). *Secrets and benefits of internal Qigong cultivation*. Malvern: Amber Leaf Press.
Yan, X., Lu, F., Jiang, H., Cao, W., Xia, Z., Shen, H., Wang, J., Dao, M., Lin, H., & Zhu, R. (2002). *Journal of Scientific Exploration, 16*(3), 381–411.
Yang, C. K. (1970). *Religion in Chinese society*. Berkeley: University of California Press.
Yolton, J. W. (1983). *Thinking matter: Materialism in eighteenth-century Britain*. Minneapolis: University of Minnesota Press.
Yoon, H.-K. (2006). *The culture of fengshui in Korea: An exploration of East Asian geomancy*. Plymouth: Rowan & Littlefield.
York, D. G., Gingerich, O., & Zhang, S.-N. (Eds.). (2012). *The astronomy revolution: 400 years of exploring the cosmos*. New York: Taylor & Francis.
Yosida, M. (1973). The Chinese concept of nature. In S. Nakayama & N. Sivin (Eds.), *Chinese science: Explorations of an ancient tradition* (pp. 71–89). Cambridge, MA: Harvard University Press.
Young, T. (1807). *A course of lectures on natural philosophy and the mechanical arts* (2 Vols.). London: J. Johnson. (University of California Digital Library).
Zarrow, P. G. (2005). *China in war and revolution, 1895–1949*. New York: Routledge.
Zarrow, P. G. (2018). *Educating China: Knowledge, society and textbooks in a modernizing world, 1902–1937*. Cambridge: Cambridge University Press.
Zebrowski, E. (1998). *A history of the circle: Mathematical reasoning and the physical universe*. Piscataway: Rutgers University Press.
Zeidler, D. L., & Sadler, T. D. (Eds.). (2008). Social and ethical issues in science education. *Special Issue of Science & Education, 17*(8–9), 799–803.
Zhang, D. (2002). *Key concepts in Chinese philosophy* (E. Ryden, Trans. & Ed.). Beijing: Foreign Languages Press.
Zhang, J. (2004). *A translation of the ancient Chinese 'the book of burial (Zang Shu)' by Guo Pu (276–324)*. Lewinston: Edwin Mellen Press.
Zilsel, E. (1942). The genesis of the concept of scientific law. *The Philosophical Review, 51*, 245–267.
Zimmerman, M. (1995). *Science, nonscience, and nonsense*. Baltimore: The Johns Hopkins University Press.

Figure Credits

Fig. 3.1 Yin-Yang seasonal variation
Google Open Source. Users.skynet.be https://www.google.com.au/search?hl=en&tbm=isch&source=hp&biw=1206&bih=708&ei=jzJqXO7gKNj5rQGRsp3wDA&q=seasonal+variation+of+yinyang&oq=seasonal+variation+of+yinyang&gs_l=img.3...2342.12985..13747...1.0..0.279.6026.0j27j4......0....1..gws-wizimg.....0..0j0i5i30j0i8i30j0i24.ZwyMzag_VX8#imgrc=JTW4gnzPLCHMWM:

Fig. 4.1 Feng shui dragon-hole buildings, Hong Kong
Wikimedia Commons: https://commons.wikimedia.org/w/index.php?curid=28593973
Wikimedia Commons: https://commons.wikimedia.org/wiki/File:HKBuilding Fengshui.jpg

Fig. 4.2 Geomantic (Lu-pan) compass
National Maritime Museum (open source) https://upload.wikimedia.org/wikipedia/commons/e/e0/Chinese_Geomantic_Compass_c._1760%2C_National_Maritime_Museum.JPG

Fig. 4.3 The yin-yang Pa Qua
The Art of Ancient Wisdom, http://www.theartofancientwisdom.com/i-ching/. Permission requested

Fig. 4.4 *I Ching* hexagram table
Polaris Open Source, http://www.polariswushu.net/

Fig. 4.5 *I Ching* hexagram #8
Wikicommons: https://commons.wikimedia.org/wiki/Category:I_Ching_hexagrams#/media/File:Ichinghexagram-08.svg

Fig. 4.6 *I Ching* hexagram 43
Wikicommons: https://commons.wikimedia.org/wiki/Category:I_Ching_hexagrams#/media/File:Ichinghexagram-64.svg

© Springer Nature Switzerland AG 2019
M. R. Matthews, *Feng Shui: Teaching About Science and Pseudoscience*,
Science: Philosophy, History and Education,
https://doi.org/10.1007/978-3-030-18822-1

Fig. 4.7 *I Ching* hexagram 64
(From: https://commons.wikimedia.org/wiki/Category:I_Ching_hexagrams#/media/File:Iching-hexagram-64.svg)

Fig. 5.1 Acupuncture meridians
Wellcome Open Access Collection https://iiif.wellcomecollection.org/image/M0014704.jpg/full/full/0/default.jpg#_ga=2.145031796.1724326645.1549110372-604797815.1549110372

Fig. 6.1 Matteo Ricci
Wikimedia commons https://upload.wikimedia.org/wikipedia/commons/a/ae/Ricciportrait.jpg

Fig. 6.2 Three Jesuit Scientists at work in China (Jean-Baptiste Du Halde 1747)
Wikimedia commons https://upload.wikimedia.org/wikipedia/commons/7/7c/Du_Halde_-_Description_de_la_Chine_-_Vol_3_feuille_110.jpg

Fig. 6.3 Zhao Youqin's fourteenth century flat-earth cosmology (*A* Yangcheng, *B* Mt. Kunlun, *C* Sihai zhi Zhong). M. R. Matthews

Fig. 7.1 Precession of the equinoxes
https://ancestorsandarchetypes.weebly.com/precession.html. Permission requested

Fig. 11.1 Galileo's Pendulum (*Two New Sciences*, 1638) public domain

Fig. 11.2 Animal magnetism therapy
Wellcome Collection: https://wellcomecollection.org/works/h5ccrkrf

Fig. 11.3 Cross-disciplinary teaching of feng shui. M. R. Matthews

Fig. 13.1 H.H. Lin Yun & H.H. Crystal C. Rinpoche
Website: http://www.yunlintemple.org/home. Permission requested

Fig. 13.2 Conceptual ecology for scientific progress (Bunge 2012a, p. 28)
Springer permission
License No 4522970686757
License date Feb 06, 2019

Fig. 13.3 Conceptual ecology for pseudoscience (Bunge 2012a, p. 33)
Springer permission
License No 4522970686757
License date Feb 06, 2019

Name Index

A
Aarnio, K., 27, 33
Acuna, C., xii
Acupuncture Evidence Project, 110
Aczel, A.D., 47, 137, 153
Aikenhead, G., 126, 165, 172
Alcock, J.E., 21, 31, 81, 226
Althusser, L., 149
American Association for the Advancement of Science (AAAS), 16, 17, 35, 36, 149, 163, 294, 295
Amorth, G., 129
Andersen, K., 26, 28, 34, 81, 267, 284
Angle, S., 5, 49, 161, 162, 201, 205, 235
Aquinas, T., 15, 28, 184
Augustine, 140

B
Bachelard, G., 149
Bacon, F., 38, 62, 178, 180
Beinfield, H., 92, 93, 106, 109, 152, 265, 288
Bellarmine, R., 14, 116
Bergson, H., 95
Berkeley, G., 52–54
Bernal, J.D., 197, 207, 242
Berthrong, J.H., 157, 161, 201
Binet, A., 226
Bird, G., 228
Birdsall, G., 68, 77
Black, M., 266
Blüh, O., 242
Bodde, D., 160, 199, 202–204, 209
Boerschmann, E., 67
Boudry, M., 28, 33, 261, 278, 285, 296

Boyle, R., 49, 53, 138, 178, 228
Brecht, B., 51
Brown, S., 7, 43
Bruun, O., 4, 7, 32, 72, 74, 75, 84, 87, 155
Bunge, M., xiii, 11, 16, 37, 50, 58, 59, 111, 144, 186, 200, 208, 209, 215, 260, 262, 263, 270, 275, 277–279, 282, 283, 285–288, 294, 295, 297

C
Campion, N., 20, 21
Capra, F., 84, 95, 96, 104
Carson, R., xii, 242
Cascioli, V., 22
Ch'en Duxiu, 11, 167, 289, 296
Chan, T., 74
Chan, W.-T., 44–46, 49, 52, 141
Chang, C. (Zhang Junmei), 173, 174
Chao, Nanming, 251
Chen, K., 66, 254, 255
Chen, Y.J., 210, 236, 237
Chinese Communist Party (CCP), 11, 30, 52, 130, 164, 167, 168, 173, 179, 182, 183, 185–188, 190, 246, 247, 256, 282
Chow, T.-T., 170
Chung, K., 59
Cixi, Empress Dowager, 166
Clark, A., 102
Clavius, C., 116, 117, 125
Cohen, R.S., xi, 33, 185
Colquhoun, D., 107, 108
Confucius (Kong Fuzi), 120, 122, 141, 142, 161, 166–169, 171, 199, 202, 205
Constantinou, C., 236, 238

Cooter, R., 277
Crombie, A.C., 139, 140, 197

D

d'Alembert, J., 53, 219
Dai, N., 160, 162, 171
Dale, R.A., 111
Davy, H., 229
de Groot, J.J.M., 32, 155
de Tocqueville, A., 126
Deng, Xiaoping, 157, 167, 179, 180, 185, 247
Descartes, R., 53, 58, 217–219
Dewey, J., 17, 18, 169, 170, 173, 182, 294, 297
Ding, W., 7, 49, 51
Doorly, J.W., 61
Driver, R., 287
du Halde, J.-B., 204
Dubs, H.H., 44
Duhem, P., 198, 257
Dukes, E.J., 47

E

Eddington, A., 273, 287
Eddy, M.B., 61
Educational Policies Commission (USA), 18
Eitel, E.J., 4, 93, 115, 133–156, 158, 195, 205, 213, 265, 296, 297
Ernst, E., 109, 110
Estrin, M., 107
Euclid, 116, 117, 159, 160, 203

F

Fan, Hongye, 185
Fang, L., 16, 59, 177, 179–181, 199, 220, 256
Fang, Yi, 185
Fara, P., 137, 211, 212
Faraday, M., 49, 233, 234
Fèrè, C., 226
Feynman, R.P., 216, 297
Fishman, Y.I., 33, 261
Franklin, B., 17, 223–225, 228, 241
Freedman, M., 32, 71, 137
Fung, Y.-L., 43, 170–173

G

Galileo, G., 49, 53, 69, 104, 116, 125, 138–140, 149, 159, 161, 196, 198, 200, 209, 211, 213, 217–219, 228, 236, 246, 257

Gallagher, L., 77, 120
Galle, J., 149
Gardner, M., 27, 33, 296
Gauld, C.F., xii, 18
Giles, H., 3
Gill, H., 29
Glymour, C., 81
Goodman, L., 150
Gould, S.J., 261, 285
Grosseteste, R., 139
Grove, J.W., 200, 233
Grove, W.R., 229–231
Guo, P., 47, 48, 68, 282
Guo, Y., 118, 163, 165, 185

H

Hall, A.R., 196, 197, 218
Hansson, S.O., 270, 279, 280, 282
Hao, S., 69, 70
Harper, D., 94, 205
Haygarth, J., 226
Hempel, C.G., 274, 275
Heraclitus, 217
Herschel, W., 149
Hesse, M.B., 266
Hessen, B.M., 197
Horton, R., 246
Hoyningen-Huene, P., 273, 280
Hsün Chhing, 158
Hsun, Tzu, 44
Hu, Shih, 173
Hua, Shipping., 123
Huff, T.E., 117, 163, 164, 199, 202, 204
Hulskramer, G., 84–88
Hume, D., 25, 37, 53, 159, 202, 272
Huxley, J., 95
Hwangbo, A.B., 77, 269

I

Ingenhousz, J., 239, 241
International Feng Shui Society, 66, 242
International Yan Xin Qigong Association, 94, 249, 252
Italian National Consumer Association, 22

J

Jammer, M., 238
Jeans, J., 287
Jefferson, T., 17, 20
Jen Hung-chün, 174, 175

Name Index 333

Jin, G., 140, 142, 173
John Paul II., 184

K
Kampourakis, K., xii
Kant, I., 53, 138, 139, 162, 286
Kaptchuk, T.J., 258, 259
Kenny, D., 25
Kepler, J., 49, 117, 149, 219
Kim, Y.S., 211
King, J., 7, 280
Knoop, A., 6, 136
Korngold, E., 92, 93, 106, 109, 152, 265, 288
Koyré, A., 138, 141, 172, 196
Kuhn, T.S., 71, 139, 234, 274, 277, 286, 296, 297

L
Lau Tzu, 44, 264
Laudan, L., 111, 274, 276–278
Lavoisier, A., 54, 59, 210, 223–225, 240
Leibniz, G.W., 53, 87, 116, 216–219
Le Verrier, U., 149
Leo XIII, 184
Li, Shan-Lan, 118
Li, Shu-chuen, 108
Li, Yaobang, 171
Liang Chí-cháo, 174
Liftig, I.F., 17
Lin, Chunyi, 101
Lin, Hui, 253
Lin, Z.X., 32, 187, 249, 251, 254
Lindeman, M., 27, 33
Liu, JeeLoo, 43, 50, 51, 53, 58, 93, 262, 263
Liu Xiaobo, 181
Locke, J., 53, 178, 182
Lodge, O., 232, 233
Losh, S.C., 21, 26, 27, 33
Low, M.F., 211
Lu, Zuyin, 249, 250, 254
Luo, Qinshun, 52

M
Mach, E., 62, 220, 242, 257, 263, 272, 297
Mackenzie, J., xii
Mahner, M., 261, 270, 278
Makeham, J., 49, 205
Mao, Zedong, 105, 107, 167, 169, 183, 184, 188, 199, 256

March, A.L., 46, 47, 68, 69, 82
Maurício, P., xii
Maxwell, J.C., 210
Mbeki, T., 24
McColl, P., xii
McIntyre, L., 18, 22, 37, 270, 286, 295, 297
Mencius, 49, 141, 163, 171
Mesmer, F.A., 221–223, 225
Miller, H.L., 168, 181
Miller, J.D., 14
Mitchell, E.D., 294
Moody, E.A., 139, 198

N
National Science Foundation (NSF), USA, 20, 21, 26
Needham, J., 52, 82, 84, 90, 94, 115, 118, 124, 126, 131, 137, 141, 150, 158, 160, 171, 172, 195–214, 247
Nehru, J.L., 17
Newman, J.H., 183
Newton, I., 37, 49, 58, 72, 117, 141, 159, 196, 198, 218–220, 225, 228, 230, 246, 257, 295
Nola, R., xii, 9, 277
Novella, S., 107, 109
Nzekwe, B., 21, 26, 27, 33

P
Papadouris, N., 236, 238
Park, Geun-hye, 30
Parkes, G., 4, 46, 67, 151, 210
Paton, M.J., 32, 47, 48, 68, 69, 189
Pennick, N., 66–68, 83, 84
Pennock, R.T., 33, 260, 261, 278
Peters, R.S., 294
Pius X, 184
Polanyi, M., 181, 182, 185
Polo, M., 115, 199
Pomeroy, D., 245
Popper, K.R., 38, 150, 161, 182, 208, 209, 250, 272–274, 297
Potter, J., 72
Priestley, J., 17, 49, 182, 185, 239–241, 257
Puro, J., 4, 32

Q
Qian, W.-Y., 53, 143, 160, 211
Qian, Xuesen, 185, 255

R

Reifler, S., 84, 87–90
Reiki Training Centre, 98
Ricci, M., 115–131, 133, 158–160, 172, 178, 188, 189, 195, 213, 215, 227, 256, 267, 289, 296, 297
Rolnick, H., 71, 72, 151, 235, 293, 296
Rossbach, S., 55, 77
Rowland, H., 163
Rule, P.A., 116, 120, 122, 129, 130
Russell, B., 167–170
Ryden, E., 58

S

Sacks, B., 98, 99
Schall, J.A., 116, 118, 119, 130
Shapin, S., 197
Shermer, M., 20, 33, 81, 231, 232, 270, 278
Shimony, A., xi, 207, 208
Shin, J., 7, 30, 98, 111
Shinji, A., 123–125, 203, 204
Shockley, W., 188
Skinner, S., 5, 43, 47, 55, 56, 71, 82, 83, 147, 148
Smith, G.H., 245
Spear, W., 67
Spinoza, B., 53, 166
Squires, N., 22
Stalker, D., 81
Starck, M., 21, 22
State Council Information Office (SCIO), 91, 103–107
Staver, J., 159, 287
Stove, D.C., 285
Suchting, W.A., xi, 52, 161, 220

T

T'áng Yüeh, 174, 175
Tang, Yijie, 141, 157, 178
Tanzer, E.J., 48, 60, 237
Taylor, R., 76, 77, 228
Thales of Miletus, 59
Thompson, B. (Count Rumford), 229, 230
Thomson, W. (Lord Kelvin), 243
Ting, Wen-chiang, 174
Tiwald, J., 5, 49, 161, 162, 201, 205, 235
Tobin, K., 96, 98, 258–260
Too, L., 66, 70, 71, 76, 77, 80, 83, 153, 258
Tresmontant, C., 15
Trump, D., 25, 36, 285, 286, 298
Tu, Y., 213

U

Urevbu, A.O., 245

V

van Helmont, J.B., 54
Verbiest, F., 116, 119
Vivekananda, S., 60–62
Volkov, A., 124, 203
von Glasersfeld, E., 52, 286
von Goethe, J.W., 143
von Helmholtz, H., 229, 231, 234, 235
von Mayer, J.R., 241
von Reichenbach, K., 227, 232

W

Wade, T., 3
Walker, W.R., 20, 21, 23, 26
Wallace, A., 139, 232
Wang, Chong, 154, 158, 260
Wang, Fuzhi, 53, 93
Wang, H.-S., 142
Wang, R.R., 7, 49, 51
Wang, Yangmin, 54
Wartofsky, M.W., xi, 208
Weber, K., 281
Wegman, M.E., 109
Wei Jingsheng, 179, 180
Weinberg, S., 138, 139, 197
Welch, H., 65
Wen, K., 6
Westfall, R.S., 196, 197, 220, 261
Whitehead, A.N., 205–208
Winchester, S., 196, 198
Wong, E., 45, 82
Wootton, D., 197, 230, 277
Wylie, A., 118

X

Xi, Jinping, 91, 167, 169, 182, 183
Xi, Zhu (Choo-he), 54, 143
Xin, Yan, 93, 94, 249–252, 254–255, 282
Xu, Guangqi, 117, 160
Xu, Liangying, 180, 181, 185

Y

Yang, Hsiung, 158
Yen, Fu, 159
Yiling Pharmaceutical Company, 105
Young, T., 216
Yun, Lin, 280–282

Z

Zhang, Danian, 56
Zhang, Xiangyu, 254
Zhang, Yunfun, 186
Zhang, Zai, 49, 51
Zhang, Zhidong, 164, 168
Zhao, Youqin, 124, 125, 203
Zhu, Xi, 54, 143
Zimmerman, M., 20, 21, 26, 31

Subject Index

A

Acupuncture
   appraisal, 32, 98, 106–111
   Chinese government support, 81, 103–106
   clinical practice, 109
   Evidence Project in Australia, 110
   meridian lines, 3, 5, 32, 43, 94, 95, 97, 98, 103, 107, 282
   Western government support, 76–78, 103, 107, 259
AIDS, 24, 25
Angels, jinn & devils, 28, 29
Animal magnetism
   Franklin Report, 215, 223, 225
   Mesmer, 221–223, 225
Aspirin, 212
Astrology
   Chinese, 145–146
   Western, 125, 145–146
Astronomy
   Chinese, 117, 118, 123–125, 130, 149, 150, 152, 220
   political control, 123, 125, 150, 284
   precession of equinoxes, 147, 148
Atomism, Greek, 206

B

Border Crossing, as multicultural education policy, 188
Buddhism, 51, 115, 122, 125, 133, 141, 201, 280

C

Chinese Communist Party (CCP)
   falun gong, opposition to, 190
   feng shui, policies, 3, 30, 52, 130, 186–188
   founding, 6, 168
   Matteo Ricci recognition, 130
Compass, geomantic, 47, 82, 137
Compass school of feng shui, 68–72, 153
Confucianism
   neo-Confucianis, 5, 49–51, 54, 70, 124, 138, 161, 168, 172, 173, 201, 205–210, 235
   science, relations to, 138
Construction
   feng shui, 3, 4, 65–67, 72–81, 134, 151, 190, 269
   Western, 76–78
Correlative thinking, 94, 95, 151
Correspondence theory of truth, 260
Corruption of traditions, 11, 12, 128, 130, 166, 283, 296
Cosmology, 6, 7, 16, 45, 48–50, 53–55, 59, 66, 82–84, 96, 124, 125, 133, 135, 136, 144, 145, 148, 151, 157, 176, 202, 203, 209, 228, 262
Creation Science (USA), 191, 283

D

Daoism (Taoism), 44, 46, 49, 65, 115, 125, 128, 178, 201, 209

Demarcation
    Bunge, Mario, 11, 270, 275, 279
    Hansson, S.O., 270, 279, 280, 282
    Hempel, Carl, 274, 275
    Laudan, Larry, 274, 276–278
    opposition to, 287
    Popper, Karl, 272–274
    pseudoscience/science, 4, 5, 9–12, 75, 150, 269, 270, 272–279
    science/non-science, 4, 10, 11, 124, 269, 270, 272–276, 278
Dewey, John
    China lectures, 169
Dialectical materialism, 16, 185, 187
Dispositional properties, 264
Divination (geomancy), 3, 4, 30, 31, 68, 78, 82–89, 126–128, 130, 137, 153, 158, 190, 195, 269
Duhem-Quine thesis, 257

E
Ecology of science, 176, 283–288
Edinburgh Strong Programme, 276, 277
Education
    feng shui in science programme, 4, 9–11, 26, 60, 146, 150, 179, 236–237, 249, 278, 293, 295
Eitel, Ernest
    astrology, 145–146, 148, 150
    astronomy, 115, 141, 147–150, 152
    education, 133, 136, 139, 150, 154–155
    education task for China, 154–155
    element theory, 209
    experiment, 137–140
    feng shui observations, 133–156
Element (phase) theory, 152, 203, 209, 211
Elements
    Chinese, 52, 58, 76, 83, 90, 117–119, 151–152, 198, 209
    Greek, 118, 161, 209
Empiricism
    Chinese thought, 54, 95, 141, 158, 166, 171, 172, 184, 207, 209
    scientific revolution, antithetical to, 20, 24, 172, 296
Energy
    animal magnetism, 220–228
    conservation law, 216, 234–236, 238
    feng shui, cross-disciplinary teaching, 10, 242, 243, 297
    Galileo, Galilei, 211, 213, 217–219, 228, 236, 246, 257
    history, 216–220
    HPS-informed teaching, 238–243
    imponderable fluids, 221, 228–231, 234
    Leibniz, 216–219
    literacy, 237–238
    Mach, Ernst, 220, 234, 242
    metaphysics, 215–216, 245
    Newton, Isaac, 218–221, 225, 228, 230
    photosynthesis, 239, 241, 242
    vis viva debate, 216, 218, 219, 231, 238
Enlightenment (European)
    Chinese adoption, 169
    education, 17, 52, 53, 165, 166, 177–179, 181
    science, 17, 52, 53, 165, 166, 177–179, 181
Ether
    chi, 6, 47, 58, 59, 111, 232, 233
    Newton, Isaac, 58–59
Euclid, translations, 116, 117, 159, 160, 203
Exorcism, 129
Experiment
    Chinese traditional science, 244, 247
    Indian, 138
    Kant, 53, 138, 139
    scientific, 137, 139, 140, 257
    scientific revolution, 138, 172, 196–198

F
Falun gong, 7, 190, 282
Form school of feng shui, 68–71, 78, 153

G
Gravity waves, 273, 274
Great Wall of China, 46, 198

H
Hinduism, 61, 62
History and philosophy of science (HPS) in education, 13, 18–19, 33, 39, 75, 111, 115, 156, 179, 191, 192, 297
Hong Kong
    feng shui construction, 7, 30, 34, 71–75, 77, 78, 134–136
    legal disputes, 73, 75, 77, 100

I
I Ching (Book of Changes)
    impact on Chinese traditional science, 7, 95, 124, 159
    interpretations, 49, 84, 87, 88

Needham, Joseph, 84, 90, 94, 124, 158, 160, 203
  origins, 84
Idealism, ontology, 187
Imperial examination system, 164, 167, 202
Imponderable fluids, 95, 221, 228–231, 234
Intervening variables, 136, 263–265

J
Jesuit mission to China, 116–119, 121

L
Landscape design
  feng shui, 68
Liberal education, 24, 36, 242, 285
Liberalism
  science, 168, 173, 179–183, 186

M
Marxism
  as official philosophy, 184–186
Mathematics, Chinese, 117–120, 159, 160
May Fourth Movement, 166–173, 180, 181
Measurement of chi, 54, 56, 76, 93, 104, 265
Meridian Energy Analysis Devices (MEAD), 54, 293
Metaphor, 68, 211, 266
Metaphysics
  Chinese, 7, 43, 47, 53, 56, 78, 79, 104, 133, 135, 140–142, 191, 195, 201, 202, 207–210
  Marxism, 16, 184–186
  religion, 15, 53, 61, 116
  science, 4, 9, 191
Modernization of China
  censorship, 182
  dissident tradition, 158, 179, 199
  Hundred Days Reform, 165
  imperial defeats, 163
  imperial examinations, 163–164, 167, 202
  liberalism, 168, 173, 179–183, 186
  Marxist philosophy, 184–186
  May Fourth Movement, 166–173, 180, 181
  national science congress (first), 185
  New Thought Movement, 166–173, 184, 247
  science, 12, 90, 157, 163, 178
  science education, 177–179
  study abroad, 171
  Westernization Movement, 162, 163
Multicultural education, 213, 244

N
Naturalism
  chi naturalism, 52, 262
  Chinese tradition, 50, 52
  methodological, 177, 260–263, 276, 283, 284
  ontological, 83, 261, 262
Needham, Joseph
  appraisal, 115, 118, 124, 126, 131, 137, 141, 150, 196, 211–213
  Cambridge University, 196
  education, 172, 195–196, 204, 211, 213, 214
  experiment, 137, 196, 197, 200, 201, 205, 208
  five element theory, 203, 209, 210
  Needham Question, 52, 126, 131, 171, 172, 195–198, 202, 205, 211
  technology, 90, 137, 160, 172, 195–202, 205, 213, 214
  universalism, 245, 286
  Whitehead and process philosophy, 195, 203, 205, 207, 208
Neo-Confucianism, 5, 138, 161, 168, 172, 201, 205–210, 235
Neptune, discovery, 149, 150, 152
New Thought Movement, 166–173, 184, 247
Newtonianism
  energy, 62, 206, 207, 218, 220
  gravitation, 58, 120, 146, 221
Non-overlapping magisteria argument (NOMA), 261, 285

P
Papal States, 128
Philosophy of Life debate, 167, 173–175
Photosynthesis, 239, 241, 242
Physicalism, 262
Placebo effects
  first study, 99, 100, 102, 107, 109–111, 224, 226
Prāṇa, 60–62, 96
President Trump, 36, 285
Priestley, Joseph
  design argument, 241
  experimentation, 141, 182, 185, 239–241, 257
Pseudoscience
  academic studies, 32, 33, 283, 285
  demarcation, 4, 9, 11, 150, 270, 272–279
  philosophical insulation of, 258–260
  Sedona (AZ), 34, 35, 38
Psychosomatic illness, 102, 226

R

Reiki therapy
   Australia, 100
   biofields, 100
Ricci, Matteo
   astronomy, 116–119, 122–126, 129, 130
   burial, 122, 130, 135
   China mission, 115–119
   education, 115, 121, 123, 125, 126
   feng shui observations, 115–131
   solar eclipse prediction, 119, 123, 125, 273
Roman Catholic Church
   philosophical controls, 10, 15, 184
Romanticism
   Chinese traditional science, 142–144
   European, 143
Russell, Bertrand
   China lectures, 169, 170
   influence of US teaching, 169, 170

S

Scientific habit of mind
   science education, 18, 19, 25, 31, 34–35, 52, 165, 188, 192, 295
   scientific temper (India), 17, 19
Scientific literacy, 35–37
Scientific revolution
   continuity/discontinuity, 197, 198
   experiment, 138, 172, 196, 197
   internalist/externalist accounts, 197, 200–205
   measurement, 138
   philosophy, 53, 173, 196, 197, 207–209
Scientific testing of chi claims
   Lu Zuyin, 249, 250, 254
   Yan Xin, 249–255, 282
   Zhang Xiangyu, 254
Scientism, 36, 37, 155, 174, 191, 259, 284, 295, 296
Sedona
   pseudoscience concentration, 34, 35, 38
Shape of Earth debate, 85, 224
Spirits, 5, 17, 18, 20, 24, 26, 29, 30, 46, 47, 49, 51, 67, 69, 72, 75, 82, 85, 94, 129, 134, 135, 137, 140–142, 153–154, 159, 166, 171, 175, 178, 195, 208, 216, 231–233, 235, 244, 245, 267, 268, 283, 294
Spiritualist science, 231–234
STEAM education, 5, 10, 242
Superstition
   Asia, 4, 30–31, 70
   educational response, 33, 186, 187, 298
   Finland, 27, 28
   Italy, 161
   Korea, 4, 30, 31, 34
   United States, 21, 298
   varieties of, 13, 26, 50, 95, 200, 205
   witch craze, 26, 56

T

Traditional Chinese Medicine
   acupuncture (*see* Acupuncture)
   components, 94, 96, 102, 103, 106, 111
   government support, 65, 81, 91, 103–106, 108, 190
   Mao Zedong, 107
   principles, 94, 96, 104, 110
   qigong exercise, 65, 95, 101, 111, 295
   school programmes, 91, 93, 102, 106
   Western adoption, 99, 103, 104, 107
Translation difficulties, 3, 5, 7, 58, 62, 88, 116–118, 120, 124, 151, 159, 162, 203, 204, 215, 272

U

University
   feng shui courses, 5, 25, 71, 77, 78, 91, 98, 106, 108, 109, 133, 138, 157, 159, 162–165, 173, 186–190, 254, 267

W

Wade-Giles system, 3, 7, 44, 45, 47, 143
Worldviews
   chi-base, 3, 19, 43, 60–62, 176, 298
   Christian, 4, 10, 15, 49, 50, 52, 60, 67, 145, 159, 207, 240
   countercultural, 4, 60, 84, 90
   education and, 14–18
   mechanical, 44, 143, 144, 206–208, 220, 228, 261
   multicultural, 19, 84, 126, 155, 188, 213, 244–246
   organicism, 205–210
   science and, 14–18, 125, 133
World Wide Web
   feng shui, 65, 78–81

Y

Yin-Yang, 8, 12, 43, 44, 48, 57, 69, 85–87, 93, 104, 107, 111, 169, 197, 209, 219, 239, 280

Printed by Printforce, the Netherlands